T0269117

ESSENTIAL METHODS FOR DESIGN BASED SAMPLE SURVEYS

Essential Methods for Design Based Sample Surveys

A derivative of Handbook of Statistics: Sample Surveys: Design, Methods and Applications, Vol. 29a

Edited by

Danny Pfeffermann

Department of Statistics, Hebrew University, Israel

C.R. Rao

The Pennsylvania State University,
University Park, USA

Amsterdam • Boston • Heidelberg • London • New York • Oxford
Paris • San Diego • San Francisco • Singapore • Sydney • Tokyo

North-Holland is an imprint of Elsevier

North-Holland is an imprint of Elsevier
30 Corporate Drive, Suite 400, Burlington, MA 01803, USA
Linacre House, Jordan Hill, Oxford OX2 8DP, UK
Radarweg 29, PO Box 211, 1000 AE Amsterdam, The Netherlands

First edition 2011

British Library Cataloguing in Publication Data
A catalogue record for this book is available from the British Library

Library of Congress Cataloging-in-Publication Data
A catalog record for this book is available from the Library of Congress

ISBN: 978-0-444-63826-7

For information on all North Holland publications
visit our web site at *books.elsevier.com*

Typeset by: diacriTech, India

Printed and bound in Great Britain
10 11 12 13   10 9 8 7 6 5 4 3 2 1

Table of Contents

List of Contributors

Brewer, Kenneth, *School of Finance and Applied Statistics, College of Business and Economics, L.F. Crisp Building (Building 26), Australian National University, A.C.T. 0200, Australia; e-mail: ken.brewer@anu.edu.au* (Ch. 1).

Chowdhury, Sadeq, *NORC, University of Chicago, 4350 East-West Highway, Suite 800, Bethesda, MD 20814; e-mail: sadeqc@yahoo.com* (Ch. 3).

Frankovic, Kathleen A., *Survey and Election Consultant, 3162 Kaiwiki Rd., Hilo, HI 96720; e-mail: kaf@cbsnews.com* (Ch. 11).

Gambino, Jack G., *Household Survey Methods Division, Statistics Canada, Ottawa, Canada K1A 0T6; e-mail: jack.gambino@statcan.gc.ca* (Ch. 5).

Glickman, Hagit, *National Authority of Measurement and Evaluation in Education (RAMA), Ministry of Education, Kiryat Hamemshala, Tel Aviv 67012, Israel; e-mail: hglickman.rama@education.gov.il* (Ch. 10).

Gregoire, Timothy, Weyerhaeuser, J.P. Jr., *Professor of Forest Management, School of Forestry and Environmental Studies, Yale University, 360 Prospect Street, New Haven, CT 06511-2189; e-mail: timothy.gregoire@yale.edu* (Ch. 1).

Hidiroglou, Michael A., *Statistical Research and Innovation Division, Statistics Canada, Canada, K1A 0T6; e-mail: Mike.Hidiroglou@statcan.gc.ca* (Ch. 6).

House, Carol C., *National Agricultural Statistics Service, U.S. Department of Agriculture, Washington, DC, USA; e-mail: Carol.House@usda.gov* (Ch. 7).

Kalton, Graham, *Westat, 1600 Research Blvd., Rockville, MD 20850; e-mail: grahamkalton@westat.com* (Ch. 2).

Kelly, Jenny, *NORC, University of Chicago, 1 North State Street, Suite 1600, Chicago, IL 60602; e-mail: Kelly-Jenny@norc.org* (Ch. 3).

Lavallée, Pierre, *Social Survey Methods Division, Statistics Canada, Canada, K1A 0T6; e-mail: pierre.lavallee@statcan.gc.ca* (Ch. 6).

Marker, David A., *Westat, 1650 Research Blvd., Rockville Maryland 20850; e-mail: DavidMarker@Westat.com* (Ch. 8).

Naidu, Gurramkonda M., *Professor Emeritus, College of Business & Economics, University of Wisconsin-Whitewater, Whitewater, WI 53190; e-mail: naidug@uww.edu* (Ch. 9).

Nirel, Ronit, *Department of Statistics, The Hebrew University of Jerusalem, Mount Scopus, Jerusalem 91905, Israel; e-mail: nirelr@cc.huji.ac.il* (Ch. 10).

Nusser, S. M., *Department of Statistics, Iowa State University, Ames, IA, USA; e-mail: nusser@iastate.edu* (Ch. 7).

Panagopoulos, Costas, *Department of Political Science, Fordham University, 441 E. Fordham Rd., Bronx, NY 10458; e-mail: costas@post.harvard.edu* (Ch. 11).

Shapiro, Robert Y., *Department of Political Science and Institute for Social and Economic Research and Policy, Columbia University, 420 West 118th Street, New York, NY 10027; e-mail: rys3@columbia.edu* (Ch. 11).

Silva, Pedro Luis do Nascimento, *Southampton Statistical Sciences Research Institute, University of Southampton, UK; e-mail: pedrolns@soton.ac.uk* (Ch. 5).

Skinner, Chris, *Southampton Statistical Sciences Research Institute, University of Southampton, Southampton SO17 1BJ, United Kingdom; e-mail: C.J.Skinner@soton.ac.uk* (Ch. 4).

Stevens, Don L. Jr., *Statistics Department, Oregon State University, 44 Kidder Hall, Corvallis, Oregon, 97331; e-mail: stevens@stat.oregonstate.edu* (Ch. 8).

Velu, Raja, *Irwin and Marjorie Guttag Professor, Department of Finance, Martin J. Whitman School of Management, Syracuse University, Syracuse, NY 13244-2450; e-mail:rpvelu@syr.edu* (Ch. 9).

Wolter, Kirk, *NORC at the University of Chicago, and Department of Statistics, University of Chicago, 55 East Monroe Street, Suite 3000, Chicago, IL 60603; e-mail: wolter-kirk@norc.uchicago.edu* (Ch. 3).

Essential Methods for Design Based Sample Surveys
ISSN: 0169-7161
DOI: 10.1016/B978-0-444-53734-8.00001-7

Introduction to Survey Sampling

Ken Brewer and Timothy G. Gregoire

1. Two alternative approaches to survey sampling inference

1.1. Laplace and his ratio estimator

At some time in the mid-1780s (the exact date is difficult to establish), the eminent mathematician Pierre Laplace started to press the ailing French government to conduct an enumeration of the population in about 700 communes scattered over the Kingdom (Bru, 1988), with a view to estimating the total population of France. He intended to use for this purpose the fact that there was already a substantially complete registration of births in all communes, of which there would then have been of the order of 10,000. He reasoned that if he also knew the populations of those sample communes, he could estimate the ratio of population to annual births, and apply that ratio to the known number of births in a given year, to arrive at what we would now describe as a ratio estimate of the total French population (Laplace, 1783[1], 1814a and 1814b). For various reasons, however, notably the ever-expanding borders of the French empire during Napoleon's early years, events militated against him obtaining a suitable total of births for the entire French population, so his estimated ratio was never used for its original purpose (Bru, 1988; Cochran, 1978; Hald, 1998; Laplace, 1814a and 1814b, p. 762). He did, however, devise an ingenious way for estimating the precision with which that ratio was measured. This was less straightforward than the manner in which it would be estimated today but, at the time, it was a very considerable contribution to the theory of survey sampling.

1.2. A prediction model frequently used in survey sampling

The method used by Laplace to estimate the precision of his estimated ratio was not dependent on the knowledge of results for the individual sample communes, which

---

[1] This paper is the text of an address given to the Academy on 30 October 1785, but appears to have been incorrectly dated back to 1783 while the Memoirs were being compiled. A virtually identical version of this address also appears in Laplace's *Oeuvres Complètes* 11 pp. 35–46. This version also contains three tables of vital statistics not provided in the Memoirs' version. They should, however, be treated with caution, as they contain several arithmetical inconsistencies.

would normally be required these days for survey sampling inference. The reason why it was not required there is chiefly that a particular model was invoked, namely one of drawing balls from an urn, each black ball representing a French citizen counted in Laplace's sample, and each white ball representing a birth within those sample communes in the average of the three preceding years. As it happens, there is another model frequently used in survey sampling these days, which leads to the same ratio estimator. That model is

$$Y_i = \beta X_i + U_i, \tag{1a}$$

which together with

$$E(U_i) = 0, \tag{1b}$$

$$E(U_i^2) = \sigma^2 X_i \tag{1c}$$

and

$$E(U_i U_j) = 0 \tag{1d}$$

for all $j \neq i$ can also be used for the same purpose.

Equation (1a) describes a survey variable value Y_i (for instance the population of commune i) as generated by a survey parameter, β, times an auxiliary value, X_i, (that commune's average annual births) plus a random variable, U_i. Equation (1b) stipulates that this random variable has zero mean, Eq. (1c) that its variance is proportional to the auxiliary variable (in this case, annual births), and Eq. (1d) that there is no correlation between any pair of those random variables.

Given this model, the minimum variance unbiased estimator of β is given by

$$\hat{\beta} = \frac{\sum_{i=1}^{n} Y_i}{\sum_{i=1}^{n} X_i}, \tag{2}$$

which in this instance is simply the ratio of black to white balls in Laplace's urn.

1.3. The prediction model approach to survey sampling inference

While, given the model of Eqns. (1), the logic behind the ratio estimator might appear to be straightforward, there are in fact two very different ways of arriving at it, one obvious and one somewhat less obvious but no less important. We will examine the obvious one first.

It is indeed obvious that there is a close relationship between births and population. To begin with, most of the small geographical areas (there are a few exceptions such as military barracks and boarding schools) have approximately equal numbers of males and females. The age distribution is not quite so stable, but with a high probability different areas within the same country are likely to have more or less the same age distribution, so the proportion of females of child-bearing age to total population is also more or less constant. So, also with a reasonable measure of assurance, one might expect the ratio of births in a given year to total population to be more or less constant, which makes the ratio estimator an attractive choice.

We may have, therefore, a notion in our minds that the number in the population in the ith commune, Y_i, is proportional to the number of births there in an average year, X_i, plus a random error, U_i. If we write that idea down in mathematical form, we arrive at a set of equations similar to (1) above, though possibly with a more general variance structure than that implied by Eqns. (1c) and (1d), and that set would enable us to predict the value of Y_i given only the value of X_i together with an estimate of the ratio β. Laplace's estimate of β was a little over 28.35.

The kind of inference that we have just used is often described as "model-based," but because it is a prediction model and because we shall meet another kind of model very shortly, it is preferable to describe it as "prediction-based," and this is the term that will be used here.

1.4. The randomization approach to survey sampling inference

As already indicated, the other modern approach to survey sampling inference is more subtle, so it will take a little longer to describe. It is convenient to use a reasonably realistic scenario to do so.

The hypothetical country of Oz (which has a great deal more in common with Australia than with Frank L. Baum's mythical Land of Oz) has a population of 20 million people geographically distributed over 10,000 postcodes. These postcodes vary greatly among themselves in population, with much larger numbers of people in a typical urban than in a typical rural postcode.

Oz has a government agency named Centrifuge, which disburses welfare payments widely over the entire country. Its beneficiaries are in various categories such as Age Pensioners, Invalid Pensioners, and University Students. One group of its beneficiaries receives what are called Discretionary Benefits. These are paid to people who do not fall into any of the regular categories but are nevertheless judged to be in need of and/or deserving of financial support.

Centrifuge staff, being human, sometimes mistakenly make payments over and above what their beneficiaries are entitled to. In the Discretionary Benefits category, it is more difficult than usual to determine when such errors (known as overpayments) have been made, so when Centrifuge wanted to arrive at a figure for the amounts of Overpayments to Discretionary Beneficiaries, it decided to do so on a sample basis. Further, since it keeps its records in postcode order, it chose to select 1000 of these at random (one tenth of the total) and to spend considerable time and effort in ensuring that the Overpayments in these sample postcodes were accurately determined. (In what follows, the number of sample postcodes, in this case 1000, will be denoted by n and the number of postcodes in total, in this case 10,000, denoted by N.)

The original intention of the Centrifuge sample designers had been to use the same kind of ratio estimator as Laplace had used in 1802, namely

$$\hat{Y} = \frac{\sum_{i=1}^{N} \delta_i Y_i}{\sum_{i=1}^{N} \delta_i X_i} \sum_{i=1}^{N} X_i, \tag{3}$$

with Y_i being the amount of overpayments in the ith postcode and X_i the corresponding postcode population. In (3), δ_i is a binary (1/0) indicator of inclusion into the sample of size n: for any particular sample, all but n of the N elements of the population will have a value of $\delta = 0$ so that the sum of $\delta_i Y_i$ over $i = 1 \ldots N$ yields the sum of just the n values of Y_i on those elements selected into the sample.

However, when this proposal came to the attention of a certain senior Centrifuge officer who had a good mathematical education, he queried the use of this ratio estimator on the grounds that the relationship between Overpayments (in this particular category) and Population in individual postcodes was so weak that the use of the model (1) to justify it was extremely precarious. He suggested that the population figures for the selected postcodes should be ignored and that the ratio estimator should be replaced by the simpler expansion estimator, which was

$$\hat{Y} = (N/n) \sum_{i=1}^{N} \delta_i Y_i. \tag{4}$$

When this suggestion was passed on to the survey designers, they saw that it was needed to be treated seriously, but they were still convinced that there was a sufficiently strong relationship between Overpayments and Population for the ratio estimator also to be a serious contender. Before long, one of them found a proof, given in several standard sampling textbooks, that without reliance on any prediction model such as Eqns. (1), the ratio estimator was more efficient than the expansion estimator provided (a) that the sample had been selected randomly from the parent population and (b) that the correlation between the Y_i and the X_i exceeded a certain value (the exact nature of which is irrelevant for the time being). The upshot was that when the sample data became available, that requirement was calculated to be met quite comfortably, and in consequence the ratio estimator was used after all.

1.5. A comparison of these two approaches

The basic lesson to be drawn from the above scenario is that there are two radically different sources of survey sampling inference. The first is prediction on the basis of a mathematical model, of which (1), or something similar to it, is the one most commonly postulated. The other is randomized sampling, which can provide a valid inference regardless of whether the prediction model is a useful one or not. Note that a model can be useful even when it is imperfect. The famous aphorism of G.E.P. Box, "All models are wrong, but some are useful." (Box, 1979), is particularly relevant here.

There are also several other lessons that can be drawn. To begin with, models such as that of Eqns. (1) have parameters. Equation (1a) has the parameter β, and Eq. (1c) has the parameter σ^2 that describes the extent of variability in the Y_i. By contrast, the randomization-based estimator (4) involves no estimation of any parameter. All the quantities on the right hand side of (4), namely N, n, and the sample Y_i, are known, if not without error, at least without the need for any separate estimation or inference.

In consequence, we may say that estimators based on prediction inference are parametric, whereas those based on randomization inference are nonparametric. Parametric estimators tend to be more accurate than nonparametric estimators when the model on which they are based is sufficiently close to the truth as to be useful, but they are also sensitive to the possibility of model breakdown. By contrast, nonparametric estimators tend to be less efficient than parametric ones, but (since there is no model to break down) they are essentially robust. If an estimator is supported by both parametric and nonparametric inference, it is likely to be both efficient and robust. When the correlation between the sample Y_i and the sample X_i is sufficiently large to meet the relevant condition, mentioned but not defined above in the Oz scenario, the estimator is also likely to be both efficient and robust, but when the correlation fails to meet that condition, another estimator has a better randomization-based support, so the ratio estimator is no longer robust, and the indications are that the expansion estimator, which does not rely upon the usefulness of the prediction model (1), would be preferable.

It could be argued, however, that the expansion estimator itself could be considered as based on the even simpler prediction model

$$Y_i = \alpha + U_i, \tag{5}$$

where the random terms U_i have zero means and zero correlations as before. In this case, the parameter to be estimated is α, and it is optimally estimated by the mean of the sample observations Y_i. However, the parametrization used here is so simple that the parametric estimator based upon it coincides with the nonparametric estimator provided by randomization inference. This coincidence appears to have occasioned some considerable confusion, especially, but not exclusively, in the early days of survey sampling.

Moreover, it is also possible to regard the randomization approach as implying its own quite different model. Suppose we had a sample in which some of the units had been selected with one chance in ten, others with one chance in two, and the remainder with certainty. (Units selected with certainty are often described as "completely enumerated.") We could then make a model of the population from which such a sample had been selected by including in it (a) the units that had been selected with one chance in ten, together with nine exact copies of each such unit, (b) the units that had been selected with one chance in two, together with a single exact copy of each such unit, and (c) the units that had been included with certainty, but in this instance without any copies. Such a model would be a "randomization model." Further, since it would be a nonparametric model, it would be intrinsically robust, even if better models could be built that did use parameters.

In summary, the distinction between parametric prediction inference and nonparametric randomization inference is quite a vital one, and it is important to bear it in mind as we consider below some of the remarkable vicissitudes that have beset the history of survey sampling from its earliest times and have still by no means come to a definitive end.

2. Historical approaches to survey sampling inference

2.1. The development of randomization-based inference

Although, as mentioned above, Laplace had made plans to use the ratio estimator as early as the mid-1780s, modern survey sampling is more usually reckoned as dating from the work of Anders Nicolai Kiaer, the first Director of the Norwegian Central Bureau of Statistics. By 1895, Kiaer, having already conducted sample surveys successfully in his own country for fifteen years or more, had found to his own satisfaction that it was not always necessary to enumerate an entire population to obtain useful information about it. He decided that it was time to convince his peers of this fact and attempted to do so first at the session of the International Statistical Institute (ISI) that was held in Berne that year. He argued there that what he called a "partial investigation," based on a subset of the population units, could indeed provide such information, provided only that the subset had been carefully chosen to reflect the whole of that population in miniature. He described this process as his "representative method," and he was able to gain some initial support for it, notably from his Scandinavian colleagues. Unfortunately, however, his idea of representation was too subjective and lacking in probabilistic rigor to make headway against the then universally held belief that only complete enumerations, "censuses," could provide any useful information (Lie, 2002; Wright, 2001).

It was nevertheless Kiaer's determined effort to overthrow that universally held belief that emboldened Lucien March, at the ISI's Berlin meeting in 1903, to suggest that randomization might provide an objective basis for such a partial investigation (Wright, 2001). This idea was further developed by Arthur Lyon Bowley, first in a theoretical paper (Bowley, 1906) and later by a practical demonstration of its feasibility in a pioneering survey conducted in Reading, England (Bowley, 1912).

By 1925, the ISI at its Rome meeting was sufficiently convinced (largely by the report of a study that it had itself commissioned) to adopt a resolution giving acceptance to the idea of sampling. However, it was left to the discretion of the investigators whether they should use randomized or purposive sampling. With the advantage of hindsight, we may conjecture that, however vague their awareness of the fact, they were intuiting that purposive sampling was under some circumstances capable of delivering accurate estimates, but that under other circumstances, the underpinning of randomization inference would be required.

In the following year, Bowley published a substantial monograph in which he presented what was then known concerning the purposive and randomizing approaches to sample selection and also made suggestions for further developments in both of them (Bowley, 1926). These included the notion of collecting similar units into groups called "strata," including the same proportion of units from each stratum in the sample, and an attempt to make purposive sampling more rigorous by taking into account the correlations between, on the one hand, the variables of interest for the survey and, on the other, any auxiliary variables that could be helpful in the estimation process.

2.2. Neyman's establishment of a randomization orthodoxy

A few years later, Corrado Gini and Luigi Galvani selected a purposive sample of 29 out of 214 districts (circondari) from the 1921 Italian Population Census (Gini and Galvani, 1929). Their sample was chosen in such a way as to reflect almost exactly the whole-of-Italy average values for seven variables chosen for their importance, but it was shown by Jerzy Neyman (1934) that it exhibited substantial differences from those averages for other important variables.

Neyman went on to attack this study with a three pronged argument. His criticisms may be summarized as follows:

(1) Because randomization had not been used, the investigators had not been able to invoke the Central Limit Theorem. Consequently, they had been unable to use

the normality of the estimates to construct the confidence intervals that Neyman himself had recently invented and which appeared in English for the first time in his 1934 paper.

(2) On the investigators' own admission, the difficulty of achieving their "purposive" requirement (that the sample match the population closely on seven variables) had caused them to limit their attention to the 214 districts rather than to the 8354 communes into which Italy had also been divided. In consequence, their 15% sample consisted of only 29 districts (instead of perhaps 1200 or 1300 communes). Neyman further showed that a considerably more accurate set of estimates could have been expected had the sample consisted of this larger number of smaller units. Regardless of whether the decision to use districts had required the use of purposive sampling, or whether the causation was the other way round, it was evident that purposive sampling and samples consisting of far too few units went hand in hand.

(3) The population model used by the investigators was demonstrably unrealistic and inappropriate. Models by their very nature were always liable to represent the actual situation inadequately. Randomization obviated the need for population modeling.[2] With randomization-based inference, the statistical properties of an estimator are reckoned with respect to the distribution of its estimates from all samples that might possibly be drawn using the design under consideration. The same estimator under different designs will admit to differing statistical properties. For example, an estimator that is unbiased under an equal probability design (see Section 3 of this chapter for an elucidation of various designs that are in common use) may well be biased under an unequal probability design.

In the event, the ideas that Neyman had presented in this paper, though relevant for their time and well presented, caught on only gradually over the course of the next decade. W. Edwards Deming heard Neyman in London in 1936 and soon arranged for him to lecture, and his approach to be taught, to U.S. government statisticians. A crucial event in its acceptance was the use in the 1940 U.S. Population and Housing Census of a one-in-twenty sample designed by Deming, along with Morris Hansen and others, to obtain answers to additional questions. Once accepted, however, Neyman's arguments swept all other considerations aside for at least two decades.

Those twenty odd years were a time of great progress. In the terms introduced by Kuhn (1996), finite population sampling had found a universally accepted "paradigm" (or "disciplinary matrix") in randomization-based inference, and an unusually fruitful period of normal science had ensued. Several influential sampling textbooks were published, including most importantly those by Hansen et al. (1953) and by Cochran (1953, 1963). Other advances included the use of self-weighting, multistage, unequal probability samples by Hansen and Hurwitz at the U.S. Bureau of the Census, Mahalanobis's invention of interpenetrating samples to simplify the estimation of variance for complex survey designs and to measure and control the incidence of nonsampling errors, and the beginnings of what later came to be described as "model-assisted survey sampling."

---

[2] The model of Eqns. (1) above had not been published at the time of Neyman's presentation. It is believed first to have appeared in Fairfield Smith (1938) in the context of a survey of agricultural crops. Another early example of its use is in Jessen (1942).

A lone challenge to this orthodoxy was voiced by Godambe (1955) with his proof of the nonexistence of any uniformly best randomization-based estimator of the population mean, but few others working in this excitingly innovative field seemed to be concerned by this result.

2.3. Model-assisted or model-based? The controversy over prediction inference

It therefore came as a considerable shock to the finite population sampling establishment when Royall (1970b) issued his highly readable call to arms for the reinstatement of purposive sampling and prediction-based inference. To read this paper was to read Neyman (1934) being stood on its head. The identical issues were being considered, but the opposite conclusions were being drawn.

By 1973, Royall had abandoned the most extreme of his recommendations. This was that the best sample to select would be the one that was optimal in terms of a model closely resembling Eqns. (1). (That sample would typically have consisted of the largest n units in the population, asking for trouble if the parameter β had not in fact been constant over the entire range of sizes of the population units.) In Royall and Herson (1973a and 1973b), the authors suggested instead that the sample should be chosen to be "balanced", in other words that the moments of the sample X_i should be as close as possible to the corresponding moments of the entire population. (This was very similar to the much earlier notion that samples should be chosen purposively to resemble the population in miniature, and the samples of Gini and Galvani (1929) had been chosen in much that same way!)

With that exception, Royall's original stand remained unshaken. The business of a sampling statistician was to make a model of the relevant population, design a sample to estimate its parameters, and make all inferences regarding that population in terms of those parameter estimates. The randomization-based concept of defining the variance of an estimator in terms of the variability of its estimates over all possible samples was to be discarded in favor of the prediction variance, which was sample-specific and based on averaging over all possible realizations of the chosen prediction model.

Sampling statisticians had at no stage been slow to take sides in this debate. Now the battle lines were drawn. The heat of the argument appears to have been exacerbated by language blocks; for instance, the words "expectation" and "variance" carried one set of connotations for randomization-based inference and quite another for prediction-based inference. Assertions made on one side would therefore have appeared as unintelligible nonsense by the other.

A major establishment counterattack was launched with Hansen et al. (1983). A small (and by most standards undetectable) divergence from Royall's model was shown nevertheless to be capable of distorting the sample inferences substantially. The obvious answer would surely have been "But this distortion would not have occurred if the sample had been drawn in a balanced fashion. Haven't you read Royall and Herson (1973a and b)?" Strangely, it does not seem to have been presented at the time.

Much later, a third position was also offered, the one held by the present authors, namely that since there were merits in both approaches, and that it was possible to combine them, the two should be used together. For the purposes of this Handbook volume, it is necessary to consider all three positions as dispassionately as possible. Much can be gained by asking the question as to whether Neyman (1934) or Royall

(1970b) provided the more credible interpretation of the facts, both as they existed in 1934 or 1970 and also at the present day (2009).

2.4. A closer look at Neyman's criticisms of Gini and Galvani

The proposition will be presented here that Neyman's criticisms and prescriptions were appropriate for his time, but that they have been overtaken by events. Consider first his contention that without randomization, it was impossible to use confidence intervals to measure the accuracy of the sample estimates.

This argument was received coolly enough at the time. In moving the vote of thanks to Neyman at the time of the paper's presentation, Bowley wondered aloud whether confidence intervals were a "confidence trick." He asked "Does [a confidence interval] really lead us to what we need—the chance that within the universe which we are sampling the proportion is within these certain limits? I think it does not. I think we are in the position of knowing that *either* an improbable event had occurred *or* the proportion in the population is within these limits... The statement of the theory is not convincing, and until I am convinced I am doubtful of its validity."

In his reply, Neyman pointed out that Bowley's question in the first quoted sentence above "contain[ed] the statement of the problem in the form of Bayes" and that in consequence its solution "*must* depend upon the probability law *a priori*." He added "In so far as we keep to the old form of the problem, any further progress is impossible." He thus concluded that there was a need to stop asking Bowley's "Bayesian" question and instead adopt the stance that the "*either...or*" statement contained in his second quoted sentence "form[ed] a basis for the practical work of a statistician concerned with problems of estimation." There can be little doubt but that Neyman's suggestion was a useful prescription for the time, and the enormous amount of valuable work that has since been done using Neyman and Pearson's confidence intervals is witness to this.

However, the fact remains that confidence intervals are not easy to understand. A confidence interval is in fact a sample-specific range of potentially true values of the parameter being estimated, which has been constructed so as to have a particular property. This property is that, over a large number of sample observations, the proportion of times that the true parameter value falls inside that range (constructed for each sample separately) is equal to a predetermined value known as the confidence level. This confidence level is conventionally written as $(1 - \alpha)$, where α is small compared with unity. Conventional choices for α are 0.05, 0.01, and sometimes 0.001. Thus, if many samples of size n are drawn independently from a normal distribution and the relevant confidence interval for $\alpha = 0.05$ is calculated for each sample, the proportion of times that the true parameter value will lie within any given sample's own confidence interval will, before that sample is selected, be 0.95, or 95%.

It is not the case, however, that the probability of this true parameter value lying within the confidence interval as calculated for any individual sample of size n will be 95%. The confidence interval calculated for any individual sample of size n will, in general, be wider or narrower than average and might be centered well away from the true parameter value, especially if n is small. It is also sometimes possible to recognize when a sample is atypical and, hence, make the informed guess that in this particular case, the probability of the true value lying in a particular 95% confidence interval differs substantially from 0.95.

If, however, an agreement is made beforehand that a long succession of wagers is to be made on the basis that (say) Fred will give Harry $1 every time the true value lies inside any random sample's properly calculated 95% confidence interval, and Harry will give Fred $19 each time it does not; then at the end of that long sequence, those two gamblers would be back close to where they started. In those circumstances, the 95% confidence interval would also be identical with the 95% Bayesian credibility interval that would be obtained with a flat prior distribution over the entire real line ranging from minus infinity to plus infinity. In that instance, Bowley's "Bayesian question" could be given an unequivocally affirmative answer.

The result of one type of classical hypothesis test is also closely related to the confidence interval. Hypothesis tests are seldom applied to data obtained from household or establishment surveys, but they are frequently used in other survey sampling contexts.

The type of classical test contemplated here is often used in medical trials. The hypothesis to be tested is that a newly devised medical treatment is superior to an existing standard treatment, for which the effectiveness is known without appreciable error. In this situation, there can never be any reason to imagine that the two treatments are identically effective so that event can unquestionably be accorded the probability zero. The probability that the alternative treatment is the better one can then legitimately be estimated by the proportion of the area under the likelihood function that corresponds to values greater than the standard treatment's effectiveness. Moreover, if that standard effectiveness happens to be lower than that at the lower end of the one-sided 95% confidence interval, it can reasonably be claimed that the new treatment is superior to the standard one "with 95% confidence."

However, in that situation, the investigators might well wish to go further and quote the proportion of the area corresponding to all values less than standard treatment's effectiveness (Fisher's p-statistic). If, for instance, that proportion were 0.015, they might wish to claim that the new treatment was superior "with 98.5% confidence." To do so might invite the objection that the language used was inappropriate because Neyman's α was an arbitrarily chosen fixed value, whereas Fisher's p was a realization of a random variable, but the close similarity between the two situations would be undeniable. For further discussions of this distinction, see Hubbard and Bayarri (2003) and Berger (2003).

The situation would have been entirely different, however, had the investigation been directed to the question as to whether an additional parameter was required for a given regression model to be realistic. Such questions often arise in contexts such as biodiversity surveys and sociological studies. It is then necessary to accord the null hypothesis value itself (which is usually but not always zero) a nonzero probability. It is becoming increasingly well recognized that in these circumstances, the face value of Fisher's p can give a grossly misleading estimate of the probability that an additional parameter is needed. A relatively new concept, the "false discovery rate" (Benjamini and Hochberg, 1995; Benjamini and Yekutieli, 2001; Efron et al., 2001; Sorić, 1989), can be used to provide useful insights. To summarize the findings in these papers very briefly, those false discovery rates observed empirically have, more often than not, been found to exceed the corresponding p-statistic by a considerable order of magnitude.

It is also relevant to mention that the populations met with in finite population sampling, and especially those encountered in establishment surveys, are often far removed from obeying a normal distribution, and that with the smaller samples often selected from them, the assumption of normality for the consequent estimators is unlikely even to produce accurate confidence intervals!

Nevertheless, and despite the misgivings presented above, it is still the case that randomization does provide a useful basis for the estimation of a sample variance. The criterion of minimizing that variance is also a useful one for determining optimum estimators. However, we should not expect randomization alone to provide anything further.

Neyman's second contention was that purposive sampling and samples consisting of fewer than an adequate number of units went hand in hand. This was undoubtedly the case in the 1930s, but a similar kind of matching of sample to population (Royall and his co-authors use the expression "balanced sampling") can now be undertaken quite rapidly using third-generation computers, provided only that the matching is not made on too many variables simultaneously. Brewer (1999a) presents a case that it might be preferable to choose a sample randomly and use calibrated estimators to compensate for any lack of balance, rather than to go to the trouble of selecting balanced samples. However, those who prefer to use balanced sampling can now select randomly from among many balanced or nearly balanced samples using the "cube method" (Deville and Tillé, 2004). This paper also contains several references to earlier methods for selecting balanced samples.

Neyman's third contention was basically that population models were not to be trusted. It is difficult here to improve on the earlier quote from George Box that "All models are wrong, but some models are useful." Equations (1) above provide a very simple model that has been in use since 1938. It relates a variable of interest in a sample survey to an auxiliary variable, all the population values of which are conveniently known.

In its simplest form, the relationship between these variables is assumed to be basically proportional but with a random term modifying that proportional relationship for each population unit. (Admittedly, in some instances, it is convenient to add an intercept term, or to have more than one regressor variable, and/or an additional equation to model the variance of that equation's random term, but nevertheless that simple model can be adequate in a remarkably wide set of circumstances.)

As previously mentioned, such models have been used quite frequently in survey sampling. However, it is one thing to use a prediction model to improve on an existing randomization-based estimator (as was done in the Oz scenario above) and it is quite another thing actually to base one's sampling inference on that model. The former, or "model-assisted" approach to survey sampling inference, is clearly distinguished from prediction-based inference proper in the following quotation, taken from the Preface to the encyclopedic book, *Model Assisted Survey Sampling* by Särndal et al. (1992, also available in paperbook 2003):

> Statistical modeling has strongly influenced survey sampling theory in recent years. In this book, sampling theory is assisted by modeling. It becomes simple to explain how the auxiliary information in a given survey will lead to a particular estimation technique. The teaching of sampling and the style of presentation in journal articles have changed a great deal by this new emphasis. Readers of this book will become familiar with this new style.

> We use the randomization theory or design-based point of view. This is the tra-
> ditional mode of inference in surveys, ever since the sampling breakthroughs in the
> 1930s and 1940s. The reasoning is familiar to survey statisticians in government and
> elsewhere.

As this quotation indicates, using a prediction model to form an estimator as Royall proposed, without regard to any justification in terms of randomization theory, is quite a different approach. It is often described as "model-based," or pejoratively as "model-dependent," but it appears preferable to use the expression, "prediction-based."

A seminal paper attacking the use of a prediction model for such purposes was that by Hansen et al. (1983), which has already been mentioned; but there can be no serious doubt attached to the proposition that this model provides a reasonable first approximation to many real situations. Once again, Neyman's contention has been overtaken by events.

2.5. Other recent developments in sample survey inference

A similarly detailed assessment of the now classic papers written by Royall and his colleagues in the 1970s and early 1980s is less necessary, since there have been fewer changes since they were written, but it is worth providing a short summary of some of them. Royall (1970b) has already been mentioned as having turned Neyman (1934) on its head. Royall (1971) took the same arguments a stage further. In Royall and Herson (1973a and 1973b), there is an implicit admission that selecting the sample that minimized the prediction-based variance (prediction variance) was not a viable strategy. The suggestion offered there is to select balanced samples instead: ones that reflect the moments of the parent population. In this recommendation, it recalls the early twentieth-century preoccupation with finding a sample that resembled the population in miniature but, as has been indicated above, this does not necessarily count against it.

Royall (1976) provides a useful and entertaining introduction to prediction-based inference, written at a time when the early criticisms of it had been fully taken into account. Joint papers by Royall and Eberhardt (1975) and Royall and Cumberland (1978, 1981a and 1981b) deal with various aspects of prediction variance estimation, whereas Cumberland and Royall (1981) offer a prediction-based consideration of unequal probability sampling. The book by Valliant et al. (2000) provides a comprehensive account of survey sampling from the prediction-based viewpoint up to that date, and that by Bolfarine and Zacks (1992) presents a Bayesian perspective on it.

Significant contributions have also been made by other authors. Bardsley and Chambers (1984) offered ridge regression as an alternative to pure calibration when the number of regressor variables was substantial. Chambers and Dunstan (1986) and Chambers et al. (1992) considered the estimation of distribution functions from a prediction-based standpoint. Chambers et al. (1993) and Chambers and Kokic (1993) deal specifically with questions of robustness against model breakdown. A more considerable bibliography of important papers relating to prediction-inference can be found in Valliant et al. (2000).

The randomization-based literature over recent years has been far too extensive to reference in the same detail, and in any case comparatively little of it deals with the question of sampling inference. However, two publications already mentioned above are of especial importance. These are the polemical paper by Hansen et al. (1983) and the highly influential text-book by Särndal et al. (1992), which sets out explicitly to indicate what can be achieved by using model-assisted methods of sample estimation without the explicit use of prediction-based inference. Other recent papers of particular interest in this field include Deville and Särndal (1992) and Deville et al. (1993).

Publications advocating or even mentioning the use of both forms of inference simultaneously are few in number. Brewer (1994) would seem to be the earliest to appear in print. It was written in anticipation of and to improve upon Brewer (1995), which faithfully records what the author was advocating at the First International Conference on Establishment Surveys in 1993, but was subsequently found not to be as efficient or even as workable as the alternative provided in Brewer (1994). A few years later, Brewer (1999a) compared stratified balanced with stratified random sampling and Brewer (1999b) provided a detailed description of how the two inferences could be used simultaneously in unequal probability sampling; also Brewer's (2002) textbook has provided yet further details on this topic, including some unsought spin-offs that follow from their simultaneous use, and an extension to multistage sampling.

All three views are still held. The establishment view is that model-assisted randomization-based inference has worked well for several decades, and there is insufficient reason to change. The prediction-based approach continues to be presented by others as the only one that can consistently be held by a well-educated statistician. And a few say "Why not use both?" Only time and experience are likely to resolve the issue, but in the meantime, all three views need to be clearly understood.

3. Some common sampling strategies

3.1. Some ground-clearing definitions

So far, we have only been broadly considering the options that the sampling statistician has when making inferences from the sample to the population from which it was drawn. It is now time to consider the specifics, and for that we will need to use certain definitions.

A *sample design* is a procedure for selecting a sample from a population in a specific fashion. These are some examples:

- simple random sampling with and without replacement;
- random sampling with unequal probabilities, again with and without replacement;
- systematic sampling with equal or unequal probabilities;
- stratified sampling, in which the population units are first classified into groups or "strata" having certain properties in common;
- two-phase sampling, in which a large sample is drawn at the first phase and a subsample from that large sample at the second phase;
- multistage sampling, usually in the context of area sampling, in which a sample of (necessarily large) first-stage units is selected first, samples within those first-stage sample units at the second stage, and so on for possibly third and fourth stages; and
- permanent random number sampling, in which each population unit is assigned a number, and the sample at any time is defined in terms of the ranges of those permanent random numbers that are to be in sample at that time.

This list is not exhaustive, and any given sample may have more than one of those characteristics. For instance, a sample could be of three stages, with stratification and unequal probability sampling at the first stage, unstratified unequal probability sampling at the second stage, and systematic random sampling with equal probabilities at the third stage. Subsequently, subsamples could be drawn from that sample, converting it into a multiphase multistage sample design.

A *sample estimate* is a statistic produced using sample data that can give users an indication as to the value of a population quantity. Special attention will be paid in this section to estimates of population total and population mean because these loom so large in the responsibilities of national statistical offices, but there are many sample surveys that have more ambitious objectives and may be set up so as to estimate small domain totals, regression and/or correlation coefficients, measures of dispersion, or even conceivably coefficients of heteroskedasticity (measures of the extent to which the variance of the U_i can itself vary with the size of the auxiliary variable X_i).

A *sample estimator* is a prescription, usually a mathematical formula, indicating how estimates of population quantities are to be obtained from the sample survey data.

An *estimation procedure* is a specification as to what sample estimators are to be used in a given sample survey.

A *sample strategy* is a combination of a sample design and an estimation procedure. Given a specific sample strategy, it is possible to work out what estimates can be produced and how accurately those estimates can be made.

One consequence of the fact that two quite disparate inferential approaches can be used to form survey estimators is that considerable care needs to be taken in the choice of notation. In statistical practice generally, random variables are represented by uppercase symbols and fixed numbers by lowercase symbols, but between the two approaches, an observed value automatically changes its status. Specifically, in both approaches, a sample value can be represented as the product of a population value and the inclusion indicator, δ, which was introduced in (3). However, in the prediction-based approach, the population value is a random variable and the inclusion indicator is a fixed number, whereas in the randomization-based approach, it is the inclusion indicator that is the random variable while the population value is a fixed number. There is no ideal way to resolve this notational problem, but we shall continue to denote population values by, say, Y_i or X_i and sample values by $\delta_i Y_i$ or $\delta_i X_i$, as we did in Eq. (3).

3.2. Equal probability sampling with the expansion estimator

In what follows, the sample strategies will first be presented in the context of randomization-based inference, then that of the nearest equivalent in prediction-based inference, and finally, wherever appropriate, there will be a note as to how they can be combined.

3.2.1. Simple random sampling with replacement using the expansion estimator
From a randomization-based standpoint, simple random sampling with replacement (srswr) is the simplest of all selection procedures. It is appropriate for use where (a) the

population consists of units whose sizes are not themselves known, but are known not to differ too greatly amongst themselves, and (b) it has no geographical or hierarchical structure that might be useful for stratification or area sampling purposes. Examples are populations of easily accessible individuals or households, administrative records relating to individuals, households, or family businesses; and franchise holders in a large franchise.

The number of population units is assumed known, say N, and a sample is selected by drawing a single unit from this population, completely at random, n times. Each time a unit is drawn, its identity is recorded, and the unit so drawn is returned to the population so that it stands exactly the same chance of being selected at any subsequent draw as it did at the first draw. At the end of the n draws, the ith population unit appears in the sample v_i times, where v_i is a number between 0 and n, and the sum of the v_i over the population is n.

The typical survey variable value on the ith population unit may be denoted by Y_i. The population total of the Y_i may be written Y. A randomization-unbiased estimator of Y is the *expansion estimator*, namely $\hat{Y} = (N/n) \sum_{i=1}^{N} v_i Y_i$. (To form the corresponding randomization-unbiased estimator of the population mean, $\bar{Y} = Y/N$, replace the expression N/n in this paragraph by $1/n$.)

The randomization variance of the estimator $\hat{Y}$ is $V(\hat{Y}) = (N^2/n) S_{wr}^2$, where $S_{wr}^2 = N^{-1} \sum_{i=1}^{N} (Y_i - \bar{Y})^2$. $V(\hat{Y})$ is in turn estimated randomization-unbiasedly by $(N^2/n) \hat{S}_{wr}^2$, where $\hat{S}_{wr}^2 = N^{-1} \sum_{i=1}^{N} v_i (Y_i - \bar{Y})^2$. (To form the corresponding expressions for the population mean, replace the expression N^2/n throughout this paragraph by $1/n$. Since these changes from population total to population mean are fairly obvious, they will not be repeated for other sampling strategies.) Full derivations of these formulae will be found in most sampling textbooks.

There is no simple prediction-based counterpart to srswr. From the point of view of prediction-based inference, multiple appearances of a population unit add no information additional to that provided by the first appearance. Even from the randomization standpoint, srswr is seldom called for, as simple random sampling without replacement (or srswor) is more efficient. Simple random sampling with replacement is considered here purely on account of its extremely simple randomization variance and variance estimator, and because (by comparison with it) both the extra efficiency of srswor and the extra complications involved in its use can be readily appreciated.

3.2.2. Simple random sampling without replacement using the expansion estimator

This sample design is identical with srswr, except that instead of allowing selected population units to be selected again at later draws, units already selected are given no subsequent probabilities of selection. In consequence, the units not yet selected have higher conditional probabilities of being selected at later draws. Because the expected number of distinct units included in sample is always n (the maximum possible number under srswr), the srswor estimators of population total and mean have smaller variances than their srswr counterparts. A randomization-unbiased estimator of Y is again $\hat{Y} = (N/n) \sum_{i=1}^{N} v_i Y_i$, but since under srswor the v_i take only the values 0 and 1, it will be convenient hereafter to use a different symbol, δ_i, in its place.

The randomization variance of the estimator $\hat{Y}$ is $V(\hat{Y}) = (N - n)(N/n)S^2$, where $S^2 = (N-1)^{-1} \sum_{i=1}^{N} (Y_i - \bar{Y})^2$. The variance estimator $V(\hat{Y})$ is in turn estimated randomization-unbiasedly by $(N - n)(N/n)\hat{S}^2$, where $\hat{S}^2 = (n - 1)^{-1} \sum_{i=1}^{N} \delta_i$ $(Y_i - \hat{\bar{Y}})^2$. The substitution of the factor N^2 (in the srswr formulae for the variance and the unbiased variance estimator) by the factor $N(N - n)$ (in the corresponding srswor formulae) is indicative of the extent to which the use of sampling without replacement reduces the variance.

Note, however, that the sampling fraction, n/N, is not particularly influential in reducing the variance, even for srswor, unless n/N is an appreciable fraction of unity. An estimate of a proportion obtained from an srswor sample of 3000 people in, say, Wales, is not appreciably any more accurate than the corresponding estimate obtained from a sample of 3000 people in the United States; and this is despite the proportion of Welsh people in the first sample being about 1 in 1000 and the proportion of Americans in the second being only 1 in 100,000. For thin samples like these, such variances are to all intents and purposes inversely proportional to the sample size, and the percentage standard errors are inversely proportional to the square root of the sample size. Full derivations of these formulae will be again be found in most sampling textbooks.

Since srswor is both more efficient and more convenient than srswr, it will be assumed, from this point on, that sampling is without replacement unless otherwise specified. One important variant on srswor, which also results in sampling without replacement, is systematic sampling with equal probabilities, and this is the next sampling design that will be considered.

3.2.3. Systematic sampling with equal probabilities, using the expansion estimator

Systematic sampling, by definition, is the selection of sample units from a comprehensive list using a constant skip interval between neighboring selections. If, for instance, the skip interval is 10, then one possible systematic sample from a population of 104 would consist of the second unit in order, then the 12th, the 22nd, etc. up to and including the 102nd unit in order. This sample would be selected if the starting point (usually chosen randomly as a number between 1 and the skip interval) was chosen to be 2. The sample size would then be 11 units with probability 0.4 and 10 units with probability 0.6, and the expected sample size would be 10.4, or more generally the population size divided by the skip interval.

There are two important subcases of such systematic selection. The first is where the population is deliberately randomized in order prior to selection. The only substantial difference between this kind of systematic selection and srswor is that in the latter case, the sample size is fixed, whereas in the former it is a random variable. Even from the strictest possible randomization standpoint, however, it is possible to consider the selection procedure as conditioned on the selection of the particular random start (in this case 2), in which case the sample size would be fixed at 10 and the srswor theory would then hold without any modification. This conditional randomization theory is used very commonly, and from a model-assisted point of view it is totally acceptable.

That is emphatically not true, however, for the second subcase, where the population is not deliberately randomized in order prior to selection. Randomization theory in that subcase is not appropriate and it could be quite dangerous to apply it. In an extreme case, the 104 units could be soldiers, and every 10th one from the 3rd onwards could be a sergeant, the remainder being privates. In that case, the sample selected above would consist entirely of privates, and if the random start had been three rather than two, the sample would have been entirely one of sergeants. This, however, is a rare and easily detectable situation within this nonrandomized subcase. A more likely situation would be one where the population had been ordered according to some informative characteristic, such as age. In that instance, the sample would in one sense be a highly desirable one, reflecting the age distribution of the population better than by chance. That would be the kind of sample that the early pioneers of survey sampling would have been seeking with their purposive sampling, one that reflected in miniature the properties of the population as a whole.

From the randomization standpoint, however, that sample would have had two defects, one obvious and one rather more subtle. Consider a sample survey aimed at estimating the level of health in the population of 104 persons as a whole. The obvious defect would be that although the obvious estimate based on the systematic sample would reflect that level considerably more accurately than one based on a random sample would have done, the randomization-based estimate of its variance would not provide an appropriate measure of its accuracy.

The more subtle defect is that the randomization-based estimate of its variance would in fact tend to overestimate even what the variance would have been if a randomized sample had been selected. So the systematic sample would tend to reduce the actual variance but slightly inflate the estimated variance! (This last point is indeed a subtle one, and most readers should not worry if they are not able to work out why this should be. It has to do with the fact that the average squared distance between sample units is slightly greater for a systematic sample than it is for a purely random sample.)

In summary, then, systematic sampling is temptingly easy to use and in most cases will yield a better estimate than a purely randomized sample of the same size, but the estimated variance would not reflect this betterment, and in some instances a systematic sample could produce a radically unsuitable and misleading sample. To be on the safe side, therefore, it would be advisable to randomize the order of the population units before selection and to use the srswor theory to analyze the sample.

3.2.4. Simple prediction inference using the expansion estimator

Simple random sampling without replacement does have a prediction-based counterpart. The appropriate prediction model is the special case of Eqns. (1) in which all the X_i take the value unity. The prediction variances of the U_i in (1c) are in this instance all the same, at σ^2. Because this very simple model is being taken as an accurate reflection of reality, it would not matter, in theory, how the sample was selected. It could (to take the extreme case) be a "convenience sample" consisting of all the people in the relevant defined category whom the survey investigator happened to know personally, but of course, in practice, the use of such a "convenience sample" would make the assumptions underlying the equality of the X_i very hard to accept. It would be much more convincing if the sample were chosen randomly from a carefully compiled list, which would then be an srswor sample, and it is not surprising that the formulae relevant to this form of prediction sampling inference should be virtually identical to those for randomization sampling srswor.

The minimum-variance prediction-unbiased estimator of Y under the simple prediction model described in the previous paragraph is identical with the randomization-unbiased estimator under srswor, namely $\hat{Y} = (N/n) \sum_{i=1}^{N} \delta_i Y_i$. Further, the prediction variance of $\hat{Y}$ is $V(\hat{Y}) = (N - n)(N/n)\sigma^2$. A prediction-unbiased estimator of $V(\hat{Y})$ is $\hat{V}(\hat{Y}) = (N - n)(N/n)\hat{\sigma}^2$, where $\hat{\sigma}^2 = (n - 1)^{-1} \sum_{i=1}^{N} (\delta_i Y_i - \hat{\bar{Y}})^2$ where $\hat{\bar{Y}} = \hat{Y}/N$. Note that although the prediction variance is typically sample-specific, in this instance it is the same for all samples. However, the estimated prediction variance does, as always, vary from sample to sample.

3.3. Equal probability sampling with the ratio estimator

So far, we have been using estimators that depend only on the sample observations Y_i themselves. More often than not, however, the sampling statistician has at hand relevant auxiliary information regarding most of the units in the population. We have already noted that Laplace, back at the turn of the 19th century, had access (at least in principle) to annual birth registration figures that were approximately proportional to the population figures that he was attempting to estimate. To take a typical modern example, the population for a Survey of Retail Establishments (shops) would typically consist mainly of shops that had already been in existence at the time of the most recent complete Census of Retail Establishments, and the principal information collected at that Census would have been the sales figures for the previous calendar or financial year. Current sales would, for most establishments and for a reasonable period, remain approximately proportional to those Census sales figures.

Returning to the model of Eqns. (1), we may equate the Y_i with the current sales of the sample establishments, the X_i with the Census sales of the sample and nonsample establishments, and the X with the total Census sales over all sample and nonsample establishments combined. It may be remembered that "Centrifuge's" ratio estimators worked well both when the model of Eqns. (1) was a useful one and also in the weaker situation when there was a comparatively modest correlation between the Y_i and the X_i. In a similar fashion, the corresponding ratio estimator for this Survey of Retail Establishments tends to outperform the corresponding expansion estimator, at least until it is time to conduct the next Census of Retail Establishments, which would typically be some time in the next 5–10 years.

It was stated above that the population for a Census of Retail Establishments would typically consist mainly of shops that had already been in existence at the time of the most recent complete Census. Such shops would make up the "Main Subuniverse" for the survey. In practice, there would usually be a substantial minority of shops of which the existence would be known, but which had not been in business at the time of that Census, and for these there would be a separate "New Business Subuniverse," which for want of a suitable auxiliary variable would need to be estimated using an expansion estimator, and in times of rapid growth there might even be an "Unlisted New Business Provision" to allow for the sales of shops that were so new that their existence was merely inferred on the basis of previous experience. Nevertheless, even then, the main core of the estimate of survey period sales would still be the sales of shops in the Main Subuniverse, these sales would be based on Ratio Estimation, and the relevant Ratio Estimator would be the product of the $\hat{\beta}$ of Eq. (2) and the Total of Census Sales X.

The modern way of estimating the variance of that ratio estimator depends on whether the relevant variance to be estimated is the randomization variance, which is based on the variability of the estimates over all possible samples, or whether it is the prediction variance, which is sample specific. (For a discussion of the difference between the randomization and prediction approaches to inference, the reader may wish to refer back to Sections 1.3 and 1.4.)

The most common practice at present is to estimate the randomization-variance, and for that the procedure is as follows: denote the population total of the Y_i by Y, its expansion estimator by $\hat{Y}$, and its ratio estimator by $\hat{Y}_R$. Then the randomization variance of $\hat{Y}_R$ is approximated by

$$V(\hat{Y}_R) \approx V(\hat{Y}) + \beta^2 V(\hat{X}) - 2\beta C(\hat{Y}, \hat{X}), \tag{6}$$

where β is the same parameter as in Eq. (1a), $V(\hat{Y})$ is the randomization variance of the expansion estimator of Y, $V(\hat{X})$ is the variance of the corresponding expansion estimator of X, based on the same sample size, and $C(\hat{Y}, \hat{X})$ is the covariance between those two estimators.

The approximate randomization-variance of $\hat{Y}_R$ can therefore be estimated by

$$\hat{V}(\hat{Y}_R) = \hat{V}(\hat{Y}) + \hat{\beta}^2 \hat{V}(\hat{X}) - 2\hat{\beta}\hat{C}(\hat{Y}, \hat{X}), \tag{7}$$

where $\hat{V}(\hat{Y})$ is the randomization-unbiased estimator of $V(\hat{Y})$, given in Subsection 3.2.2, $\hat{V}(\hat{X})$ is the corresponding expression in the X-variable, $\hat{C}(\hat{Y}, \hat{X})$ is the corresponding expression for the randomization-unbiased estimator of covariance between them, namely $(N - n)(N/n) \sum_{i=1}^{N} \delta_i (Y_i - \bar{Y})(X_i - \bar{X})$, and $\hat{\beta}$ is the sample estimator of β, as given in Eq. (2).

3.4. Simple balanced sampling with the expansion estimator

An alternative to simple random sampling is simple balanced sampling, which has already been referred to in Section 2.3. When the sample has been selected in such a way as to be balanced on the auxiliary variables X_i, in the way described in that section, the expansion estimator is comparable in accuracy to that section's ratio estimator itself. This is because the expansion estimator based on the balanced sample is then "calibrated" on those X_i. That is to say, the expansion estimate of the total X is necessarily without error; it is exactly equal to X. It is easy to see that in the situation described in the previous subsection, $\hat{Y}_R$ was similarly "calibrated" on the X_i, that is, $\hat{X}_R$ would have been exactly equal to X.

It is a matter of some contention as to whether it is preferable to use simple random sampling and the ratio estimator or simple balanced sampling and the expansion estimator. The choice is basically between a simple selection procedure and a relatively complex estimator on the one hand and a simple estimator with a relatively complex selection procedure on the other. The choice is considered at length in Brewer (1999a). It depends crucially on the prior choice of sampling inference. Those who hold exclusively to randomization for this purpose would necessarily prefer the ratio estimation option. It is only those who are prepared to accept prediction inference, either as an alternative or exclusively, for whom the choice between the two strategies described above would be a matter of taste.

For a further discussion of balanced sampling, see Sections 2.3 and 2.4.

3.5. Stratified random sampling with equal inclusion probabilities within strata

If any kind of supplementary information is available that enables population units to be grouped together in such a way that they are reasonably similar within their groups and reasonably different from group to group, it will usually pay to treat these groups as separate subpopulations, or *strata*, and obtain estimates from each stratum separately. Examples of such groups include males and females, different descriptions of retail outlets (grocers, butchers, other food and drink, clothing, footwear, hardware, etc.), industries of nonretail businesses, dwellings in urban and in rural areas, or in metropolitan and nonmetropolitan areas.

It takes a great deal of similarity to obtain a poorer estimate by stratification, and the resulting increase in variance is almost always trivial, so the default rule is "Use all the relevant information that you have. When in doubt, stratify." There are, however, several exceptions to this rule.

The first is that if there are many such groups, and all the differences between all possible pairs of groups are known to be small, there is little to gain by stratification, and the business of dealing with lots of little strata might itself amount to an appreciable increase in effort. However, this is an extreme situation, so in most cases, it is safer to stick with the default rule. (In any case, do not worry. Experience will gradually give you the feel as to when to stratify and when not to do so.)

The remaining exceptions all relate to stratification by size. Size is an awkward criterion to stratify on because the boundaries between size strata are so obviously arbitrary. If stratification by size has already been decided upon, one useful rule of thumb is that size boundaries such as "under 10,000," "10,000–19,999," "20,000–49,999," "50,000–99,999," "100,000–199,999," and "over 200,000" (with appropriate adjustments to take account of the scale in which the units are measured) are difficult to improve on appreciably. Moreover, there is unlikely to be much gain in forming more than about six size strata.

Another useful rule of thumb is that each stratum should be of about the same order of magnitude in its total measure of size. This rule can be particularly helpful in choosing the boundary between the lowest two and that between the highest two strata. Dalenius (1957) does give formulae that enable optimum boundaries between size strata to be determined, but they are not recommended for general use, partly because they are complicated to apply and partly because rules of thumb and common sense will get sufficiently close to a very flat optimum. A more modern approach may be found in Lavallée and Hidiroglou (1988).

Finally, there is one situation where it might very well pay not to stratify by size at all, and that is where PRN sampling is being used. This situation will be seen later (in Section 3.9).

3.5.1. Neyman and optimal allocations of sample units to strata

Another important feature of stratification is that once the strata themselves have been defined, there are some simple rules for allocating the sample size efficiently among them. One is "Neyman allocation," which is another piece of sampling methodology recommended by Neyman in his famous 1934 paper that has already been mentioned several times. The other, usually known as "Optimum allocation," is similar to Neyman allocation but also allows for the possibility that the cost of observing the value of a sample unit can differ from stratum to stratum.

Neyman allocation minimizes the variance of a sample estimate subject to a given total sample size.[3] Basically, the allocation of sample units to a stratum h should be proportional to $N_h S_h$, where N_h is the number of population units in the hth stratum and S_h is the relevant population standard deviation in that stratum.[4]

Optimum allocation is not very different. It minimizes the variance of a sample estimate subject to a given total cost and consequently allocates units in a stratum to sample proportionally to $N_h S_h / \sqrt{C_h}$, where C_h is the cost of obtaining the value Y_i for a single sample unit in the hth stratum. Since, however, it is typically more difficult to gather data from small businesses than from large ones, the effect of using Optimal rather than Neyman allocation for business surveys is to concentrate the sample toward the larger units.

Strangely, Optimum allocation seems seldom to have been used in survey practice. This is partly, perhaps, because it complicates the sample design, partly because (for any given level of accuracy) it results in the selection of a larger sample, and partly because it is not often known how much more expensive it is to collect data from smaller businesses.

3.5.2. Stratification with ratio estimation

Since the effect of stratification is effectively to divide the population into a number of subpopulations, each of which can be sampled from and estimated for separately, it is theoretically possible to choose a different selection procedure and a different estimator for each stratum. However, the arguments for using a particular selection procedure and a particular estimator are usually much the same for each stratum, so this complication seldom arises.

A more important question that does frequently arise is whether or not there is any point in combining strata for estimation purposes. This leads to the distinction between "stratum-by-stratum estimation" (also known as "separate stratum estimation") and "across-stratum estimation" (also known as "combined stratum estimation"), which will be the principal topic of this subsection.

The more straightforward of these two options is stratum-by-stratum estimation, in which each stratum is regarded as a separate subpopulation, to which the observations in other strata are irrelevant. The problem with this approach, however, is that in the randomization approach the ratio estimator is biased, and the importance of that bias, relative to the corresponding standard error, can be large when the sample size is small. It is customary in some statistical offices to set a minimum (say six) to the sample size for any stratum, but even for samples of six, it is possible for the randomization bias

---

[3] We are indebted to Gad Nathan for his discovery that Tschuprow (or Chuprov) had actually published the same result in 1923, but his result was buried in a heap of less useful mathematics. Also, it was Neyman who brought it into prominence, and he would presumably have devised it independently of Tschuprow in any case.

[4] A special allowance has then to be made for those population units that need to be completely enumerated, and the question as to what is the relevant population standard deviation cannot be answered fully at this point, but readers already familiar with the basics of stratification are referred forward to Subsection 3.5.2

to be appreciable, so the assumption is made that the estimation of the parameter β in Eq. (1a) should be carried out over all size strata combined. That is to say, the value of β is estimated as the ratio of the sum over the strata of the expansion estimates of the survey variable y to the sum over the strata of the expansion estimates of the auxiliary variable x. This is termed the across-stratum ratio estimator of β, and the product of this with the known sum over all sampled size strata of the auxiliary variable X is termed the across-stratum estimator of the total Y of the survey variable y.

This across-stratum ratio estimator, being based on a larger effective sample size than that of any individual stratum, has a smaller randomization bias than the stratum-by-stratum ratio estimator, but because the ratio of y to x is being estimated over all size strata instead of separately for each, there is the strong probability that the randomization variance of the across-stratum ratio estimator will be greater than that of the stratum-by-stratum ratio estimator. Certainly, the estimators of variance yield larger estimates for the former than the latter. So there is a trade-off between unestimated (but undoubtedly real) randomization bias, and estimated randomization variance.

When looked at from the prediction approach, however, the conclusion is quite different. If the prediction models used for the individual size strata have different parameters β_h, say, where h is a stratum indicator, then it is the across-stratum ratio estimator that is now biased (since it is estimating a nonexistent common parameter β) while the stratum-by-stratum ratio estimator (since it relies on small sample sizes for each) may have the larger prediction variance. If however, the prediction models for the different size strata have the same parameter β in common, the stratum-by-stratum ratio estimator is manifestly imprecise, since it is not using all the relevant data for its inferences, and even the across-stratum ratio estimator, while prediction-unbiased, is not using the prediction-optimal weights to estimate the common parameter β.

It therefore appears that looked at from either approach, the choice between these two estimators is suboptimal, and if viewed from both approaches simultaneously, it would usually appear to be inconclusive. The underlying fact is that stratification by size is at best a suboptimal solution to the need for probabilities of inclusion in sample to increase with the size of the population unit. We shall see later (Section 3.9) that a more logical approach would be to avoid using size as an axis of stratification entirely and to use unequal probabilities of inclusion in sample instead. While this does involve certain complications, they are nothing that high-speed computers cannot cope with, whereas the complications brought about by frequent transitions from one size stratum to another within the framework of PRN sampling are distinctly less tractable.

3.6. Sampling with probabilities proportional to size with replacement

As we have just seen, there are now serious arguments for using Unequal Probability Sampling within the context of surveys (chiefly establishment surveys) for which the norm has long been stratification by size and equal inclusion probabilities within strata. However, the genesis of unequal probability sampling, dating from Hansen and Hurwitz (1943), occurred in the very different context of area sampling for household surveys. The objective of Hansen and Hurwitz was to establish a master sample for the conduct of household surveys within the continental United States. It was unreasonable to contemplate the construction of a framework that included every household in the United States.[5]

Because of this difficulty, Hansen and Hurwitz instead constructed a multistaged framework. They started by dividing the United States into geographical strata, each containing roughly the same number of households. Within each stratum, each household was to have the same probability of inclusion in sample and to make this possible the selection was carried out in stages. The first stage of selection was of Primary Sampling Units (PSUs), which were relatively large geographical and administrative areas. These were sometimes counties, sometimes amalgamations of small counties, and sometimes major portions of large counties.

The important fact was that it was relatively easy to make a complete list of the PSUs within each stratum. However, it was not easy to construct a complete list of PSUs that were of more or less equal size in terms of numbers of households (or dwellings or individuals, whatever was the most accessible measure of size). Some were appreciably larger than others, but the intention remained that in the final sample, each household in the stratum would have the same probability of inclusion as every other household. So Hansen and Hurwitz decided that they would assign each PSU in a given stratum a measure of size; that the sum of those measures of size would be the product of the sample interval (or "spacing interval" or "skip interval") i and the number of PSUs to be selected from that stratum, say n, which number was to be chosen beforehand. Then, a random number r would be chosen between one and the sample interval, and the PSUs selected would be those containing the size measures numbered $r, r + i, r + 2i \ldots r + (n - 1)i$ (see Table 1).

Clearly, the larger the size of a PSU, the larger would be its probability of inclusion in sample. To ensure that the larger probability of selection at the first stage did not translate into a larger probability of inclusion of households at the final stage, Hansen and Hurwitz then required that the product of the probabilities of inclusion at all subsequent stages was to be inversely proportional to the probability of selection at the first stage. So at the final stage of selection (Hansen and Hurwitz contemplated up to three such stages), the population units were individual households and each had the same eventual probability of inclusion in sample as every other household in the stratum.

To ease the estimation of variance, both overall and at each stage, Hansen and Hurwitz allowed it to proceed as though selection had been with replacement at each stage. Since the inclusion probabilities, even at each stage, were comparatively small, this was a reasonable approximation. One of the great simplifications was that the overall variance, the components from all stages combined, could be estimated as though there had been only a single stage of selection. Before the introduction of computers, this was a brilliant simplification, and even today the exact estimation of variance when sampling is without replacement still involves certain complications, considered in Section 3.7.

---

[5] Conceptually, it might be easier to think of this as a list of every dwelling. In fact, the two would have been identical since the definition of a dwelling was whatever a household was occupying, which might for instance be a share of a private house. A household in turn was defined as a group of people sharing meals on a regular basis.

Table 1
Example of PSU selection with randomized listing

Sample fraction 1/147		Number of sample PSUs 2		Cluster size 32.8	
PSU No.	No. of Dwellings	No. of Clusters	Cumulated Clusters	Selection Number	Within-PSU Sample Fraction
1	1550	47	47		
10	639	20	67		
7	728	22	89		
5	1055	32	121	103	1/32
9	732	22	143		
2	911	28	171		
6	553	17	188		
3	1153	35	223		
4	1457	44	267	250	1/44
8	873	27	294		
Total	9651	294			

Note: The number of clusters in PSU number 10 has been rounded up from 19.48 to 20 in order for the total number of clusters to be divisible by 147. Note also that the selection number 103 lies in the interval between 90 and 121 while the selection number 250 lies in the interval between 224 and 267.

3.7. Sampling with unequal probabilities without replacement

The transition from sampling with replacement to sampling without replacement was reasonably simple for simple random sampling but that was far from the case for sampling with unequal probabilities. The first into the field were Horvitz and Thompson (1952). Their estimator is appropriately named after them as the Horvitz-Thompson Estimator or HTE. It is simply the sum over the sample of the ratios of each unit's survey variable value (y_i for the ith unit) to its probability of inclusion in sample (π_i). The authors showed that this estimator was randomization unbiased. They also produced a formula for its variance and a (usually unbiased) estimator of that variance. These last two formulae were functions of the "second-order inclusion probabilities," that is, the probabilities of inclusion in sample of all possible pairs of population units. If the number of units in the population is denoted by N, then the number of possible pairs is $N(N-1)/2$, so the variance formula involved a summation over $N(N-1)/2$ terms, and even the variance estimation formula required a sum over $n(n-1)/2$ pairs of sample units.

Papers by Sen (1953) and by Yates and Grundy (1953) soon followed. Both of these made use of the fact that when the selection procedure ensured a sample of predetermined size (n units), the variance was both minimized in itself and capable of being estimated much more accurately than when the sample size was not fixed. Both papers arrived at the same formulae for the fixed-sample-size variance and for an estimator of that variance that was randomization unbiased, provided that the joint inclusion probabilities, π_{ij}, for all possible pairs of units were greater than zero. However, this Sen–Yates–Grundy variance estimator still depended on the $n(n-1)/2$ values of the π_{ij} so that the variance could not be estimated randomization-unbiasedly without evaluating this large number of joint inclusion probabilities.

Many without-replacement selection schemes have been devised in attempts to minimize these problems. One of the earliest and simplest was randomized systematic sampling, or "RANSYS," originally described by Goodman and Kish (1950). It involved randomizing the population units and selecting systematically with a skip interval that was constant in terms of the size measures. After 1953, dozens of other methods followed in rapid succession. For descriptions of these early methods, see Brewer and Hanif (1982) and Chaudhury and Vos (1988). However, it seemed to be generally true that if the sample was easy to select, then the inclusion probabilities were difficult to evaluate, and the converse also holds.

Poisson sampling (Hájek, 1964) is one such method that deserves a special mention. Although in its original specification, it did not ensure samples of fixed size, it did have other interesting properties. To select a Poisson sample, each population in turn is subjected to a Bernoulli trial, with the probability of "success" (inclusion in sample) being π_i, and the selection procedure continues until the last population unit has been subjected to its trial. The achieved sample sizes are, however, highly variable, and consequently, Poisson sampling in its original form was not an immediately popular choice. However, several modified versions were later formulated; several of these and also the original version are still in current use.

One of the most important of these modified versions was Conditional Poisson Sampling or CPS, also found in Hájek (1964) and discussed in detail by Chen et al. (1994). For CPS, Poisson samples with a particular expected sample size are repeatedly selected, but only to be immediately rejected once it is certain that the eventual sample will not have exactly that expected sample size. One notable feature of CPS is that it has the maximum entropy attainable for any population of units having a given set of first-order inclusion probabilities π_i.[6] Several fast algorithms for using CPS are now available, in which the second-order inclusion probabilities are also computed exactly. See Tillé (2006).

In the meantime, however, another path of investigation had also been pioneered by Hájek (1964). He was concerned that the estimation of variance for the HTE was unduly complicated by the fact that both the Sen–Yates–Grundy formula for the randomization variance and their estimator of that variance required knowledge of the second-order inclusion probabilities. In this instance, Hájek (and eventually others) approximated the fixed sample size variance of the HTE by an expression that depended only on the first-order inclusion probabilities. However, initially these approximations were taken to be specific to particular selection procedures. For instance, Hájek's 1964 approximation was originally taken to be specific to CPS.

In time, however, it was noted that very different selection procedures could have almost identical values of the π_{ij}. The first two for which this was noticed were RANSYS, for which the π_{ij} had been approximated by Hartley and Rao (1962), and the Rao–Sampford selection procedure (J.N.K. Rao, 1965; Sampford, 1967), for which

---

[6] Entropy is a measure of unpredictability or randomness. If a population is deliberately arranged in order of size and a sample is selected from it systematically, that sample will have low entropy. If however (as with RANSYS) the units are arranged in random order before selection, the sample will have high entropy, only a few percentage points smaller than that of CPS itself. While low entropy sample designs may have very high or very low randomization variances, high entropy designs with the same set of first-order inclusion probabilities all have more or less the same randomization variance. For a discussion of the role of entropy in survey sampling, see Chen et al. (1994).

they had been approximated by Asok and Sukhatme (1976). These were radically different selection procedures, but the two sets of approximations to the π_{ij} were identical to order n^3/N^3. Although both procedures produced fixed size samples, and the population units had inclusion probabilities that were exactly proportional to their given measures of size, it appeared that the only other thing that the two selection procedures had in common was that they both involved a large measure of randomization. Entropy, defined as $\sum_{k=1}^{M} [P_k - log(P_k)]$, where P_k is the probability of selecting the kth out of the M possible samples, is a measure of the randomness of the selection. It therefore appeared plausible that all high-entropy sampling procedures would have much the same sets of π_{ij}, and hence much the same randomization variance. If so, it followed that approximate variance formulae produced on the basis of any of these methods would be valid approximations for them all, and that useful estimators of these approximate variances would be likely also to be useful estimators of the variances of the HTE for all high-entropy selection procedures.

Whether this is the case or not is currently a matter of some contention, but Preston and Henderson (2007) provide evidence to the effect that the several randomization variance estimators provided along these lines are all reasonably similar in precision and smallness of bias, all at least as efficient as the Sen–Yates–Grundy variance estimator (as measured by their randomization mean squared errors MSEs), and all a great deal less cumbersome to use.

In addition, they can be divided into two families, the members of each family having both a noticeable similarity in structure and a detectable difference in entropy level from the members of the other family. The first family includes those estimators provided by Hájek (1964), by Deville (1993, 1999, 2000; see also Chen et al., 1994) initially for CPS, and by Rosn for Pareto πps (Rosén, 1997a, 1997b). The second family, described in Brewer and Donadio (2003), is based on the π_{ij} values associated with RANSYS and with the Rao–Sampford selection procedure. These two procedures have slightly smaller entropies and slightly higher randomization variance than CPS, but both Preston and Henderson (2007) and Henderson (2006) indicate that the Hájek-Deville family of estimators should be used for CPS, Pareto πps and similar selection procedures—thus probably including Tillé (1996)—while the Brewer-Donadio family estimators would be appropriate for use with RANSYS and Rao-Sampford.

It is also possible to use replication methods, such as the jackknife and the bootstrap, to estimate the HTE's randomization variance. The same Preston and Henderson paper provides evidence that a particular version of the bootstrap can provide adequate, though somewhat less accurate, estimates of that variance than can be obtained using the two families just described.

Finally, it is of interest that the "anticipated variance" of the HTE (that is to say the randomization expectation of its prediction variance, or equivalently the prediction expectation of its randomization variance; see Isaki and Fuller, 1982) is a simple function of the π_i and independent of the π_{ij}. Hence, for any population that obeys the model of Eqns. (1), both the randomization variance and the anticipated variance of the HTE can be estimated without any reference to the π_{ij}.

3.8. The generalized regression estimator

Up to this point, it has been assumed that only a single auxiliary variable has been available for improving the estimation of the mean or total of a survey variable. It has also been assumed that the appropriate way to use that auxiliary variable was by using Eq. (1a), which implies a ratio relationship between those two variables. More generally, the survey variable could depend on a constant term as well, or on more than a single auxiliary variable, or both. However, that relationship is seldom likely to be well represented by a model that implies the relevance of ordinary least squares (OLS).

One case where OLS might be appropriate is where the survey variable is Expenditure and the auxiliary variable is Income. The relationship between Income and Expenditure (the Consumption Function) is well known to involve an approximately linear dependence with a large positive intercept on the Expenditure axis. But OLS assumes homoskedasticity (the variance of Expenditure remaining constant as Income increases) while it is more than likely that the variance of Expenditure increases with Income, and in fact the data from the majority of sample surveys do indicate the existence of a measure of heteroskedaticity. This in itself is enough to make the use of OLS questionable. Eq. (1c) allows for the variance of the survey variable to increase linearly with the auxiliary variable, and in fact it is common for this variance to increase somewhat faster than this, and occasionally as fast as the square of the auxiliary variable.

A commonly used estimator of total in these more general circumstances is the generalized regression estimator or GREG (Cassel et al., 1976), which may be written as follows:

$$\hat{Y}_{\text{GREG}} = \hat{Y}_{\text{HTE}} + \sum_{k=1}^{p} (X_k - \hat{X}_{\text{HTE}k})\hat{\beta}_k, \tag{8}$$

or alternatively as

$$\hat{Y}_{\text{GREG}} = \sum_{k=1}^{p} X_k \hat{\beta}_k + \left(\hat{Y}_{\text{HTE}} - \sum_{k=1}^{p} \hat{X}_{\text{HTE}k} \hat{\beta}_k \right). \tag{9}$$

In these two equations, $\hat{Y}_{\text{HTE}}$ is the HTE of the survey variable, $\hat{X}_{\text{HTE}k}$ is the HTE of the kth auxiliary variable and $\hat{\beta}_k$ is an estimator of the regression coefficient of the survey variable on the kth auxiliary variable, where the regression is on p auxiliary variables simultaneously. One of those auxiliary variables may be a constant term, in which case there is an intercept estimated in the equation. (In that original paper, $\hat{\beta}_k$ was a generalized least squares estimator, but this was not a necessary choice. For instance, Brewer (1999b) defined $\hat{\beta}_k$ in such a way as to ensure that the GREG was simultaneously interpretable in the randomization and prediction approaches to sampling inference, and also showed that this could be achieved with only trivial increments to its randomization and prediction variances).

In the second of these two equations, the first term on the right-hand side is a prediction estimator of the survey variable total, but one that ignores the extent to which the HTE of survey variable total differs from the sum of the p products of the individual auxiliary variable HTEs with their corresponding regression estimates. Särndal et al. (1992) noted that the first term (the prediction estimator) had a randomization variance that was of a lower order of magnitude than the corresponding variance of the second term and therefore suggested that the randomization variance of the GREG estimator be estimated by estimating only that of the second term. It is true that as the sample size increases, the randomization variance of the prediction estimator becomes small with respect to that of the second term, but when the sample size is small, this can lead to a substantial underestimate of the GREG's randomization variance.

This is not an easy problem to solve wholly within the randomization approach, and in Chapter 8 of Brewer (2002, p. 136), there is a recommendation to estimate the anticipated variance as a substitute. (The anticipated variance is the randomization expectation of the prediction variance.). This is obviously not a fully satisfactory solution, except in the special case considered by Brewer, where the GREG had been devised to be simultaneously a randomization estimator and a prediction estimator, so more work on it seems to be called for. Another alternative would be to estimate the GREG's randomization variance using a replication method such as the jackknife or the bootstrap, but again this alternative appears to need further study. For more information regarding the GREG, see Särndal et al. (1992).

3.9. Permanent random number (PRN) sampling

One of the important but less obvious objectives of survey sampling is to be able to control intelligently the manner in which the sample for a repeating survey is allowed to change over time. It is appropriate for a large sample unit that is contributing substantially to the estimate of total to remain in sample for fairly long periods, but it is not so appropriate for small population units to do the same, so it is sensible to rotate the sample around the population in such a way that the larger the unit is, the longer it remains in sample. One of the ways of doing this is to assign each unit a PRN, say between zero and unity, and define the sample as consisting of those population units that occupy certain regions of that PRN space. Units in a large-size stratum might initially be in sample if they had PRNs between zero and 0.2 for the initial survey, between 0.02 and 0.22 for the second, 0.04 and 0.24 for the third, and so on. In this way, each unit would remain in sample for up to 10 occasions but then be "rested" for the next 40. Those in a small-size stratum would remain occupy a smaller region of the PRN space, say initially between zero and 0.04, but the sample PRN space would be rotated just as fast so that units would remain in sample for no more than two occasions before being "rested."

From the data supplier's point of view, however, it is particularly inappropriate to be removed from the sample and then included again shortly afterwards. This can easily happen, however, if a population unit changes its size stratum, particularly if the change is upward. Consequently, it is inconvenient to use PRN sampling and size stratification together. Moreover, as has already been indicated in Section 3.5, stratification by size is a suboptimal way of satisfying the requirement that the larger the unit, the greater should be its probability of inclusion in sample.

Hence, when attempting to control and rotate samples using the PRN technique, it becomes highly desirable, if not indeed necessary, to find a better solution than stratification by size. Brewer (2002) (Chapter 13, pp. 260–265), provides a suggestion as to how this could be done. It involves the use of a selection procedure known as Pareto πps sampling, which is due to Rosén (1997a, 1997b). This is a particular form of what is known as *order sampling*, and is very similar in its π_{ij} values to CPS sampling, so it is a high-entropy sample selection procedure. It is, however, somewhat complicated to describe and therefore inappropriate to pursue further in this introductory chapter. Those who wish to pursue the possibility of using PRN sampling without stratification by size are referred to those two papers by Rosén and to Chapter 13 of Brewer (2002).

4. Conclusion

From the very early days of survey sampling, there have been sharp disagreements as to the relative importance of the randomization and prediction approaches to survey sampling inference. These disagreements are less severe now than they were in the 1970s and 1980s but to some extent they have persisted into the 21st century. What is incontrovertible, however, is that prediction inference is parametric and randomization nonparametric. Hence the prediction approach is appropriate to the extent that the prediction models are useful, whereas the randomization approach provides a robust alternative where they are not useful. It would therefore seem that ideally both should be used together, but there are many who sincerely believe the one or the other to be irrelevant. The dialogue therefore continues.

Both the randomization and the prediction approaches offer a wide range of manners in which the sample can or should be selected, and an equally wide range of manners in which the survey values (usually, but not exclusively consisting of population totals, population means, and ratios between them) can be estimated. The choices among them depend to a large extent on the natures of the populations (in particular, whether they consist of individuals and households, of establishments and enterprizes, or of some other units entirely) but also on the experience and the views of the survey investigators. However, there are some questions that frequently need to be asked, and these are the ones that have been focussed on in this chapter. They include, "What are the units that constitute the population?" "Into what groups or strata do they naturally fall?" "What characteristics of the population need to be estimated?" "How large a sample is appropriate?" (or alternatively, "How precise are the estimates required to be?") "How should the sample units be selected?" and "How should the population characteristics be estimated?"

In addition, there are many questions that need to be answered that fall outside the scope of the discipline of survey sampling. A few of them would be as follows: "What information are we seeking, and for what reasons?" "What authority, if any, do we have to ask for this information?" "In what format should it be collected?" "What organizational structure is required?" "What training needs to be given and to whom?" and not least, "How will it all be paid for?"

So those questions that specifically relate to survey sampling always need to be considered in this wider framework. The aim of this Chapter will have been achieved if the person who has read it has emerged with some feeling for the way in which the discipline of survey sampling can be used to fit within this wider framework.

Essential Methods for Design Based Sample Surveys
ISSN: 0169-7161
© 2009 Elsevier B.V. All rights reserved
DOI: 10.1016/B978-0-444-53734-8.00002-9

Designs for Surveys over Time

Graham Kalton

1. Introduction

Most of the literature on survey methodology focuses on surveys that are designed to produce a snapshot of the population at one point in time. However, in practice, researchers are often interested in obtaining a video of the changes that occur over time. The time dimension can be introduced by repeating the survey at different time points or by using some form of panel design. This chapter reviews the *design* choices available for surveys over time and provides a brief overview of many additional methodological complexities that arise. References are provided to direct readers to more detailed treatments of the various topics discussed.

Many surveys aim to estimate characteristics of a population at a specific point in time. For example, the U.S. Census of Population provides a snapshot of the population for April 1 at the beginning of each decade. In practice, of course, data collection for the majority of surveys cannot be conducted on a single day, but the goal is still to represent the population as of that date. Also, often some of the data collected will not relate to that specific date; some retrospective data are generally collected, such as employment status in a given earlier week, illnesses experienced in the past month, and expenditures over the past six months. Nevertheless, the objective of these *cross-sectional surveys* is to collect the data needed for describing and analyzing characteristics of the population as it exists at a point in time.

This focus of cross-sectional surveys on a particular point in time is important because both the characteristics and the composition of a population change over time. The survey estimates are therefore time specific, a feature that is particularly important in some contexts. For example, the unemployment rate is a key economic indicator that varies over time; the rate may change from one month to the next because of a change in the economy (with businesses laying off or recruiting new employees) and/or because of a change in the labor force (as occurs, e.g., at the end of the school year, when school leavers start to seek employment).

Changes in population characteristics over time raise many important issues for study. At one level, policymakers need to estimate population characteristics repeatedly over time to obtain estimates that are as current as possible. They are also interested in the change in the estimates across time: Has the unemployment rate increased or decreased since the previous survey? This change is termed the *net change* and reflects changes in both the characteristics and composition of the population. A more detailed analysis would involve understanding the components of change. To what extent is the change (or lack of change) due to population dynamics, with people entering the population through "births" (e.g., people reaching age 15, for surveys of adults; people who immigrate; or people who leave institutions, for surveys of the noninstitutional population) and leaving the population through "deaths" (e.g., people who die, emigrate, or enter an institution)? To what extent is the change due to changes in the statuses of the persons in the population? Furthermore, how does change operate in cases where there is a change in status? For example, assuming no population dynamics, if the unemployment rate undergoes a net increase of 1 percent, is that because 1 percent of previously employed persons lost their jobs or because, say, 10 percent lost their jobs and 9 percent of the previously unemployed found work? The decomposition of net change into its two components leads to a measure of *gross change*. While net change can be measured from separate samples for the two occasions, measuring gross change requires repeated measurements on the same sample, or at least a representative subsample.

There are two broad classes of objectives for surveys across time, and these give rise to different approaches to survey design. In many cases, the objectives are restricted to estimating population parameters at different time points and to estimating net changes and trends. This class of objectives also includes the estimation of average values of population parameters over a period of time. None of these objectives requires repeated measurements on the same sample. They can all be achieved by collecting the survey data from representative cross-sectional samples of the survey population at different time points. Satisfying these objectives imposes no restrictions on the relationships between the samples at different time points. In particular, these objectives can be met with samples selected entirely independently at each time point. They can also be met with samples that are constructed to minimize sample overlap across time to spread the respondent burden over different sample elements. Such *repeated surveys* are discussed in Section 2.

This first class of objectives can also be met with panel designs that include some or all of the sample members at different time points. In fact, the precision of cross-sectional and net change estimates can be improved using a *rotating panel* sample design that creates some degree of sample overlap over time. Rotating panel designs may also be used to eliminate telescoping effects that occur when respondents erroneously report an event as occurring in a given interval of time. Rotating panel designs are discussed in Section 3.

The second class of objectives focuses primarily on the estimation of gross change and other components of individual change, and on the aggregation of responses (e.g., expenditures) for individuals over time. These objectives can be satisfied only by some form of panel survey that collects data from the same individuals for the period of interest. Issues involved in conducting various types of *panel*—or *longitudinal*—*surveys* are discussed in Section 4.

Other objectives for surveys over time relate to the production of estimates for rare populations (i.e., a subset of the general population that has a rare characteristic). One such objective is to accumulate a sample of cases with the rare characteristic over time. If the characteristic is an event, such as getting divorced, then this objective can be satisfied by any of the designs. However, if the characteristic is stable, such as being a member of a rare racial group, accumulation works only when fresh samples are added over time. In either case, analysts need to recognize that the characteristics of members of the rare population may vary over time. For example, in a survey of recent divorcees, the economic consequences of the divorce may change over the period of sample aggregation.

A different objective with a rare population is to produce estimates for that population at various points in time. If the rare characteristic is a stable one, it may be economical to identify a sample of members of that population at an initial time point and then return to that sample repeatedly in a panel design. This approach has been used, for instance, in sampling graduate scientists and engineers in the U.S. Scientists and Engineers Statistical Data System (SESTAT) (Fecso et al., 2007). An initial sample was created based on data collected in the latest decennial Census of Population, and that sample was treated as a panel to be resurveyed at intervals during the next decade. While this scheme covered all those who were already scientists and engineers and living in the United States at the time of the census, there was a need to add supplemental samples of new U.S. graduates—"births"—as the decade progressed (with a remaining gap for scientists and engineers entering the United States after the census).

The final section of the chapter (Section 5) briefly summarizes the issues to be considered in making a choice of the type of design to adopt for surveying a population over time. It also summarizes the methodological challenges that are to be faced in producing valid findings from surveys over time.

2. Repeated surveys

This section discusses a range of issues and designs for surveys over time when the analytic focus is on the production of a series of cross-sectional estimates that can be used in analyses of net changes and trends at the aggregate level. The designs considered here are not structured to enable longitudinal analyses at the element level.

A common form of repeated survey is one in which separate samples of the ultimate sampling units are selected on each occasion. When the interval between rounds of a repeated survey is long (say, 5–10 years), the selection of entirely independent samples may well be an effective strategy. However, in repeated surveys with multistage sample designs and with shorter intervals between rounds, sizeable benefits may be achieved by retaining the same primary sampling units (PSUs), and perhaps also units at later stages (but not the ultimate units), at each round. Master samples of PSUs are widely used for national household survey programs because of the fieldwork and statistical efficiencies they provide, both for repeated surveys on a given topic and surveys that range over different topics (U.N. Department of Economic and Social Affairs Statistics Division, 2005, Chapter V). Overlapping higher level sampling units also leads to more precise estimates of net change over time. However, a master sample becomes increasingly less statistically efficient over time, as the population changes. When updated frame information becomes available and indicates that substantial population changes have occurred, the need arises to modify the PSUs' measures of size and revise the stratification. (In national household surveys, the availability of new population census results is usually the basis for an update). To address these issues, a variety of methods have been developed to retain as many sampled PSUs as possible in the new sample while updating the measures of size and strata (e.g., Ernst, 1999; Keyfitz, 1951; Kish and Scott, 1971).

With repeated surveys of businesses, a methodology based on some form of permanent random numbers (PRNs) is often used (see Ernst et al., 2000; Ohlsson, 1995; and Chapter 6 in this volume). In essence, the methodology consists of assigning a random number between 0 and 1 to each population element on a list frame. Then a disproportionate stratified sample can be readily selected by including all elements with random numbers less than the sampling fraction in each stratum. The random numbers assigned remain with the elements over time, with the result that an element selected in the sample at a given round of a repeated survey will also certainly be in the samples at all other rounds for which its selection probability is no lower than that of the given round. This flexible procedure automatically covers changes in overall and stratum sample sizes across rounds, elements that change strata, and births and deaths. It is primarily intended to improve the precision of estimates of change across rounds and sometimes to facilitate data collection, but it can also be used to generate a panel sample for a given period. The elements in the sample for all rounds of a given period constitute a probability sample of elements that exist throughout the period, with an element's probability of being in the panel given by the minimum of its selection probabilities across the rounds (Hughes and Hinkins, 1995). The Statistics of Income Division of the U.S. Internal Revenue Service uses a PRN methodology to sample both individual and corporate tax returns and has created panel files from the cross-sectional samples for longitudinal analysis (Dalton and Gangi, 2007). The PRN methodology can also be modified to provide sample rotation to limit respondent burden on sampled businesses, particularly small businesses for which selection probabilities are low. For example, the PRN for each business can be increased by, say, 0.1 on each round and taking the fractional part if the result exceeds 1.

A critical objective for repeated surveys is the production of valid estimates of trends throughout the period of interest, particularly change from one round to the next. However, changes in the survey design are often desirable, and unfortunately even small design changes may affect the estimates. Thus, changes in question wording or questionnaire content, mode of data collection, interviewer training, interviewer field force, sampling frame, coding procedures, and imputation and weighting procedures can all threaten the validity of trend estimates. It is, for example, well documented that changing the questionnaire content can lead to context effects that can distort trend estimates (see, e.g., Biemer et al., 1991; Tourangeau et al., 2000), and even the meanings of identical questions may change over time (see, e.g., Kulka, 1982). Even a major increase in sample size alone can affect the survey estimates because of the need to recruit new interviewers and perhaps because of a reduced level of effort to obtain responses. Those conducting repeated surveys—particularly a long-running series of repeated surveys—are frequently confronted with the dilemma of whether to improve the survey procedures based on experience gained in past rounds, general methodological research, and changes in the population and topics of current interest, or to stay with past methods to maintain valid trend estimates.

When a significant methodological change is found to be necessary, a common practice is to carry out a bridging survey for one or more time periods, that is, to conduct one part of the survey using the old methods and another part using the new methods simultaneously. For example, in preparation for a conversion of the monthly

U.S. Current Population Survey (CPS) from a combination of face-to-face paper-and-pencil interviewing (PAPI) and computer-assisted telephone interviewing (CATI) to a combination of computer-assisted personal interviewing (CAPI) and CATI, the U.S. Census Bureau and Bureau of Labor Statistics (2000) conducted CATI and CAPI overlap experiments in 1992, in which a sample of 12,000 households were interviewed with the CAPI/CATI combination using a revised CPS questionnaire, to provide estimates with the revised methodology that could be compared with those produced by the official CPS. The experiments found that the new questionnaire in combination with the computer-assisted data collection did not significantly affect the estimate of the overall unemployment rate but did affect a number of other estimates, such as the duration of unemployment for the unemployed and the proportion of employees working part-time.

As another example, the U.S. National Household Survey on Drug Abuse (NHSDA, now the National Survey on Drug Use and Health) underwent a major redesign in 1999 to adopt a new method of data collection (a change from PAPI to computer-assisted interviewing, including the use of an electronic screener for respondent selection within sampled households), a different sample design with a much larger sample (and hence the need for a larger interviewer pool), and some other changes. Interviews were conducted with a national sample of approximately 14,000 households using the old PAPI methodology to evaluate the effects of the redesign and to maintain comparable data for analyses of trends. Using this sample for comparison, Gfroerer et al. (2002) provide a detailed analysis of the various effects of the NHSDA redesign.

The data collected in a series of rounds of a repeated survey are sometimes combined to produce larger samples and hence reduce sampling errors, particularly for estimates pertaining to small population subgroups (Kish, 1999). Also, the data collection for a survey may be spread over time to facilitate the fieldwork. In this case, the survey's sample may be built up from a set of replicates, each of which can produce estimates—albeit less precise—for the entire survey population. The U.S. National Health and Nutrition Examination Survey (NHANES), for example, conducts medical examinations in mobile examination centers that travel around the country from one PSU to the next (Mohadjer and Curtin, 2008). The examination centers visit 15 PSUs per year. Each yearly sample is a replicate sample that can be used to produce national estimates, but the estimates generally have low precision. For most analyses, the samples are aggregated over three or six years.

Some repeated surveys are specifically designed to be combined to produce average estimates for a period of time for characteristics that change over time. The U.S. National Health Interview Survey (NHIS) is one such survey (Botman et al., 2000). The NHIS collects health-related data from weekly samples of persons, with the data being aggregated to produce annual estimates. The annual estimate of the two-week prevalence of a seasonal illness, such as influenza, is thus an average prevalence estimate across the year. The NHIS may also be aggregated over longer periods to produce estimates for rare subgroups or over shorter periods to produce estimates for a heavy outbreak of a disease such as measles.

The complexities involved in aggregating data from a series of rounds of a repeated survey are well illustrated by the U.S. Census Bureau's American Community Survey (ACS) (U.S. Census Bureau, 2006c; Citro and Kalton, 2007). A single-stage sample of households is selected each month for the ACS, accumulating to approximately two million responding households per year. The data are aggregated to produce

1-year period estimates for governmental units with populations of 65,000 or more, 3-year period estimates for governmental units with populations of 20,000 or more, and 5-year period estimates for all governmental units, including school districts and census tracts. A user needs to understand the difference between these period estimates and the usual point-in-time estimates, and to carefully assess their applicability for his or her needs.

An ACS period estimate reflects both any changes in the characteristic under study over the period and any changes in an area's population size and composition. Thus, period estimates will differ from point-in-time estimates for characteristics that can change markedly over a period (e.g., unemployment rates). A particular issue here relates to estimates that involve monetary amounts, such as income during the past 12 months, rent or value of the accommodation, or fuel costs last month. When aggregating data over, say, 5 years, should the effect of inflation on changes in such amounts be taken into account? If so, how should this be done? Point-in-time and period estimates will also differ for areas with highly seasonal populations and for areas of rapid growth or decline. The use of period estimates of totals (e.g., the number of persons in poverty) is particularly unclear for areas that experience substantial changes in population size during the period.

3. Rotating panel surveys

Repeated surveys of households may retain the same set of PSUs, second-stage sampling units, and units at other higher stages of sampling from one round to the next, but they select fresh samples of reporting units on each occasion.[1] In contrast, rotating panel surveys are designed to ensure some degree of overlap in the final sample units at specified rounds. However, unlike full panel surveys, not all the same final sample units are retained over all rounds. With a rotating design, each final sample unit remains in the sample for only a limited period of time.

Rotating panel designs are widely used for labor force surveys. In the monthly Canadian Labour Force Survey (LFS), for example, each sampled housing unit is included in the sample for six consecutive months. Each month, one-sixth of the housing units enter the sample and one-sixth of them rotate out of the sample (Statistics Canada, 1998). Table 1 illustrates this rotation scheme over a period of 12 months. The sample in month 1 is made up of six rotation groups, each comprising one-sixth of the overall sample. Rotation group A has been in sample for the previous five months and rotates out in month 2, rotation group B has been in sample for four previous months and will remain in sample in month 2 before rotating out in month 3, and so on.

Various rotating panel designs are used for labor force surveys in different countries. For example, the quarterly U.K. LFS employs a five-wave rotating panel design, with one-fifth of the sample entering each quarter and one-fifth rotating out (U.K. National Statistics, 2007). The U.S. monthly labor force survey—CPS—employs a more complex rotation scheme. Each sampled housing unit is in sample for eight months, but not

---

[1] Indeed, repeated surveys may be designed to minimize the chance that a reporting unit is selected for rounds that are close in time (as is the case with the U.S. American Community Survey, which ensures that housing units are not selected more than once in a five-year period).

Table 1
An illustration of the Canadian LFS six-month rotation scheme for a 12-month period

Month											
1	2	3	4	5	6	7	8	9	10	11	12
A	G	G	G	G	G	G	M	M	M	M	M
B	B	H	H	H	H	H	H	N	N	N	N
C	C	C	I	I	I	I	I	I	O	O	O
D	D	D	D	J	J	J	J	J	J	P	P
E	E	E	E	E	K	K	K	K	K	K	Q
F	F	F	F	F	F	L	L	L	L	L	L

consecutive months (U.S. Census Bureau and Bureau of Labor Statistics, 2000). A housing unit enters the sample in a given month and stays in sample for the next three months, drops out of sample for 8 months, and then returns for four consecutive months. This scheme is termed a 4-8-4 rotation scheme.

The primary analytic objectives of these labor force surveys are the same as those of repeated surveys: to produce cross-sectional estimates at each time point and to measure net changes over time. As compared with independent samples for each round, a rotating design induces a correlation between estimates at rounds in which there is some sample overlap. Since this correlation is almost always positive, the overlap results in a reduction in the sampling error of estimates of net change. The 4-8-4 rotation pattern in the U.S. CPS, for example, is fashioned to provide substantial overlap from one month to the next and also from a given month in one year to the same month in the next year.

In fact, with a rotating panel design, the precision of estimates of current level and net change can "borrow strength" from the data collected during all previous rounds of the survey, using the technique of composite estimation. See U.S. Census Bureau and Bureau of Labor Statistics (2000) for the application of composite estimation in the CPS and Fuller and Rao (2001) for an investigation into the use of regression composite estimation in the Canadian LFS. See also Binder and Hidiroglou (1988) for more detailed discussions of composite estimation and optimal rotation designs.

As well as improvements in the precision of survey estimates, a rotating design can yield important cost savings because returning to the same housing units is often less expensive than starting afresh. In particular, while the initial interview may need to be conducted face-to-face, subsequent interviews may be conducted by telephone where possible. This procedure is used in the Canadian LFS and the U.K. LFS. In the U.S. CPS, face-to-face interviewing is required for the first and fifth interviews (when a sampled housing unit returns to the sample after a gap of eight months), but other interviews may be conducted by telephone. There is, however, a concern that responses obtained by telephone may not be comparable with those obtained by face-to-face interviewing. See, for example, de Leeuw (2005) on the effects of mixed-mode data collections in general and Dillman and Christian (2005) on the issues of mixing modes across waves of a panel survey.

The LFS conducted by the Australian Bureau of Statistics uses a rotating panel survey design, with sampled dwellings remaining in sample for eight consecutive months. Until 1996, face-to-face interviewing was used on each wave, but then telephone

interviewing was introduced for all interviews except the first. This change was introduced over time by rotation group. Bell (1998) used this balanced feature to analyze
the effects of the change while also taking account of rotation group bias (see below).
He concluded that the change in data collection methods resulted in a transitory effect
on estimates by labor force status, but the effect had almost disappeared by the end of
the phase-in period.

Even apart from a change in mode, there is concern that the responses obtained in
repeated interviews with the same respondents may not be comparable. This effect,
which is termed *panel conditioning*, occurs when responses at later waves of interviewing are affected by the respondents' participation in earlier waves of the survey.
For example, respondents may change their behavior as a result of being sensitized to
the survey's subject-matter and perhaps learning something from the interview (such
as the existence of a welfare program). Some respondents may change their response
behaviors in later interviews, perhaps demonstrating better recall after learning more
about the survey contents, being more motivated to give accurate responses, giving
less-considered responses because they have lost interest, or responding to filter questions in a manner that will avoid lengthy sets of follow-up questions. Respondents
may also seek to be overly consistent in responses to attitude items. See Waterton and
Lievesley (1989) and Sturgis et al. (2009) for a discussion of possible reasons for panel
conditioning effects and Cantor (2007) for an extensive review of the research on panel
conditioning.

Yet another issue with panel designs, in general, is *panel attrition* (see also Section 4). While both cross-sectional and panel surveys are subject to total nonresponse
at the initial wave of data collection, a panel survey also suffers losses at later waves.
Although nonresponse weighting adjustments may help to compensate for panel attrition, the estimates derived from a rotation group that has been in sample for several
waves, with associated panel attrition, may differ for this reason from those derived
from rotation groups that have been in sample for shorter durations.

A well-documented finding in the U.S. CPS and the Canadian LFS is that the survey results for the same time point differ across rotation groups (Bailar, 1975, 1989;
Ghangurde, 1982; U.S. Census Bureau and Bureau of Labor Statistics, 2000). This
effect, which reflects some combination of panel conditioning and attrition effects,
is variously known as *rotation group bias*, *time-in-sample bias*, and *month-in-sample
bias*. The existence of this bias for a given monthly estimate almost certainly implies
that the estimate is biased but, under an additive model of the bias by rotation group,
estimates of month-to-month changes are unbiased provided that the rotation group pattern is balanced at each time point (Bailar, 1975). This balance is not always achieved;
in any case, it does not hold during the start-up period for a rotating panel design.
Also, as Solon (1986) shows, change estimates are biased under a model that includes
a multiplicative bias term.

Rotating panel designs are not restricted to labor force surveys. They are also used
for purposes other than efficiency of fieldwork and improved precision of estimates of
change and level. An important reason for using a rotating panel design in the U.S.
National Crime Survey (now the National Crime Victimization Survey) is for the purpose of *bounding*, to deal with telescoping effects (Cantor, 1989). These effects can
occur when respondents are asked to report events that happened in a given period,
such as victimizations that they have experienced in the past six months or any cars
purchased in the past year. Telescoping occurs when a respondent reports an event as

occurring within the reference period when, in fact, it fell outside the period. See Neter and Waksberg (1964a,b) for a classic study of both panel conditioning and telescoping effects. Telescoping effects can be addressed with a rotating panel survey—indeed, any panel survey—in which the sample is reinterviewed at intervals corresponding to the reference period: events reported at the current wave that were also reported at a previous wave can be discounted because they occurred prior to the current reference period. (See also Section 4.2 for a discussion of dependent interviewing, in which respondents are reminded of their answers to the previous wave.)

Another application of rotating panel designs is when respondents cannot be expected to recall accurately all the information required for a given reference period. They may then be reinterviewed at set intervals to report the information for shorter reference periods, with the information then being aggregated to provide the information required for the full period. The household component of the U.S. Medical Expenditure Panel Survey (MEPS), for example, uses a rotating panel approach to collect data on health status and health care access, utilization, and expenditures; data on panel members are collected over five waves covering 30 months, with a new panel being introduced every year (Ezzati-Rice and Cohen, 2004). The data from two consecutive rotation groups (panels) can be pooled for a given year, and the data aggregated to produce annual estimates. The Survey of Income and Program Participation, described in Section 4, has used a similar approach (Kalton et al., 1998).

A rotating panel design can sometimes provide the data needed for longitudinal analyses for a limited time span. For example, with the ultimate sampling units retained in the MEPS rotating panel for five waves, the data for one rotation group (panel) can be analyzed to study gross changes over a specific 30-month period (Ezzati-Rice and Cohen, 2004). However, in labor force surveys, the basis of the rotation scheme is usually housing units rather than household members. While this choice eliminates the need to trace and interview households or household members that move between waves, it fails to provide the data needed for longitudinal analyses of households or persons.

A general issue with cross-sectional estimation from panel surveys is that, unless special steps are taken, the sample at later waves does not represent elements that have entered the population after the sample was selected for the initial wave. This is seldom a very serious concern for rotating panel surveys because the duration of each rotation group in the panel is fairly short. Moreover, at any point in time, new entrants can appear in the sample through the more recent rotation groups provided that the sample coverage is updated for each group (such as by updating dwelling lists in sampled segments in a multistage design). They can therefore be properly represented in cross-sectional estimation by the use of an appropriate weighting scheme that reflects the fact that they had chances of selection in only some of the rotation groups on which the cross-sectional estimates are based.

4. Panel surveys

This section considers survey designs in which the same elements are followed over time. Such designs are commonly known as either panel surveys or longitudinal surveys. The term "panel survey" is used in this chapter for all such designs, with the term "longitudinal" being used to describe the data such designs produce.

The distinction between panel surveys and rotating panel surveys becomes blurred in the case of panel surveys of fixed duration, when fresh panels are introduced periodically; the new panel may overlap with the current panel or it may start only after the current panel has been terminated. The distinction made here is between surveys that focus primarily on cross-sectional estimates and estimates of net change, as with the rotating panels in labor force surveys, and panel surveys that are concerned primarily with longitudinal analyses at the individual unit level (e.g., gross change).

The first part of this section (Section 4.1) describes some types of panel surveys that are in widespread use, and the second part (Section 4.2) reviews a range of methodological issues that arise with panel surveys.

4.1. Types of panel survey

The benefit of a full panel survey is that it produces the data needed for longitudinal analysis, thus greatly expanding on the analytic potential of a cross-sectional survey. Although panel surveys have a long history, interest in them has greatly increased in recent years, especially with developments in the computing power needed to handle complex longitudinal data files and the software needed for conducting longitudinal analyses. There are many panel surveys currently in operation and a sizeable literature on the conduct of such surveys (e.g., Duncan and Kalton, 1987; Kalton and Citro, 1993; Kasprzyk et al., 1989; Lynn, 2009; Trivellato, 1999).

Longitudinal data obtained from panel surveys provide the opportunity for a wide variety of analyses not possible with cross-sectional data. For example, longitudinal data are needed for analyses of gross change; durations of spells (e.g., of poverty); growth trajectories with growth curve modeling (e.g., children's physical and cognitive development); early indicators that predict later outcomes (e.g., environmental exposures in childhood and health outcomes in later years); and causal temporal pathways between "causes" and effects, using longitudinal structural equation modeling (e.g., self-efficacy as a mediator between stressful life events and depressive symptoms; see Maciejewski et al., 2000). Sometimes, the longitudinal data needed can be obtained by retrospective recall or from administrative records. Indeed, some panel surveys are based entirely on administrative records; for example, the U.K. Office for National Statistics Longitudinal Study links data across censuses together with vital event data on a sample basis (Blackwell et al., 2005). However, when the quality of retrospective recall is inadequate and administrative data are unavailable or insufficient, direct data collection in a panel survey is required.

Martin et al. (2006) provide brief descriptions of a sizeable number of large-scale panel surveys of social conditions conducted in several countries. These surveys cover topics such as physical and mental health and disability; physical, social, and educational development; employment history; family dynamics; effects of divorce; dynamics of income and assets; transitions to retirement and effects of aging; and social and cultural integration of immigrants. Most focus initially on specific areas but, over time, they are likely to cover a wide range of subject-matter. (Indeed, one of the advantages of the panel design is the opportunity for data collection on many topics at different waves of the survey.) A long-term panel survey may change its areas of inquiry over time in response to changing societal concerns and, in the case of panel studies of age cohorts, to examine changes in age-related topics of interest. Panel designs have also long been used in fields such as epidemiological research on specific illnesses (e.g., Doll and Hill, 1964), studies of voting behavior (e.g., American National Election Study, 2007), and studies to evaluate the effects of intervention programs (Piesse et al., 2009).

Although panel surveys are primarily concerned with producing the data needed for longitudinal analysis at the element level, in most cases they are also analyzed cross-sectionally for each wave of data collection. An important issue for cross-sectional analysis is the adequacy of the sample coverage at each wave: unless steps are regularly taken to "freshen" the cross-sectional samples by adding samples of new entrants to the population since the last sample update, the new entrants will not be represented in the cross-sectional analyses. New entrants are generally not included, and hence are not needed, for those longitudinal analyses that start with data from the first wave of the panel.

4.1.1. Cohort studies

One class of panel survey is often known as a cohort study (Bynner, 2004). Many cohort studies take samples of persons of a particular age and follow them through important periods of transition in their lives. A birth cohort may be followed throughout the life course, and, indeed, the study can be extended to follow the offspring of the original cohort. Four British national birth cohorts well illustrate this type of design in a sequential form: the first was the 1946 National Birth Cohort, also known as the National Survey of Health and Development (Wadsworth et al., 2005); the subsequent studies are the 1958 National Child Development Study, the 1970 British Cohort Study, and the Millennium Cohort Study (Centre for Longitudinal Studies, 2007). A new U.S. birth cohort study, the National Children's Study, will enroll women early in their pregnancies, and even prior to pregnancy, to study the effects of environmental exposures and other factors in early life on the children's health and development till they are 21 years old (NCS, 2007).

Although birth cohorts provide extremely valuable longitudinal data for examining the effects of early childhood experiences on health and other factors much later in life, this great strength is accompanied by some limitations. The sample members of all age cohort studies are subject to the same historical events, or period effects (e.g., wars and environmental disasters) that affect the entire population at that time. They may also be affected by such events differentially because of their susceptibility at the ages at which the events were experienced (cohort effects). Data from a single cohort confound age, period, and cohort effects, and the results must be interpreted accordingly (Bynner, 2004; Yang, 2007). For instance, the effects of early life experiences on later outcomes must be viewed in the context of the period and cohort studied; the magnitude, or even the existence, of the effects may not apply to the current or later generations.

Another limitation of a birth cohort study is that its members would have to be followed for a very long time before it would be possible to investigate, say, the effects of retirement on health outcomes. For this reason, many age cohort studies start at later ages. For instance, the U.S. Health and Retirement Study (HRS) (Juster and Suzman, 1995), the English Longitudinal Study of Ageing (ELSA, 2007), and the Survey of Health, Ageing and Retirement in Europe (SHARE, 2007) follow cohort members from later middle age into later life. The U.S. Bureau of Labor Statistics' National Longitudinal Surveys (NLS) comprise seven cohorts defined by starting age group and sex (e.g., men and women aged 12–17 years, women aged 30–44, and men aged 45–59)

(U.S. Bureau of Labor Statistics, 2005). The U. S. National Center for Education Statistics has conducted many cohort studies, including the Early Childhood Longitudinal Studies, Birth and Kindergarten Cohorts; the Education Longitudinal Study of 2002 (which is following a 10th grade cohort through high school to postsecondary education and/or work); the Beginning Postsecondary Students Longitudinal Study; the Baccalaureate and Beyond Longitudinal Study; High School and Beyond; and the National Longitudinal Study of the High School Class of 1972 (U.S. Institute of Education Sciences, National Center for Education Statistics, 2007).

4.1.2. Household panel surveys

A second major class of panel survey is the household panel survey (Rose, 2000). The ongoing U.S. Panel Study of Income Dynamics (PSID) (Hill, 1992), started in 1968, provided a major impetus for this type of study. The PSID design has been adopted for studies in many countries, including the ongoing British Household Panel Survey (Taylor et al., 2007) (currently being expanded into the Understanding Society study), the ongoing German Socio-Economic Panel (German Institute for Economic Research, 2007), and the European Community Household Panel (ECHP, 2007). These surveys start with a sample of households and then follow household members for the duration of the panel. To reflect the economic and social conditions of the households of panel members at each wave, the surveys also collect data on persons with whom panel members are living at later waves, termed variously as *cohabitants*, *nonsample persons*, or *associated persons*. The surveys are thus, in reality, samples of persons rather than households. Households are constantly changing, with members joining and leaving, new households coming into existence, and others ceasing to exist. For this reason, the definition of a longitudinal household is extremely problematic unless the time period is very short (see, e.g., Duncan and Hill, 1985, and McMillen and Herriot, 1985). Person-level analyses with wave-specific household characteristics attributed to panel members are generally preferred for longitudinal analyses.

Other household panel surveys modeled along the lines of the PSID are the U.S. Survey of Income and Program Participation (SIPP) (Kalton et al., 1998), the Canadian Survey of Labour and Income Dynamics (SLID) (Statistics Canada, 2008), and the European surveys conducted under the European Union Statistics on Income and Living Conditions (EU-SILC) regulations (replacing the ECHP surveys) (Eurostat, 2005). A distinctive feature of all these surveys is that they are designed to last for a fixed duration. The SIPP panels, which collect data every four months, have varied between 2½ and 4 years in length; the SIPP design has varied between overlapping panels and abutting panels where a new panel starts only as the last panel ends. The SLID panels, which collect data twice a year, last for six years; the SLID uses a rotating design with a fresh panel starting every three years. The EU-SILC surveys follow a four-year rotation design.

An advantage of abutting panels is that the full sample size is available for longitudinal analyses for the period of the panel. However, abutting panels cannot handle longitudinal analyses for a period that spans two panels. Also, valid estimates of trends from cross-sectional estimates, such as the annual estimates that are important for these surveys, cannot be produced because of variable time-in-sample biases across years. Rotating designs with an annual rotation can produce acceptable trend estimates because of the constant balance of time-in-sample across waves.

The choice of the length of time for household panel surveys of limited duration depends on a combination of analytic objectives and practical data collection considerations, particularly respondent burden and its effects on response rates at later waves. In the SIPP, for example, the duration was extended from the original goal of eight waves (2⅔ years) to 12 waves (4 years) in 1996 to provide longer periods of observation for use in longitudinal analyses, such as durations of spells of poverty. (See Citro and Kalton, 1993, for a review of the SIPP program and a discussion of the SIPP panel length.) However, with three waves of data collection each year, the extension to four years was accompanied by lower response rates. Response rate and other practical implementation issues influenced the decision to move from the ECHP to the EU-SILC rotating design (Eurostat, 2005).

4.1.3. Cross-national panel surveys

An important recent development in general survey research is the use of survey data for cross-national comparisons, enabling the investigation of the effects of differing societal conditions on the populations involved. This development applies equally with some panel surveys. Notable examples are the aging panel surveys (HRS, ELSA, and SHARE) that are coordinated across many countries and the household panel income surveys. While coordination is valuable, there remains a need for harmonization of the data collected in surveys in different countries. To facilitate cross-national research on economic and health issues in Australia, Canada, Germany, Great Britain, and the United States, Burkhauser and Lillard (2007) have created a Cross-National Equivalent File from the data collected in the household panel surveys in these countries.

4.2. Methodological issues in panel surveys

Many of the same methodological issues apply to panel surveys and rotating panel surveys, although the importance of an issue may be different. For example, all forms of panel surveys must take into consideration the issues of sample attrition, panel conditioning, time-in-sample bias, and the need to cover new entrants to the population for cross-sectional estimation. However, concerns about sample attrition and covering new entrants increase with panels of longer duration, and conditioning effects are a serious concern for many forms of longitudinal analysis.

4.2.1. Maintaining panel participation

Maintaining participation of panel members throughout the life of the study is a critical issue with a panel survey. As in a cross-sectional survey, a panel survey is subject to *total nonresponse*, which occurs when a sampled unit fails to participate in any wave of the panel. In addition, a panel survey is subject to *wave nonresponse*, which occurs when a sampled unit participates in some but not all waves of the survey.

Wave nonresponse may consist of *attrition nonresponse* (when the unit drops out of the panel at one wave and never returns to it) or *nonattrition nonresponse* (when a sampled unit misses a wave but responds at one or more later waves). The potential patterns of wave nonresponse depend on the following rules adopted for the panel. For instance, for practical reasons, many surveys make no attempt to convert initial nonrespondents into respondents at the next or later waves. Thus, initial nonrespondents are all total nonrespondents. Also, nonrespondents who adamantly refuse to participate or cannot be located at one wave may not be followed in subsequent waves and, hence, become attrition cases. Frequently, no attempt is made to follow up those who have missed two consecutive waves.

In general, panel surveys encounter the highest loss rate at the initial wave, after which high proportions of those responding at each successive wave also respond at the following wave. However, the accumulation of nonrespondents over time frequently results in a high overall nonresponse rate. With long-term panels, this situation raises the dilemma of whether to continue with the existing panel, with its increasing analytic potential, or terminate it and start afresh. Some panels—particularly those with high respondent burden—are designed to be of limited duration because of concerns about attrition. The possible biasing effects of accumulating nonresponse are a serious concern in nearly all panel surveys; see, for example, the special issue on attrition in longitudinal surveys in the *Journal of Human Resources* (Volume 33, Number 2, 1998).

The primary causes of wave nonresponse are loss to follow-up for panel members who move and refusals resulting from the repeated burden of being a panel member. A variety of methods are used to minimize loss to follow-up, particularly in panels with lengthy intervals between waves. One approach is to institute methods for tracking panel members so they can be located if they move. For instance, mailings such as birthday cards or newsletters with key findings from the last wave can be sent to panel members, using delivery methods that require the post office to forward mail and inform the sender of the new address. When the tracking methods fail, panel members must be traced to their new addresses. Collecting contact information (e.g., telephone numbers) for neighbors and relatives who are not likely to be mobile (e.g., parents of young people) can be helpful. Otherwise, a variety of web-based and other search procedures may be used. With sufficient effort, loss to follow-up can be limited, even for panels with long intervals between waves (Couper and Ofstedal, 2009).

Preventing the loss of panel members who are no longer willing to participate is a challenge. As with a cross-sectional survey, incentives may be offered to increase participation, but there are additional factors to be considered in a panel survey. Should incentives be offered at every wave? If not, will panel members paid an incentive at one wave refuse to participate when they are not offered an incentive at the next wave? The results of a 1996 U.S. SIPP incentive experiment, in which a monetary incentive was offered only at the first wave, do not support that concern. This experiment found that the sample loss at that and the following five waves was lower with an incentive of $20 than with an incentive of $10 and that both rates of loss were lower than the sample loss in the control group, which received no incentive (Kalton et al., 1998; Mack et al., 1999). With panel surveys, the opportunity exists to target incentives to respondents who demonstrated reluctance at the previous wave, such as those who failed to answer many questions. There are debatable equity issues with this procedure because it serves to reward behaviors that are undesirable from the survey organization's perspective. Another form of targeting is to offer increased incentives to those who facilitate the interview at a given wave. This form of targeting has been found to be cost-effective in the U.S. NLS, where larger incentives are offered to those who call in for a telephone interview (Olson, 2005). Incentives are used in one form or another in many panel surveys.

Minimizing respondent burden is another approach to limiting the loss to follow-up from refusals at later waves of a panel survey. One way to reduce respondent burden is via linkages to administrative data; such linkages can also provide data that panel members are unable to report accurately. In the Canadian SLID, income tax records are used to reduce the reporting burden on many respondents. The survey includes a January interview to collect data on labor market experiences, educational activity, and family relationships and a May interview to collect income data; more than 80 percent of the respondents grant Statistics Canada permission to collect income data from their tax files, and thereby avoid the burden of a May interview (Statistics Canada, 2008). Linkages to administrative data can also extend the period of observation of a panel survey without extending the period of the survey data collection, with its attendant nonresponse losses. A good example is provided by the U.S. HRS, a panel survey that starts with samples of persons aged 51–61. The analytic value of the survey data is greatly enhanced by linkages to lifetime earnings and benefits records in Social Security files and to health insurance and pension data from employers (Juster and Suzman, 1995). See Calderwood and Lessof (2009) for a discussion of linkages implemented in panel surveys in the U.K. For ethical reasons, panel members must be asked for their permission to make the linkages, and procedures must be put in place to ensure that the linkages cannot harm the panel members.

4.2.2. Measurement error

Measurement errors are a concern in all surveys, but they are particularly problematic for longitudinal analyses of panel survey data. To illustrate this point, consider the estimation of the stability of an attitude score across two waves of a panel survey and assume that the scores are subject to random measurement errors that are independent between waves. While the cross-sectional and net change estimates are unbiased in this case, the stability of the estimates over time is underestimated. Another example is the important case of estimating gross change in employment status from a labor force survey. Even if the cross-sectional and net change estimates are considered acceptable, measurement errors can lead to serious overestimates of gross change (see, e.g., Chua and Fuller, 1987). Kalton et al. (1989) list a variety of sources of differential measurement errors that can distort the estimation of gross change: panel conditioning effects, change in mode of data collection, change in respondents between waves (including the possibility of proxy respondents when the sampled person cannot be contacted), changes in personnel (e.g., interviewers, coders of responses to open questions), changes in the questionnaire (with possible context effects even when the questions involved remain the same), changes in the questionnaire content, changes in interpretations of a question, imputation of missing responses, matching errors in linking the files for the two waves, and keying errors. The kinds of effects that measurement errors have on analyses of gross change also apply for many other forms of longitudinal analysis.

One way to try to reduce the overestimation of gross change is to use dependent interviewing, in which respondents are reminded of their responses on the previous wave of the panel. A risk with dependent interviewing is, of course, the generation of false consistency in the responses. Many studies have been conducted to evaluate the effect of dependent interviewing (e.g., Hill, 1994; Hoogendoorn, 2004; Jäckle, 2009; Lynn and Sala, 2006); see Mathiowetz and McGonagle (2000) and Jäckle (2009) for reviews of the technique. In general, dependent interviewing is thought to reduce response errors and the overestimation of gross change. Dependent interviewing may be applied in a proactive form by reminding respondents about their reported status at the previous wave (e.g., they were employed) and then asking for their current status, or in a reactive form by asking about their current status and any discrepancy from their previous response. Proactive-dependent interviewing can also be useful to reduce respondent burden, whereas reactive-dependent interviewing may be less susceptible to false consistency effects. In situations where household respondents may change between waves of a panel, dependent interviewing may result in the disclosure of one household respondent's responses to a different household respondent in the next wave. This feature raises a confidentiality issue that may need to be addressed with a consent form that permits sharing of responses across household members (Pascale and Mayer, 2004).

A particular aspect of excessive change between waves becomes evident in panels that ask respondents to report their statuses for subperiods within the interval between waves. Thus, for example, the U.S. SIPP collects data on a monthly basis within each four-month interval between waves. A common finding in this situation is that the amounts of gross change between adjacent months are much greater for pairs of months for which the data are collected in different waves than for pairs of months for which the data are collected in the same wave. This effect, which is termed the *seam effect*, has been investigated in many studies (see, e.g., Cotton and Giles, 1998; Kalton and Miller, 1991; Kalton et al., 1998; Moore and Kasprzyk, 1984; Rips et al., 2003). The effect is likely to be a combination of false consistency within a wave and overstatement of change across waves. Dependent interviewing can also be used to address this problem.

4.2.3. Weighting and imputation

The standard weighting adjustment methods used to compensate for total nonresponse in cross-sectional surveys can be applied for total nonresponse in panel surveys. However, compensating for wave nonresponse (particularly, nonattrition nonresponse) and item nonresponse is much more challenging.

Apart from total nonrespondents, a good deal of information is known about the other cases with missing data based on the responses they provided in the wave(s) of data collection in which they have participated. One approach for handling missing data is to impute all the missing items, including all the items in waves in which the sampled unit is a nonrespondent. This approach has the advantage of retaining in the analysis file all the information that the sampled units have reported. However, the large amount of imputation involved raises concerns about distortions that the imputed values may introduce into an analysis. An alternative approach is to use nonresponse weighting adjustments to handle some or all of the missing waves. This approach limits the analysis file to sampled units that responded in all the relevant waves. It uses a limited number of the survey responses in making the adjustments, but the responses to other items are lost (Kalton, 1986; Lepkowski, 1989). Imputation is the natural solution when a survey unit fails to respond to just a few items, but the choice between imputation and weighting adjustments is less straightforward for wave nonresponse.

To retain the covariance structure in the dataset, imputation requires that all the other variables associated with the variable to be imputed should be used as auxiliary variables in the imputation model. Satisfying that requirement adequately is difficult enough with a cross-sectional survey, but it is much more challenging with a panel survey because the model has to incorporate variables from other waves of the survey and variables from the given wave. In particular, it is important to include the responses to the same variable at other waves, or gross change will be overestimated. A practical issue is that, in many cases, a completed data set is produced after each wave of a panel survey to enable analyses of all the data collected up to that point. Imputations for the current wave can use current and previous wave data as auxiliary variables but not data from later waves. One solution is to produce preliminary imputations for each wave and then produce final imputations when the next wave's data can be incorporated into the imputation scheme. Although somewhat laborious, this solution is used in the British Household Panel Survey (Taylor et al., 2007).

Concerns that mass imputation for wave nonresponse may introduce distortion into analyses have led to a general, but not universal, preference for weighting adjustments over imputation for handling this type of missing data. In the case of attrition non-response, a common practice is to develop weights for all those responding at each wave, based on data collected in previous waves (by the attrition definition, all those who responded at the current wave have responded at all previous waves). With the large amount of data available for respondents and attrition nonrespondents at a given wave, the development of the weighting adjustments is more complex than in most cross-sectional surveys, but the process is essentially the same. Procedures such as Chi-squared Automatic Interaction Detector (CHAID), propensity score weighting, and raking can be used in developing the weighting adjustments (see, e.g., Kalton and Brick, 2000; Rizzo et al., 1996). Analysts select the set of weights for the latest wave in which they are interested and they can then use that set of weights to conduct any longitudinal analyses they desire.

With many possible patterns of response/nonresponse for nonattrition cases across waves, the use of weighting adjustments in this case can result in the production of a multitude of sets of weights if all the data for responding waves are to be retained for analyses involving data from any given set of waves. Including attrition nonresponse, there is, in fact, a maximum of $2^H - 1$ patterns of response/nonresponse across H waves, but this number is often reduced somewhat by the following rules that are used. Rather than compute sets of weights for each of the potential patterns, analysts often reduce the sets of weights by discarding data for some reported waves. For instance, discarding data from all waves after the first nonresponding wave converts all nonattrition cases to attrition cases and reduces the number of sets of weights to H. Starting from that basis, a restricted number of additional sets of weights may be added based on a review of the combinations of waves that are of analytic importance together with an assessment of the extent of data loss for these combinations resulting from a weighting scheme that treats all wave nonrespondents as attrition cases.

An alternative approach for handling certain nonattrition patterns is to use imputation rather than weighting. For example, in the 1991, 1992, and 1993 U.S. SIPP panels, a longitudinal imputation procedure was used for persons who missed one wave that fell between two waves in which they responded; in the 1996 panel, this procedure was extended to include two consecutive missing waves (see Kalton et al., 1998, for research on longitudinal imputation and the missing wave imputation procedures adopted for the SIPP).

A unique weighting issue arises with cross-sectional estimation in household panel surveys as a result of the collection of survey data for the cohabitants with whom panel members are living at each wave. To take advantage of all the data collected at a given wave in producing cross-sectional estimates for that wave requires a weighting scheme that takes account of the multiple routes of selection by which the household and household members can be selected for the sample at that wave. For example, if a member of an originally sampled household leaves to join another household, that person's new household could be selected either via that person's original household or via the original households of the other members of the new household. A weighting scheme that takes the multiple routes of selection into account and that depends only on the selection probabilities of originally sampled households is described by Ernst (1989), Kalton and Brick (1995), and Lavallée (1995, 2007b). Verma et al. (2007) discuss the application of this scheme to rotating panel designs, with particular reference to the EU-SILC surveys.

4.2.4. Sampling issues

There are many special sampling considerations that arise with panel surveys. One concerns the degree of clustering to be used in selecting the sample for the first wave of a panel. The effectiveness of clustering in reducing interviewers' travel time and facilitating face-to-face callbacks dissipates over time as some panel members move to new addresses. Also, once a panel sample has been enrolled with face-to-face interviews, other methods of data collection that do not benefit from clustering (telephone, mail, and web) may be used in later waves. These considerations argue for less clustering in the first wave of a panel survey than in a cross-sectional survey, thereby cutting back the increases in the variances of survey estimates resulting from clustering.

Most surveys aim to produce estimates for certain subgroups of the population as well as for the total population. Smaller subgroups are often oversampled to generate sample sizes that produce adequate levels of precision for the subgroup estimates. The use of oversampling in a panel survey must be carefully assessed. With a long-term panel, survey designers should take into account that the survey objectives and subgroups of interest may change over time so that the initial oversampling may be detrimental later on. Also, the type of subgroup must be considered. When the defining characteristic of the subgroup is a static one, as with a racial/ethnic subgroup, oversampling can be particularly advantageous in a panel survey. In this case, the benefits of the oversampling apply throughout the life of the panel, whereas any additional costs associated with the oversampling are incurred only in the initial wave. However, when the defining characteristic is liable to change over time (e.g., being in poverty or living in a particular province), oversampling based on the initial state can be problematic, particularly when a high degree of oversampling is used. Over time, panel members will move into and out of the subgroup. As a result, subgroup members at later waves will have markedly different weights, leading to a serious loss in precision of subgroup estimates, even to the point that they may not be useful. An extreme example occurs with panels of businesses; often, highly disproportionate samples of businesses are used, but over time, some small businesses that were sampled at very low rates may grow substantially. These high-growth businesses retain the large weights associated with their initial selection probabilities, a feature that gives rise to a serious loss of precision in the survey estimates. If oversampling is to be used with nonpermanent subgroups in a panel survey, consideration should be given to keeping the variability in sampling rates within reasonable bounds to avoid the loss of precision associated with movement across subgroups.

In the types of panel survey described here, the primary focus is on providing the data needed for longitudinal analyses. However, panel survey data are also widely used for cross-sectional analyses of the data produced at each wave. An important issue for these analyses is representation of the full population at the time of the wave in question, that is, covering units that entered the population after the sample for the initial wave was selected. The same issue arises for longitudinal analyses, where the starting point for the analyses is later in the life of the panel. New samples may be added to give representation to new entrants at later waves. They may also be added to counteract the sample loss from initial wave and attrition nonresponse, in response to an expanded definition of the population of inference, to increase sample sizes for certain subgroups that have become of analytic interest, or just to expand overall sample size.

A number of panel surveys use methods to add sample at later waves as, for example was described in Section 1 for the SESTAT (Fecso et al., 2007). As another example, the U.S. National Education Longitudinal Study of 1988 (NELS:88) started with a sample of 8th grade students in 1988. At the first follow-up wave at 10th grade, the sample was freshened by adding a sample of 10th graders who were not in 8th grade in the 1987–1988 school year. The ongoing U.S. PSID, started in 1968, added a sample of post-1968 immigrants in 1997 (PSID, 2007). The German Socio-Economic Panel, started in West Germany in 1984, added a sample in East Germany in 1990, a sample of immigrants in 1994–1995, and further new samples since then to increase sample size and to provide an oversample of high-income households (German Institute for Economic Research, 2007). The weighting schemes for the various longitudinal data files can become complex when sample additions are introduced at later waves of a panel survey.

4.2.5. Ethical and data disclosure issues

To conclude the discussion in this section, the special ethical issues and data disclosure risks associated with panel surveys deserve comment (Lessof, 2009). The requirement that sampled persons be informed about the purposes of the study at the outset can be difficult to satisfy in a long-term panel study, the purposes of which may change during the life of the panel. Also, those directing the study and conducting the data collections may change over time. Researchers should pay attention to these issues in designing consent forms.

Panel surveys are expensive to conduct, but they produce extremely rich data that can be valuable for analyses of many different subjects. To capitalize on the investment, the data should be made available to many researchers. However, the rich longitudinal data often pose a high disclosure risk (see, e.g., Béland, 1999). Although standard techniques such as data suppression (particularly of detailed geography), top coding, data swapping, and subsampling may provide adequate protection to enable the release of a public use data set for one wave, these techniques often do not afford sufficient protection for a public use panel file, which includes much more data. In this case, alternative methods may be needed to make the data available to researchers, such as restricted use files, secure data enclaves, and remote analyses. Another aspect of making panel data available for analysts is that full documentation must be maintained on an ongoing basis, both to advise analysts on the contents of the complex panel data and to record the survey details for use by those analyzing the data years later.

5. Conclusions

Those planning survey data collections to provide data across time have a choice between repeated cross-sectional, rotating panel designs, and full panel designs. If the data are to be used for longitudinal analyses, then only a panel design will serve the purpose. However, if the data are to be used only for overall trend analyses, any of the designs can be used, provided that the sample is freshened at each wave to give representation to new entrants to the population.

The design considerations for a series of repeated cross-sectional surveys would appear to be the same as those for a single cross-sectional survey, but in fact there are differences. Those planning a series of repeated cross-sectional surveys need to reflect on what the data needs might be in the future in order to cover them from the outset. During the course of the series, they will likely also face difficult decisions about changing aspects of the design to meet current conditions and conform to current best survey practice, or whether to stay with the existing design to maintain valid estimates of trends. Analysts of repeated surveys must be cognizant of any changes made to the design that may distort trend estimates.

Panel surveys are far more complex to design and analyze than cross-sectional surveys. In addition to general issues of survey design, designers of panel must pay a great deal of attention to such issues as maintaining the cooperation of panel members, tracking and tracing methods, introducing sample freshening to be able to provide valid cross-sectional estimates, the use of dependent interviewing, and the use of incentives. Analysts must be cognizant of the effects of measurement errors and panel conditioning on their analyses, as well as the likely deterioration in the representative nature of the sample over time.

Essential Methods for Design Based Sample Surveys
ISSN: 0169-7161

3

DOI: 10.1016/B978-0-444-53734-8.00003-0

Design, Conduct, and Analysis of Random-Digit Dialing Surveys

Kirk Wolter, Sadeq Chowdhury and Jenny Kelly

1. Introduction

Random-digit dialing (RDD) is a method of probability sampling that provides a sample of households, families, or persons via a random selection of their telephone numbers. For simplicity of explication, we use the person as the final unit of analysis in this article; yet, virtually, all our comments and methods extend naturally to the household or family.

In this chapter, we discuss the design, conduct, and analysis of RDD surveys primarily in the context of large-scale work performed in the United States. We believe that the material generalizes to other countries with an established landline infrastructure. In the United States, there is generally no sampling frame that enables a direct sampling of persons. RDD changes the sampling unit from the person to the telephone number, for which sampling frames do exist. Then, people can be sampled indirectly through their telephone numbers, enabling valid inferences to populations of people.

In the modern era, the RDD survey has come to embody the following three elements: (1) random sampling of telephone numbers from a listing of all (or most) assigned telephone numbers; (2) dialing the selected numbers from a central call center(s); and (3) administering the survey questionnaire to residential respondents via a system of computer-assisted telephone interviewing (CATI). RDD surveys became an accepted form of survey research in the 1970s, and their prevalence increased considerably in the 1980s and 1990s.

Today, the RDD survey stands as one of the dominant survey forms for social-science and market research. It has attained this position because it offers important advantages in cost, timing, and accuracy. RDD surveys eliminate travel costs and, thus, are far less costly than surveys that use face-to-face interviewing methods. They may be more expensive, however, than mail or web surveys that shed labor costs by eliminating or reducing the use of human interviews. RDD surveys have the capacity to deliver survey information very quickly relative to surveys that use other modes of enumeration. They may be launched and their interviewing operations completed quickly in a matter of days or weeks. The survey questionnaire—possibly even with elaborate skip patterns, large lookup tables, and other complications—can be entered quickly into the CATI software. Since the interview data are already in machine-readable form, entered by the interviewer into the CATI system, there is no need for a data conversion from paper to computer format. Mail and web surveys may match RDD surveys in terms of start-up time, whereas only web surveys can match them in terms of data collection and delivery speed. Face-to-face surveys are much slower to launch, to complete data collection, and to deliver the survey data.

RDD surveys, and generally CATI surveys, also possess features that enhance data quality. Because interviewing operations are usually centralized in a small number of call centers, it is possible to achieve specified standards relating to the hiring, qualifications, and training of the interviewers and their supervisors. Supervisors can monitor interviews and the general performance of the interviewers; they can take corrective actions in real time through retraining or replacement of underperforming interviewers. A work force that undergoes continuous improvement has the capacity to produce better and better interview data. Computer edits can be built into the CATI instrument, thus limiting missing values and the possible entry of out-of-range values, erroneous skip patterns, and the like. Face-to-face surveys that use a system of computer-assisted personal interviewing (CAPI) can incorporate the advantages of online edits, yet such surveys, with or without CAPI, cannot match CATI surveys in terms of close, real-time monitoring of interviewers. Face-to-face surveys, however, can generally achieve higher response rates than CATI surveys; they may achieve greater data completeness relative to certain types of questions or subject matter (e.g., it is obviously not possible to collect biomarkers in a pure CATI survey) and they usually have the capacity to handle a longer interview. Mail surveys generally experience lower response rates than CATI surveys; they cannot offer the benefits of online edits and cannot accommodate complicated skip patterns. Web surveys may offer online edits, yet their response rates will likely be lower than those of RDD surveys. Depending on the target population, there may be no acceptable sampling frame to support the use of a web-interviewing approach, and thus, web surveys would tend to have lower coverage than RDD surveys.

Survey planning and the selection of a mode of interview require consideration of many complex trade-offs between cost, speed, and accuracy. The foregoing discussion reveals many circumstances in which the RDD survey will be preferred and demonstrates why RDD surveys have reached a dominant position in the survey research market-place.

Before proceeding, it is also important to observe that modern surveys increasingly use multiple modes of enumeration to collect acceptable data in the fastest time feasible at an affordable price. While we do not explicitly treat mixed-mode surveys in this chapter, the methods we do present show how the telephone component of a mixed-mode survey may operate.

2. Design of RDD surveys

2.1. Structure of telephone numbers

The telephone industry in the United States consists of many individual telephone companies. Each company manages a block of telephone numbers and assigns those numbers to their subscribers. Subscribers typically maintain their numbers through time, regardless of switching telephone companies. The telephone number itself consists of eleven digits as depicted in the following diagram:

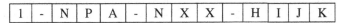

The first digit in the structure (i.e., 1) is a constant, which is the international county code for the United States, and is required to be dialed first for calling outside the local calling area and in some cases even within the local area. The three-digit area code is represented by the symbol NPA, the three-digit exchange code or prefix is represented by the symbol NXX, and the four-digit suffix is represented by the symbol HIJK. The symbol NPA-NXX-H represents a 1000-block of numbers.

Area codes represent compact geographic areas; they are generally non-overlapping and exhaust the land area of the United States. Not all three-digit combinations are in service at this time, and in some cases, area codes overlay one another. The area codes are generally nested within states, but they do not generally correspond to political, postal, or census geography. Exchange codes within area codes or 1000-blocks within exchange are designated for landline telephones, wireless (or cell) telephones, or some other type of use. Historically, the exchange codes for cell telephones have been excluded from sampling for RDD surveys. Exchange codes for landline telephones are not geographically compact nor are boundaries defined in a useful way. The area covered by an exchange can cross city or county boundaries, making it difficult to use the exchange directly for any geographic stratification. The four-digit suffix within an exchange also contains no useful geographic information. However, since the telephone companies activate banks of consecutive numbers and assign numbers to residential and nonresidential subscribers in such a way that the consecutive suffixes within an exchange may be clustered in terms of working or nonworking or residential or nonresidential status, banks of suffixes have sometimes been used as clusters in the sample selection process.

In what follows, we describe sampling frames and sampling designs for telephone numbers. We refer to 100 consecutive telephone numbers—from 1-NPA-NXX-HI00 to 1-NPA-NXX-HI99—as a *100-bank* or simply a *bank* of numbers. At any given point in time, some banks have not been assigned and placed into service, some are assigned to landline telephone subscribers, some to cell-telephone subscribers, some to other types of use, and a few to mixed use.

2.2. Sampling frames

The main premise of RDD is that each eligible person in the survey target population can be linked to—that is, reached and interviewed at—one or more residential telephone numbers in the population of landline telephone numbers. Sampling telephone numbers with known probabilities of selection means that people are selected with calculable probabilities, and thus, that valid inferences can be made to the population of eligible people. Another key premise of RDD is that each telephone number is linked to a specific, identifiable (approximate) geographic location. This feature makes it possible to select representative samples of people in defined geographic areas such as states, counties, or cities.

At the outset, it is clear that RDD surveys provide no coverage of people who do not have ready access to a landline telephone in their home, including those who have access to a cell telephone only and those who have no access to any telephone. We discuss this undercoverage and approximate ways of adjusting for it later.

A major challenge for RDD surveys is the development of a complete list of telephone numbers that covers all the remaining persons in the target population, that is, all persons in the population with access to a landline telephone. A natural but naive approach would be to sample telephone numbers at random from residential telephone directories. This method is nearly always unsatisfactory because not all residential numbers are listed in a directory and not all listed numbers actually connect to an occupied residence. The former problem is the more acute since it would result in an additional bias of undercoverage to the extent that people with an unlisted telephone are different from people with a listed telephone with respect to the issues under study in the survey. Various authors have shown that the sociodemographic characteristics of the persons in households with unlisted telephone numbers are different from those of persons in households with listed telephone numbers (see, e.g., Brunner and Brunner, 1971; Fletcher and Thompson, 1974; Glasser and Metzger, 1975; Leuthold and Scheele, 1971; Roslow and Roslow, 1972; Shih, 1980). The latter problem leads to inefficiency but not bias because some telephone numbers will screen out as nonresidential.

As an alternative to the directory-based sampling frame, one could consider the use of the conceptual list of all telephone numbers in assigned landline telephone banks. Currently, there are over 718 million numbers in such banks in the United States. Although this frame covers both listed and unlisted numbers, it can be highly inefficient to work with. The major problem is that this frame includes a large percentage of nonworking and nonresidential telephone numbers. Dialing purely at random from such a frame would result in relatively few residential calls, many unproductive calls, and the consumption of excessive cost and time.

To make the conceptual list viable as a sampling frame, some method of sampling is needed to diminish the rate of out-of-scope calls and increase the rate of productive residential calls. In fact, such methods have been developed and the conceptual list of telephone numbers does provide the sampling frame for most RDD surveys conducted today. We provide a brief account of useful sampling methods in the next section.

2.3. Sampling procedures

Various sampling procedures have been developed and used over the years for random digit dialing (Glasser and Metzger, 1972; Hauck and Cox, 1976; Sudman, 1973; Tucker et al., 1992; Waksberg, 1978). Lepkowski (1988) provides an extensive discussion of telephone sampling methods used in the United States until the late 1980s. We have already noted that pure random sampling of telephone numbers is unworkable on grounds of high cost and time. We proceed to describe three of many methods of sample selection that have found favor in RDD surveys over the years through increases in cost-effectiveness.

Sudman (1973) described a procedure in which directory-based sampling is combined with RDD sampling. The procedure considers each block of 1000 consecutive telephone numbers as a cluster, and the clusters are selected by obtaining a simple random (or systematic) sample of numbers from a telephone directory and the corresponding clusters of all selected numbers are included in the sample. Then, calls within each selected cluster are made using RDD sampling to reach a predetermined number of households with listed telephone numbers. Because the sampling of clusters is based on the directory-listed numbers, the clusters are selected with probability proportional to the (unknown) count of listed telephone numbers in the cluster. The procedure improves over the unrestricted method by concentrating on the banks with one or more listed telephone numbers, but it still requires making a large number of calls to reach the predetermined number of listed telephone numbers. Also, since the initial sample is selected based on a telephone directory, the procedure introduces a small bias of undercoverage by omitting the clusters with recently activated numbers or with no listed numbers.

During the 1970s and 1980s, the *Mitofsky–Waksberg method* (Waksberg, 1978) was used widely in RDD surveys. It involves a two-stage sampling procedure, consisting of (i) a selection of banks at the first stage, using Lahiri's (1951) rejective method for probability proportional to size (PPS) sampling, and (ii) a random selection of telephone numbers within selected banks at the second stage. The idea is to select banks with probability proportional to the number of working residential numbers (WRNs) they contain. Then, telephone numbers are sampled in the selected banks, which tend to be rich in WRNs. In this manner, dialing of unproductive numbers (nonworking or nonresidential) is greatly reduced. Each bank is considered as a cluster or primary sampling unit (PSU). A random bank is first selected and then a random telephone number is dialed within the selected bank. If the dialed number is determined to be nonworking or nonresidential, the bank is rejected and excluded from the sample. However, if the number dialed is determined to be a WRN, then the bank is accepted and additional telephone numbers are selected from the bank at random and dialed, until a fixed number, k, of residential numbers is reached. The two-stage process is continued until a predetermined number of banks, m, is selected. The total sample size of WRNs is, therefore, $n = m (k + 1)$. If M denotes the number of banks in the sampling frame and K_i is the number of WRNs in the population within the ith bank, then the probability that a given bank i is selected and accepted at a given draw is

$$\pi_i = p_i \frac{1}{1 - \ddot{p}} = \frac{K_i}{M\overline{K}}, \tag{1}$$

where

$$p_i = \frac{1}{M} \frac{K_i}{100}, \tag{2}$$

$$\ddot{p} = \frac{1}{M} \sum_{i'=1}^{M} \left(1 - \frac{K_{i'}}{100} \right), \tag{3}$$

and $\overline{K} = \sum_{i'=1}^{M} K_{i'}/M$. Thus, the rejective selection method is a PPS sampling method, despite the fact that the measure of size K_i is unknown at the time of sampling! Only banks with one or more WRNs have a nonzero probability of selection, thus reducing the number of unproductive calls. The conditional probability that a given WRN j is selected, given that its bank i is selected and accepted on the given draw, is simply $\pi_{j|i} = \min\{k + 1, K_i\}/K_i$. Thus, the unconditional probability of selecting a given

WRN on the given draw is

$$\pi_{ij} = \pi_i \pi_{j|i} = \frac{\min\{k+1, K_i\}}{M\overline{K}} = \frac{k+1}{M\overline{K}}. \tag{4}$$

Typically, k is specified to be a small number, usually no more than 4 so that it is usually less than the number of WRNs in the bank. As long as $k < K_i$ for all i, all WRNs in the population have an equal probability of selection. The values of m and k can be optimized for a given ratio of the costs of an unproductive call to the costs of a productive call (including the cost of calling, interviewing, and processing).

While the Mitofsky–Waksberg sampling scheme makes considerable improvements over unrestricted random sampling, it runs into severe operational problems. Because of potential nonresponse – or at least a lag in response between the initial release of the number and its resolution – for the first selected telephone number, the critical determination of whether a given bank is accepted or rejected may be unacceptably delayed. In addition, because of the risk of subsequent nonresponse within the bank or because of small K_i, it is sometimes difficult to locate k residential numbers. Due to these operational problems and since another method of sampling has proved to be successful, rejective selection has now sharply declined in use for modern, large-scale RDD surveys.

List-assisted sampling is a term used to describe a class of methods that select the sample from the conceptual list of all telephone numbers in assigned landline banks while exploiting information in residential telephone directories to improve the efficiency of the selection. In 1+*sampling*, one restricts the sampling frame to all assigned landline banks in which one or more telephone numbers are listed in the residential telephone directory. Banks containing zero listed numbers are dropped. Then, a probability sample of telephone numbers is selected from the remaining banks. The method provides complete coverage of all listed and unlisted numbers in banks with at least one listed number, omitting only unlisted numbers in banks with no listed number. The scheme covers around 98% of the universe of landline telephone households and excludes only the approximately 2% of households with unlisted numbers in zero banks (Giesbrecht et al., 1996; Fahimi et al., 2008; Boyle et al., 2009). The bias due to the noncoverage of these unlisted telephone households in zero banks is thought to be small in many surveys (Brick et al., 1995). The method easily extends to $p+$ sampling, where p is a small integer, say between 1 and 5. The larger the p is, the greater the rate of productive calls, saving time and money. The downside of a larger p is a reduced population coverage rate and an increased risk of bias. Ultimately, $p+$ sampling seems to have the following three virtues:

- Easy to implement
- Acceptably small undercoverage bias, especially for 1+ designs
- Yields an unclustered sample with a smaller design effect and larger effective sample size than the aforementioned clustered sampling designs

As a result, $p+$ sampling has emerged as the dominant form of RDD sampling today.

Table 1 describes our recent experience at NORC in calling 1+ samples. Approximately 24% of the telephone numbers may be classified as WRNs and 59% as something else, such as disconnected lines or businesses. Almost 17% of telephone numbers cannot be classified either way, meaning that information is incomplete and it is not

Table 1
Directory-listed status and working residential number status for U.S. telephone numbers in 1 + banks: 2005

Directory-Listed Telephone Status	Working Residential Number Status			
	No	Yes	Not Resolved	Total
No	49.1%	4.4%	8.0%	61.5%
Yes	10.0%	19.9%	8.6%	38.5%
Total	59.0%	24.3%	16.7%	100.0%

Note: The proportion of unresolved numbers depends in part on the calling protocol and on the length of the data-collection period. The proportions we cite are in the context of social-science surveys conducted by NORC with ample periods for data collection and relatively high response rates.

possible to resolve whether the numbers are residential or not, despite repeated call-backs.

Almost 39% of telephone numbers are listed in a directory, whereas 61% are not. In all, 10% of numbers are directory-listed but turn out not to be WRNs and, on the other hand, 4% of numbers are not directory listed yet turn out to be WRNs. Lags in the development of directories and the timing of the field period explain the apparent misclassifications. From these data, one can calculate that 82% of resolved WRNs are listed in a directory. The remaining 18% of resolved WRNs are the unlisted. This unlisted percentage varies considerably from one part of the country to another. To restrict sampling to directory-listed numbers only would risk a bias in survey statistics, and the bias could be differential from one area to another, clouding comparisons.

2.4. Stratification

Little auxiliary information is typically available on the RDD sampling frame and opportunities for stratification of the sample are limited. Stratification is possible by directory-listed status or by whether a mailing address can be linked to the telephone number. Some broad geographic information is embedded within the structure of telephone numbers, which can be used for coarse geographic stratification. Since area codes are nested within states, stratification by state is feasible. Finer geographic stratification is difficult because exchanges may cross area boundaries.

It is possible to make an approximate assignment of telephone exchanges to finer census-defined geographic areas by geocoding the addresses of the listed telephone numbers within exchanges. Given a census-defined area of interest, one calculates the *hit rate* and the *coverage rate* for each exchange, where the hit rate is the proportion of listed telephone numbers in the exchange that belongs to the designated area, and the coverage rate is the proportion of listed telephone numbers in the designated area that is covered by the exchange. The sampling statistician may implement a rule involving these two factors to stratify exchanges by finer geographic areas, but this is an imperfect process. For example, each exchange may be classified to one of the set of geographically defined strata, spanning the target population, according to a majority rule; that is, one may assign the exchange to the stratum for which its hit rate is the maximum over the set of strata. As a second example, one may classify an exchange to a designated area if the hit rate exceeds a threshold, such as 0.05, and the cumulative coverage rate over all such exchanges exceeds another threshold, such as 0.95.

Stratification by socioeconomic status becomes possible by mapping census tracts onto telephone exchanges based upon the geocoded addresses of listed telephone numbers. Again, such mapping is necessarily approximate. Census variables at the tract level—such as race/ethnicity, age, sex, income, poverty, education, and housing tenure variables—can then be donated to the exchange and, in turn, exchanges can be assigned to socioeconomic strata. Among other things, such stratification enables one to over-sample subpopulations of interest (Mohadjer, 1988; Wolter and Porras, 2002). Variables of this kind are sometimes called contextual or environmental variables.

2.5. Determination of sample size

The sample of telephone numbers selected for an RDD survey will need to be many times larger than the required number of completed interviews. During data-collection operations, there will be a number of losses to the sample, and the initial sample of telephone numbers must be large enough to offset the losses.

As usual, the statistician must begin by determining the effective number of complete interviews needed to achieve the survey's resource constraints and goals for statistical precision and power. Multiplying by the anticipated design effect—due to any clustering, differential sampling rates, and differential weighting effects—gives the target number of completed interviews.

The next step is to inflate the sample size to account for sample attrition due to nonresolution of telephone numbers, resolution of nonresidential telephone numbers, failure to complete the survey screening interview for some WRNs, the survey eligibility rate, and failure to complete the main interview among some eligible respondents.

The attrition starts with the nonresolution of many numbers regarding their WRN status. Despite repeated callbacks, it will be impossible to resolve many telephone numbers as to whether they are WRNs or something else. Then, among the resolved numbers, a large percentage will be non-WRNs, such as business numbers, computer modems, or disconnected lines. Once a WRN is identified, the next step is to conduct a brief screening interview to determine eligibility for the main survey. Screening for the eligible population is usually not possible beforehand because eligibility is not known at the time of sampling. (A special case is where all persons are eligible for the survey. In this case, there is in effect no screening interview.) Some screening interviews will not be completed because the respondent is never home or refuses to cooperate. Among completed screeners, households containing no eligible people are omitted (or "screened out") from the main interview. Finally, some main interviews will be missing because the eligible respondent is not at home or refuses to participate.

To determine the appropriate inflation of the sample size, the statistician must make assumptions about the foregoing factors based on general experience or information available from prior surveys. Ultimately, the target sample size in terms of telephone numbers in the hth sampling stratum is given by

$$n_h'' = \frac{n_h'}{\pi_{h1}\,\pi_{h2}\,\pi_{h3}\,\pi_{h4}\,\pi_{h5}} = \frac{n_h D_h}{\pi_{h1}\,\pi_{h2}\,\pi_{h3}\,\pi_{h4}\,\pi_{h5}}, \tag{5}$$

where n_h' is the target number of completed interviews, n_h is the effective sample size required from stratum h, D_h is the design effect assumed for stratum h, π_{h1} is the resolution completion rate assumed in stratum h, π_{h2} is the WRN rate among resolved numbers assumed in stratum h, π_{h3} is the screener completion rate assumed in stratum h, π_{h4} is the eligibility rate among screened households assumed in stratum h, and π_{h5} is the interview completion rate among eligible people assumed in stratum h.

2.6. Emerging problems and solutions

The RDD sampling frames and sampling designs discussed so far cover the population of people who reside in households with at least one landline telephone and fail to cover people who live in cell-phone-only households and nontelephone households. This undercoverage was of little concern in the early cell-telephone era. Yet at this writing, concern among survey researchers is growing. Using data from the National Health Interview Survey, Blumberg et al. (2006) show that 9.6% of adults (in the period January to June 2006) live in households with only cell-telephone service. Three years earlier (in the period January to June 2003), this population stood at only 2.8% of adults. Throughout this three-year period, adults in nontelephone households remained steady at 1.8–2.0% of the adult population. Some characteristics of the population with no telephone or only a cell telephone tend to be different from those of the population with a landline telephone. Thornberry and Massey (1988) discuss the patterns of landline telephone coverage across time and subgroups in the United States for the period of 1963–1986. Adults living with unrelated roommates tend to be cell-telephone-only at a higher rate than other adults. Other domains displaying a higher rate of cell-phone-only status include renters, young adults (age 18–24), males, adults in poverty, and residents of the South and Midwest regions. The population with no telephone service is likely to be unemployed, less well educated, below the poverty line, or in older age groups (Blumberg et al., 2006, 2005; Khare and Chowdhury, 2006; Tucker et al., 2007). Tucker et al. (2007) also show the rapid growth of the cell-telephone population using data from the Consumer Expenditure Survey and the Current Population Survey. Because of the current size of the populations without a landline telephone and the likelihood of their continued growth, there is increasing concern about potential bias in standard RDD surveys that omit these populations. What is not known for sure is whether cell-only and nontelephone populations differ from landline populations with respect to the main characteristics under study in surveys.

Until now, the main approach used to compensate for the undercoverage in RDD survey statistics has been the use of various calibration adjustments in the weighting process (Brick et al., 1996; Frankel et al., 2003; Keeter, 1995; Khare and Chowdhury, 2006). With the rapidly increasing cell-phone-only population, consideration must now be given to direct interviewing of this population. A dual-frame approach with supplementation of the RDD frame by an area-probability frame has received consideration in the past, but the approach is too expensive for general use. Dual frame designs using the traditional RDD frame of landline telephone numbers and a supplementary frame of numbers in cell-telephone banks must be considered.

Studies are being conducted to investigate the viability of interviewing respondents via cell telephones (Brick et al., 2007; Wolter, 2007). Sampling frames for cell-telephone numbers can be constructed from the Telcordia® TPM™ Data Source or Telemarketing Data Source. Both can be used to identify prefixes or 1000-blocks that are likely to be used for cell telephones. In addition, many cell-telephone numbers in use today are the result of "porting" the numbers from landline use to cell-telephone use. Such numbers are, by definition, located in landline blocks and are not covered by the type of sampling frame just described.

A number of additional problems for RDD surveys are emerging due to the recent rapid advancement in telephone technology. Voice over internet protocol (VoIP), which offers routing of telephone conversations over the internet or through any other IP-based network, is penetrating the mass market of telephony. Under the VoIP technology, a user may have a telephone number assigned under an exchange code in one city but can make and receive calls from another city or indeed anywhere in the world, just like a local call. The possibility of a universal telephone numbering system is also on the horizon, whereby a subscriber may have a number under any exchange code but can live anywhere in the country or even anywhere in the world. Call forwarding is another problem where a landline WRN may be forwarded to a business number, a cell-telephone number, a VoIP number, or a landline number out of area. Since geographic stratification for an RDD survey is usually based on the telephone area and exchange codes, extra screening efforts to identify the location of a number will be required with the increasing use of these systems. Also, increasing use of these systems will increase the rate of out-of-scope residential numbers or differences between frame and actual locations of in-scope sampling units, in turn increasing costs.

Sometimes, two telephone numbers are linked to the same landline telephone without the knowledge of a subscriber. The hidden number is called a *ghost number*, and calls made to this number will reach the subscriber just as will calls made to the real telephone number. The respondent's selection probability increases to an unknown extent and the statistician is left with no information to allow adjustment in the survey weighting.

A final set of problems has arisen from the rapid deployment and now widespread use of caller id and voice mail/answering machines. More and more people are using these technologies to screen their calls, and it is becoming more and more difficult to reach them or even resolve whether the telephone number is a WRN or not. In our recent experience with large studies in the United States, approximately 17% of all unresolved telephone numbers are due to answering machine systems and the lack of sufficiently detailed recorded messages or other markers that would allow us to otherwise classify the cases as WRNs, businesses, or other nonresidential. We also find that approximately 45% of our unresolved telephone numbers are due to repeated non-contact on all call attempts. Some unknown portion of this percentage is assumed to be due to people using caller-id to screen their calls. While caller-id and answering machines do not alter sampling units or their probabilities of selection, they do substantially increase survey costs by inflating the sample size necessary to achieve a specified number of completed interviews. They correspondingly depress response rates and thus heighten concern among statisticians about potential bias in RDD survey statistics.

3. Conduct of RDD surveys

The advent of CATI systems and automated dialers has made RDD surveys, and telephone surveys generally, a workhorse of social-science and market research. Scheduling, working, and manipulating the sample of telephone numbers to achieve a targeted number of completed interviews by stratum, on a timely basis, at low cost, and with a high response rate, is the task of the sample-management function.

3.1. Technology of data collection

CATI is a term used to cover all computer-aided aspects of telephone interviewing. It covers both hardware requirements (including telephony systems) and software. Some CATI systems use a single integrated piece of software that controls the sample, the questionnaire, and the dialing; other systems combine elements from multiple vendors to take advantage of some specializations.

A good CATI system can connect dozens or hundreds of workstations in multiple locations and can offer interviewers and their supervisors the facility to work simultaneously sharing the same system. CATI was first introduced in the early 1970s and is now commonplace in market and social-science research.

A typical system offers customized tools for survey instrument development, call scheduling, display of survey items, recording of survey responses, monitoring and supervision of interviewers' work, keeping a record of calls for each case in the released sample, online data editing and processing, preparation and export of data sets, and other automatic record keeping. A typical system allows one quickly to develop a survey questionnaire in multiple languages, thus enabling the conduct of multilingual interviews. The system can automatically record call outcomes such as no answer, answering machine, disconnect, busy, or fax/modem; it also can dispatch only the connected calls to interviewers or schedule callbacks where required. During the interview itself, the system can automatically execute simple or complex skip patterns without interviewer intervention, and conduct subsampling, if required. It offers the facility to view the call history of a case and to add to it. Integrated CATI systems offer the facility to produce frequency tabulations, survey statistics, response rates, and productivity reports, which are useful for ongoing monitoring of progress and of key indicators of quality. They also admit data exports in a wide range of formats, thus enabling external analysis of the data and reporting of progress.

The two components of a CATI system which are of most relevance to RDD surveys are the sample management component and the dialer technology. Survey costs are largely driven by interviewer time, and in a RDD survey, it is not unusual for many dozens of dials to be required for each completed interview. Dialer technology provides efficiencies by reducing the amount of time the interviewer is involved in each dial, whereas the case management system (see Section 3.3) can provide efficiencies by reducing the number of dials needed to obtain the same result.

Automated dialers can assist the dialing process at both ends of the phone call. For example, at dial initiation, they can automatically deliver and dial the next number when an interviewer comes free, and at call outcome, they can detect outcome tones such as a busy line or a fax machine, and automatically apply the appropriate disposition to the case and file it away. Under a fully manual system, it takes approximately 40 seconds for an interviewer to dial and disposition an engaged or disconnected number; yet an automatic dialer can do this in less than a quarter of this time.

Dialers can be classified into two distinct groups, based on whether they are capable of predictive dialing or not. Predictive dialers are distinguished from nonpredictive (or preview) dialers in that they do not connect cases to an interviewer until after a connection is established. This means they can dial more numbers than there are interviewers available and deliver to the interviewers only those calls which need an interviewer (i.e., those that connect to an answering machine or a live person) while handling in the background a comparable number of dials which do not need interviewer attention at all. In this manner, interviewer time is eliminated from handling unproductive calls, increasing efficiency and lowering cost.

Different predictive dialers use different algorithms, but all rely essentially on predicting two probabilities: the probability of a number being answered and the probability of an interviewer being free to take a call. One can assess the optimal speed of the dialer by using a "supply and demand" analysis, where the demand side is due to the dialer dialing numbers and creating a demand for interviewer labor, whereas the supply side is created by the pool of interviewers who are free to accept a call. When the dialer runs faster than its optimal speed and demand exceeds supply, calls will be abandoned, that is, the dialer will need to hang up on a connection because it cannot find a free interviewer to whom to pass the connection. When the dialer runs slower than the optimal speed and supply exceeds demand, interviewers will sit idle and efficiencies will be lost. The fastest dialing will occur under the following situations:

- Relatively few connections occurring among the dials made immediately preceding the current dial
- Tolerance setting for the risk of abandonment set relatively high
- Many interviewers logged into the system
- Fairly short average connection length and overall relatively little variation in the connection lengths.

However, there are two distinct drawbacks to predictive dialing. One is the risk of abandoning calls, which can be a nuisance for respondents and can ultimately translate into respondent complaints or additional refusals. The other drawback is that since the dialer will not assign a case to an interviewer until after the connection is established, the first opportunity an interviewer will have to review call notes associated with the case is when they hear someone saying "hello." Particularly when trying to convert refusals, it is important that interviewers be informed on the details of previous refusals, but predictive dialers prevent this transfer of information and thus diminish response rates.

Hybrid dialing delivers most of the efficiencies of predictive dialing while eliminating these drawbacks. The dialer starts in predictive mode for virgin cases and continues subsequent callbacks in that mode, but then shifts to preview mode for all callbacks once contact has been established. Given that it is only upon contact that a household is identified and call notes written, the risk of abandoning a call to a known household is eliminated, and the interviewer can prepare properly for the call by reading the call notes left by the interviewer who handled the previous contact.

It is important to note that hybrid dialing mixes the two modes of dialing – predictive and preview – within a single CATI survey in real time. This is quite different from splitting the sample to start it in predictive mode (to clear out disconnects) and then switching the numbers once connected into a separate survey to run in preview mode, or vice versa. Splitting samples in such a manner is inherently less efficient because it means fewer interviewers logged in for the predictive algorithm to maximize its speed and substantial time required for sample management and call history reconstruction.

Hybrid also enhances the interviewer's task, which in turn translates to an improved respondent experience. With a traditional preview dialer, interviewers can spend at least one-third of their time listening to dials being placed or ringing out, which can lead to loss of attention and preparedness for those few crucial seconds when an interviewer first interacts with a potential respondent. The hybrid dialer enables interviewers to spend a much larger proportion of their time actively engaged in respondent interactions.

3.2. Sample preparation and release

Once the sample is selected for an RDD survey, three preparatory steps are applied before releasing the sample to the telephone center for calling. The steps involve (a) subdividing the sample into replicates, (b) identifying the unproductive cases in the sample through prescreening of the telephone numbers for nonworking, nonresidential, and cellular telephone numbers, and (c) mailing advance letters to the households in the sample for which address information is available to increase response rates. These steps provide for good management and control of the sample and efficient dialing.

3.2.1. Forming replicates

In an RDD survey, the sample is usually not released and loaded into the CATI system as a single monolithic batch. Instead, the specified sample in each stratum is divided into a number of replicates that are formed randomly so that each replicate is a random subsample of the full sample. The replicates are then released to the telephone center as needed to spread the interviews for each stratum evenly across the duration of the interview period. Careful release of replicates allows for both efficient use of the interviewer pool and tight control over the number of completed interviews achieved. Recognizing the uncertainty in the level of survey response that will be achieved in an RDD survey, the statistician will often select a larger sample of telephone numbers than is expected to be consumed in the survey. Release of replicates is terminated when the difference between the (real-time) projected number of completed interviews and the target sample size determined at the planning stage is deemed to be acceptable. Unused replicates are not considered released and only released replicates constitute the actual sample for purposes of weighting and the calculation of response rates.

3.2.2. Prescreening replicates

Because approximately 75% (This percentage, derived from Table 1, assumes an ample data collection period with the survey protocol configured to achieve a relatively high response rate. Different survey conditions could yield a somewhat larger percentage.) of all selected telephone numbers are nonworking, nonresidential, or unresolved, a large part of the interviewers' efforts may potentially go into simply identifying the status of these numbers. To reduce the size of the task and allow interviewers to focus more fully on household interviews, replicates are prescreened before they are loaded into the CATI system to identify as many unproductive telephone numbers as possible. Telephone numbers can be classified as not eligible for interviewing in three ways.

First, the replicates are prescreened for businesses. Typically, this means the replicates are matched against business directories, and any telephone number that matches a directory listing is classified as a business number and is made ineligible to be called in the survey. If the survey topic might be sensitive to the exclusion of households which share a business number, additional matching of the numbers classified as businesses might be made against residential listings and the ineligible status retained only if no match is found in this additional process.

Second, the remaining numbers are matched against residential directories and the matches are classified as eligible to be called in the survey. All remaining nonmatches (did not match the business directories and the residential directories) are run through a predictive dialer, which automatically detects nonworking numbers by unique signal tones issued by the telephony system, by extended periods of silence, or by continuous noise on the telephone line. The dialer also detects fax and modem numbers. Such numbers are classified as ineligible to be called in the survey.

Third, in the United States, a telephone number originally assigned as a landline number can now be ported to a cell telephone at the request of the subscriber. This means that even if the RDD sample is selected from landline telephone exchanges, it may, by the time of the RDD survey, contain a few cell-telephone numbers. This porting of numbers creates a legal problem that must be solved before dialing can commence in the survey. Except for emergency calls or calls made with the prior express consent of the person being called, the Telephone Consumer Protection Act of 1991 prohibits the use of automatic dialers in calling telephone numbers assigned to a cell telephones. (The act does not bar the manual dialing of cell-telephone numbers.) Because the typical RDD survey uses an automatic dialer, the replicates to be released are matched to a commercial database that contains all ported numbers in the nation. Matched numbers are made ineligible to be called in the RDD survey.

Finally, all telephone numbers within the prescreened replicates that are not designated as ineligible are loaded into the CATI system and are made ready for the launch of interviewing operations. In our recent experience, approximately 56–58% of the selected sample of telephone numbers are loaded to the CATI system and are eventually called. The remaining (42–44% of the numbers) are prescreened as ineligible. Clearly, prescreening saves interviewers a lot of work.

3.2.3. Sending the advance letter
When time and resources permit, an advance letter is mailed to the subscriber(s) of all selected telephone numbers for whom a mailing address is available and not removed in the prescreening process. The addresses are obtained by matching the replicates to be released to commercially available databases that contain telephone numbers and corresponding names and addresses. (These databases are populated with information from credit histories and other sources.) The advance letter explains the purpose of the survey and its importance, and identifies the sponsor and other pertinent facts about the task of completing the survey. The letters are usually mailed a fixed number of days before the expected release of a replicate, timed to arrive two to five days before the first dial is anticipated. The choice of mailing class (e.g., express, first class, or bulk) will depend on budget and available lead time. At the time of writing, an advance letter can be sent to about one-third of the cases in an overall sample in the United States. Approximately 50% of the cases loaded to the CATI system are sent an advance letter.

A well written and presented advance letter increases the rate of cooperation in a RDD survey. De Leeuw et al. (2007) in a meta-analysis found that cooperation was on average 11 percentage points higher among cases that receive an advance letter than among cases that do not receive a letter. The size of increase for any one study will vary depending on a range of factors including topic, letter, and timing, and because there is self-selection involved in which cases present themselves with a complete address for mailing the advance letter, we cannot conclude with certainty that the higher rate is "caused" entirely by the letter.

3.3. Case management

Once all sample preparation steps have been completed, the sample is released to CATI for actual interviewing operations. Usually, the sample is released by replicate or batches of replicates and the release may occur on a daily basis or according to some other schedule. The size of the release of virgin replicates is coordinated with the number of interviewer hours available for interviewing, the distribution of those hours by shift and time zone, and in consideration of the number of pending cases already scheduled for an additional call attempt. The number of virgin cases to be released is determined by taking the difference between the total number of calls that can be handled by the telephone center in a given time period and the number of pending cases already scheduled for a call in that period. If staffing exceeds the amount of work currently in the system, a release of virgin replicates is made. The size of the release may vary by stratum, with relatively more virgin replicates released in any strata that are lagging in achieving their target numbers of completed interviews.

Once replicates are released, the next tasks are to schedule the individual cases to be called, to track and report on their status, to re-schedule them for additional calls, as needed, and generally to bring as many cases to a completed status as possible.

3.3.1. Tracking of case status

To effectively manage the sample throughout the data collection period, it is essential to be able to track the status of cases in real time, each and every day. Such management requires four distinct types of codes assigned at the case level, for every case in the released sample, and the survey manager must have the capability to tabulate frequencies and cross-tabulations of these codes at any moment. The codes are updated with each call attempt. The four types of codes are the Life Cycle Stage (LCS) code, the call-outcome code, the finalization code, and the disposition code.

The *LCS code*, presented in Table 2, describes the overall status of a case in the CATI system. It describes the key stages that a case goes through as it progresses through the telephone interviewing system. It can be used to determine the scheduling and intensity of future call attempts. Before any live contact is established with a case, there is a good chance that no household exists at the end of that line, so the objective is to be able to clear that telephone number out of the system as fast as possible. Once live contact is established, it is usually possible to classify the case as a WRN or as a nonresidential number. The WRNs will likely be worthy of additional call attempts, and once a

Table 2
Life cycle stage codes

Code	Label
0	Virgin (fresh) cases
1	No contact
2	No live contact but possible household
3	Live contact and likely household
4	Screened household

household has completed the screening interview and is determined to be eligible, it becomes the subject of still more call attempts, with the option to restrict these cases so that they are handled only by interviewers with advanced refusal conversion skills. The marginal cost of obtaining a completed interview from an already-screened household is generally relatively low. Thus, cases in LCS 4 receive the highest and most focused level of effort.

The LCS code follows a ratchet system that can move forward but not backward. For example, a case that was busy on the last call and on all previous calls will be in LCS 1. However, a case that was busy on the last call but was previously a refusal at the interview level will be in LCS 4, which captures the fact that the case has been identified as an eligible household.

The *call-outcome code* describes the state of a case based on the outcome of the last call attempt. Table 3 gives some examples of possible call-outcome codes. The combination of the LCS and call-outcome codes suggest what rule should be used to determine the schedule for the next call attempt, if any. If the call outcome is a busy signal, it is evident that someone is at home, and it is probably worthwhile to callback within the hour or half-hour. On the other hand, if the call outcome is a refusal, it is probably worthwhile to delay the next callback for several days, assuming the data collection period is long enough, to allow for a cooling-off period.

The *finalization code* is a simple indicator of whether a case has been finalized or not. Completed interviews, ineligible households, and nonworking or nonresidential telephone numbers are considered finalized, meaning no additional callbacks are planned for them. Cases not finalized require additional callbacks. In an RDD survey, the statistician may develop and use alternative rules to designate when a case is finalized. For example, a case may become final after

- a specified total number of call attempts
- a specified number of call attempts within each of several day-parts or shifts (weekday evenings, weekend days, and so forth)
- a specified number of call attempts after reaching LCS 4 status.

The *disposition code* for a case summarizes the current evidence with respect to its resolution as a WRN or not, its screening interview, and (if eligible) its main interview. The disposition code is a synthesis of the current LCS code, the sequence of call-outcome codes for all call attempts to date, and the current finalization code. We defer further discussion of disposition codes until Section 3.4.

3.3.2. Shift and resource attributes

Call scheduling is driven by a number of factors, such as shift and resource attributes, time zone, and current LCS and call-outcome codes. There may be several shift types used in the RDD survey, including weekday days, evenings, and nights, and weekend days, evenings, and nights. A shift attribute is assigned to the next call for each case in the active sample.

There may be several resource types employed in the RDD survey, including regular interviewers, interviewers with refusal conversion skills, special language (e.g., Spanish) interviewers, and special language refusal converters. A resource attribute is assigned to the next call for each case in the active sample.

Table 3
Illustrative call-outcome codes

Code	Label	Usage Notes
01	Engaged/busy	Non-autodisposition engaged/busy
02	No reply [no answer]	Non-autodisposition ring no answer/no reply
03	New phase	Virgin sample
04	Refusal HUDI	Used when a respondent hangs up during the introduction (HUDI). Respondent does not speak to interviewer at all.
05	Refusal (gatekeeper)	Used when an adult, noneligible respondent refuses. Only available if specific respondent has already been selected.
06	Refusal (soft)	Adult refuses to participate in the study after full introduction.
07	Refusal (hostile)	Used when respondents use profanity towards interviewer in a threatening manner or threatens any legal or governmental action. Reviewed by supervisors before being finalized.
08	Made soft appointment	Used when a household member gives callback time for a respondent.
09	Made hard appointment	Used when respondent gives you a specific callback time for them.
10	Unable to proceed: general call back	Used when no specific appointment information was given but recontact is viable.
11	Unable to proceed: supervisor review requested	Used when Supervisor review is required.
12	Privacy manager (known housing unit)	Used if unable to successfully bypass a privacy manager machine - known housing unit.
13	Privacy manager (unknown if housing unit)	Used if unable to successfully bypass a privacy manager machine - unknown if housing unit.
16	Answering machine: message left (known housing unit)	Used when a call machine is reached and a message is left: known housing unit.
17	Answering machine: message left (unknown if housing unit)	Used when a call machine is reached and a message is left - unknown if housing unit.
18	Business/government	Used when a business or home business line (not used for personal calls) or government office is reached.
19	Dorm/prison/hostel	Used when a dorm is reached.
20	Cell phone/mobile/GPS phone	Used when number dialed belongs to a cell phone or other mobile device.
21	Call forwarding	Used when number dialed is permanently forwarded to a different number.
22	Fast busy	Used when a fast busy signal is received.
23	Disconnect/temporarily disconnected	Used when a number is permanently disconnected, temporarily disconnected, or has been changed.
25	Fax/modem/data line	Used when a fax/modem signal is received.
32	HH ineligible (no eligible person in HH)	Screener complete, no household members eligible

The next call for a case in the active sample is scheduled in light of its current LCS and call-outcome codes, the number of times the case was already called in the various shift types, its time zone, and the availability of an appropriate resource in the interviewer pool.

3.3.3. Three phases of the data-collection period

Interviewing operations proceed through three distinct phases during the data-collection period. During the initial start-up phase, the active sample will be dominated by virgin replicates. At this time, few refusal converters are needed. During the longer middle phase, the active sample is characterized by a mixture of virgin and pending cases. The final phase, also called the close-down period, begins at the point in time when the last replicate is released. Because no virgin replicates are introduced during this period, and as noncontact cases are retired, the active sample will come to be dominated more and more by difficult and eligible cases. At this time, a larger number of specialized interviewers are needed to deal with partially completed interviews, refusals and hidden refusals, appointments, and cases that require a specialized language.

3.3.4. Staffing and staff scheduling

Once the middle phase of data collection is reached, analysis of the most productive times to call and the dynamics of the call scheduling rules (particularly if shift types are used) will produce a distinct pattern of dials that must be performed at different times of the day and week. For example, many companies concentrate RDD dialing into weekday evenings, when households are more likely to be at home, leaving the day only lightly staffed.

This pattern is further complicated by the number of time zones that a study is spanning from a single site. For example, if two-thirds of the sample is in the Eastern time zone and the remainder is in the Central time zone, and if the sample size requires 10 booths to be active at 9:00 am Eastern time, then by 10:00 am Eastern time another 5 will need to be added for a total of 15. If a sharp distinction is drawn between midweek evening and midweek days with volumes increasing four-fold, the booth requirements at 5:00 pm will jump to 45 once the Eastern evening dialing starts, and up to 60 at 6:00 pm when the Central evening dialing starts, dropping to only 20 at 9:00 pm Eastern time when the Eastern dialing closes for the day. If only five interviewers were dialing at 9:00 am, a proportion of the Eastern sample would not be dialed in that first hour and would back up for later dialing at a less optimal time of day or even to the next day. Conversely, if 20 interviewers started dialing at 9:00 am, then the day work would be exhausted by mid-afternoon, and either the interviewers would sit idle, or cases requiring a night time call would be dialed far too early with a much lower chance of success. While the patterns of calling by time zone cited here are merely illustrative, call center managers need to understand their own survey-specific sample sizes and interviewer capacity to schedule and manage the staff.

3.4. Case disposition

For an RDD survey, just as for most surveys, the statistician must provide an explicit definition of what constitutes a completed interview. The definition may hinge on such factors as whether key items were completed, whether a majority of items were completed, or whether a certain range of sections of the questionnaire were completed before break-off, if any.

Given this definition, telephone numbers in the released sample should move through the CATI system until they finalize as resolved-nonhousehold, ineligible-household, or eligible-household-completed-interview. The cases that are not finalized directly are eventually finalized as unresolved, unscreened-households, or eligible-household-incomplete-interview.

After going through the prescreening, calling, and interviewing process, all cases in an RDD survey must end up receiving final disposition codes based on their prescreening statuses, life cycle, and call-outcome codes. The final disposition codes must be defined at a sufficient level of detail so that the cases can be treated appropriately in estimation, analysis, and reporting of response rates.

The American Association for Public Opinion Research has developed standard definitions for disposition categories for RDD surveys (AAPOR, 2006). At the highest level, the AAPOR codes are:

(1) Interview includes fully complete interviews or partial interviews with all necessary questions answered.

(2) Eligible, noninterview includes cases that are resolved as WRNs and are screened as eligible for the survey but have not completed the interview.

(3) Unknown eligibility, noninterview includes cases that are unresolved for WRN status or are resolved as WRN but the screener interview to determine eligibility was incomplete.

(4) Not eligible includes cases that are not WRNs and cases that are WRNs but are screened as ineligible.

For a listing of the detailed AAPOR codes, see Table 1 in the document www.aapor.org/uploads/standarddefs_4.pdf.

While the detailed categories of AAPOR can be very helpful to data-collection managers in diagnosing problems and opportunities and to survey statisticians in planning sample sizes and data-collection operations for future surveys, they provide more detail than is necessary to conduct estimation and analysis for the current survey. For the latter purposes, we find that a simpler set of final disposition categories is sufficient and fully acceptable. We present our set of final disposition categories in Table 4 along with a cross-walk that demonstrates the link between the AAPOR detailed categories and our final disposition categories. The next subsection shows how these final categories support the calculation of response rates for an RDD survey, and Section 4 demonstrates how the final categories support the survey estimation procedure.

To give an illustration of the distribution of an RDD sample by final disposition categories, we have analyzed recent RDD surveys conducted at NORC. Figure 1 presents the recent trend in case dispositions found in our surveys. WRNs comprise categories U1, J, ER, and C. Categories V and UN are unresolved, whereas all other categories are resolved. Notice the downward trend in the proportion of the sample classified as resolved WRNs. This trend is due to the evolving telephony infrastructure in the United States.

Table 4
Cross-walk between final disposition categories and AAPOR categories

Final Categories	Meaning	AAPOR Categories
V	Virgin, cases released to telephone center but never dialed	3.11
UN	Unresolved telephone number	3.10 (except 3.11)
D	Nonworking, out-of-scope	4.20 + 4.30 + 4.40
NR	Nonresidential, out-of-scope	4.50
U1	Known household, screening incomplete	3.20
J	Screened household, not eligible	4.70
ER	Eligible respondent, incomplete interview, or refusal	2.0
C	Completed interviews	1.0

Note: AAPOR categories 3.90, 4.10, and 4.80 are not used. 3.90 is ambiguous and should be classified to either UN or U1, as the case may be; 4.10 should not exist in a properly managed survey; 4.80 does not apply in the case of strict probability sampling discussed here.

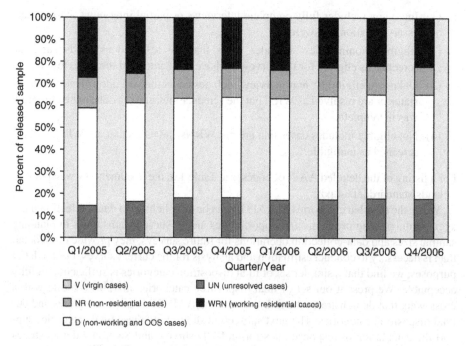

Fig. 1. Recent trends in the classification of cases in RDD surveys

3.5. Measures of response rates

It is important to report the response rate for an RDD survey, just as it is for any other survey. This rate stands as an indicator of the risk of nonresponse bias that may be present in survey estimators. Features of RDD surveys make the calculation of the response rate less than completely straightforward. To illustrate the principles, we take the situation in which a brief screening interview is administered to households to ascertain eligibility for the survey, followed by a main interview administered only to eligible households. Given this situation, what is the appropriate response rate?

The *response rate* is a summary measure used to designate the ratio of the number of completed interviews to the number of eligible units in the sample. It is a measure of the result of all efforts, properly carried out, to execute the RDD survey. Since this definition is supported by the Council of American Survey Research Organizations (CASRO), we often call it the CASRO response rate. Assumptions or estimation is usually required to work out the denominator of the response rate.

In general, the response rate is defined by

$$r = \frac{C}{C + ER + \lambda_1 U1 + \lambda_2 \phi UN} = \frac{\text{completed interviews}}{\text{eligible cases in the released sample}}, \quad (6)$$

where λ_1 is the unknown proportion of unscreened households that are in fact eligible for the survey, ϕ is the unknown proportion of unresolved telephone numbers that are in fact WRNs, and λ_2 is the unknown proportion of unresolved WRNs that are in fact eligible for the survey. Equation (6) also agrees with response rate 3 defined in AAPOR (2006). Because the rates in the denominator of the response rate are unknown, we estimate them from the sample itself by

$$\hat{\lambda}_1 = \frac{C + ER}{C + ER + J} = \frac{\text{eligible cases}}{\text{eligible and ineligible cases}}, \quad (7)$$

$$\hat{\phi} = \frac{C + ER + J + U1}{C + ER + J + U1 + D + NR} = \frac{\text{WRNs}}{\text{resolved cases}}, \quad (8)$$

and

$$\hat{\lambda}_2 = \hat{\lambda}_1. \quad (9)$$

For simplicity, we are using a system of notation wherein the same symbol is used to represent both the final disposition category, the set of survey cases classified in the category, and the cardinality of the set. For example, C designates both the category of completed interviews, the set of cases with a completed interview, and the number of cases that achieved a completed interview.

Given assumptions (6)–(8), the estimated response rate becomes

$$\hat{r} = \frac{C}{C + ER + \hat{\lambda}_1 U1 + \hat{\lambda}_2 \hat{\phi} UN} = c_R c_S c_I, \quad (10)$$

where

$$c_R = \frac{C + ER + J + U1 + D + NR}{C + ER + J + U1 + D + NR + UN}$$

$$= \frac{\text{resolved telephone numbers}}{\text{total telephone numbers in released sample}} \quad (11)$$

is the *resolution completion rate*;

$$c_S = \frac{C + ER + J}{C + ER + J + U1} = \frac{\text{screening interviews}}{\text{WRNs}} \quad (12)$$

is the *screener completion rate*; and

$$c_I = \frac{C}{C + ER} = \frac{\text{completed interviews}}{\text{eligible households}} \tag{13}$$

is the *interview completion rate*. The completion rates are useful not only for computing the response rate but also for planning, monitoring, and managing the RDD survey.

Massey (1995) and Ezzati-Rice et al. (2000) propose an alternative to assumptions (6)–(8) that takes account of both nonresponse and undercoverage, if any, in the number of identified eligible households in the survey. Some authors prefer to quote weighted response rates. To implement this idea, one could let C, ER, and so on be the sum of the base weights of the members of the corresponding set, instead of simply the cardinality of the set. The use of the weights could be important for an RDD survey that employs differential sampling rates from stratum to stratum.

4. Analysis of RDD surveys

In this section, we discuss the procedures used to develop *weights* for the survey respondents. Properly calculated weights are needed to provide essentially unbiased estimators of population parameters. Given a large-scale RDD survey, an estimator of the population total, Y, is of the general form

$$\hat{Y} = \sum_{i \in C} W_i Y_i, \tag{14}$$

where C is the set of completed interviews, Y_i is the characteristic of interest for the ith completed interview, and W_i is the survey weight for the ith completed interview. Estimators of proportions, ratios, regression coefficients, and the like are calculated as functions of estimated totals.

Weighting begins with the construction of the probabilities of selection for the telephone numbers in the released sample. The *base weights*, or the reciprocals of the probabilities of selection, implement the Horvitz and Thompson (1952) estimator of the population total. Several adjustments to the weights are required, each being designed to compensate for missing data at the several steps in the survey response process. The steps usually include nonresolution of telephone numbers as to their residential or nonresidential status; nonresponse of households to the screening interview, if any, designed to determine the persons eligible for the survey; and nonresponse of eligible respondents to the main interview. In addition, weights should be adjusted to correct for any multiple probabilities of selection and to compensate for any households missing from the sampling frame because they do not have landline telephones. Finally, weights are usually benchmarked to external population control totals. In what follows, we define all these steps in weighting. Some steps may not apply to all RDD surveys.

4.1. Base weights

The list-assisted RDD survey design is essentially a simple random sample without replacement of telephone numbers within stratum, with independent sampling from one stratum to the next. Let N_h be the size of the population of all landline telephone numbers in stratum h and let n_h be the size of the sample selected and released. Then, the base weight for the kth telephone number in the set of released telephone numbers, A, is defined by

$$W_{1k} = 1/\pi_{1k} = N_h/n_h, \quad \text{if } k \in A, \tag{15}$$

where π_{1k} is the probability of selection given the sampling design. The base weight is a constant for all released telephone numbers in a stratum.

4.2. Adjustment for nonresolution of telephone numbers

As we have seen, the RDD frame is highly inefficient with more than 70% of the telephone numbers being out-of-scope, either nonworking or nonresidential. The first step in the survey response process is to identify the WRN status of all released telephone numbers. Even after repeated call attempts, some telephone numbers will inevitably remain unresolved, meaning that there was not enough evidence collected to classify them as residential, nonresidential, or nonworking. To compensate for the missing classifications, a nonresponse adjustment is conducted within cells. The adjustment is applied by forming adjustment cells within each stratum and assuming that the WRN rate within a cell is the same for both the resolved and unresolved cases. The weights of the unresolved telephone numbers are distributed to the weights of the resolved cases in the same adjustment cell. For the kth resolved telephone number within the ℓth adjustment cell, the nonresolution adjusted weight is defined by

$$W_{2k} = W_{1k}/\pi_{2\ell}, \quad \text{if } k \in B \cap \ell, \tag{16}$$

where B is the subset in A of resolved telephone numbers, and

$$\pi_{2\ell} = \sum_{k \in B \cap \ell} W_{1k} \Bigg/ \sum_{k \in A \cap \ell} W_{1k} \tag{17}$$

is the weighted resolution completion rate for the ℓth adjustment cell.

4.3. Adjustment for nonresponse of households to the screening interview

Once a WRN is identified, it is screened for eligibility. The target population of many surveys does not include all persons within the household. A given RDD survey may only target persons of a specific age, sex, education level, or health condition. The brief screening interview, conducted prior to the main interview, is intended to identify the members of the target population living in households. Despite repeated callbacks, the screener is sometimes left missing due to refusal or noncontact. To account for such nonresponse, the weights of the WRNs with completed screening interviews are adjusted to account for the nonresponding WRNs within cells. The screener nonresponse adjusted weight of the kth screener complete in the mth nonresponse adjustment cell is defined by

$$W_{3k} = W_{2k}/\pi_{3m}, \quad \text{if } k \in S \cap m, \tag{18}$$

where S is the subset in B_1 consisting of screener completes, B_1 is the subset in B consisting of the resolved WRNs, and

$$\pi_{3m} = \sum_{k \in S \cap m} W_{2k} \Big/ \sum_{k \in B_1 \cap m} W_{2k} \qquad (19)$$

is the weighted screener completion rate for the mth adjustment cell.

4.4. Adjustment for selection of eligible respondents

In some surveys, once the eligible persons are identified within the household, one or more of them are selected for participation in the survey. Let E be the subset in S of households with eligible persons, P be the set of eligible persons identified in completed screening interviews, and P_1 be the subset in P of persons selected for the survey interview. Then, an adjustment for subsampling is defined by

$$W_{4i} = W_{3k}/\pi_{4i}, \quad \text{if } i \in k \text{ and } i \in P_1, \qquad (20)$$

where i designates the selected, eligible person in household k, π_{4i} is the conditional probability of selecting the ith eligible person given that their household screening interview was completed, and W_{3k} is the screener-nonresponse adjusted weight from the previous weighting step. For example, if one eligible person was selected at random from a household containing three eligible persons, then $\pi_{4i} = 1/3$. If the survey calls for interviewing all eligible persons within the household, then $\pi_{4i} = 1$ and $W_{4i} \equiv W_{3k}, i \in k$.

4.5. Adjustment for nonresponse of eligible persons to the main interview

Following screening and subsampling, the survey's main interview is administered to the selected eligible persons. At this juncture, further nonresponse is likely due to refusals and not-at-homes. An interview nonresponse adjustment is applied within cells to account for the eligible cases that fail to provide a completed interview. The interview nonresponse-adjusted weight for the ith eligible person with a completed interview in the qth adjustment cell is defined by

$$W_{5i} = W_{4i}/\pi_{5q}, \quad \text{if } i \in C \cap q, \qquad (21)$$

where C is the subset in P_1 of eligible persons who completed the main interview, and

$$\pi_{5q} = \sum_{i \in C \cap q} W_{4i} \Big/ \sum_{i \in P_1 \cap q} W_{4i} \qquad (22)$$

is the interview completion rate within the qth cell.

Table 5 presents a summary of a typical response pattern in an RDD survey, showing how the units with various disposition codes are treated in different weighting steps.

Table 5
Summary of hierarchical response pattern and mix of units for a typical RDD survey

Disposition Codes	Released Sample of Telephone Numbers	Resolution Status (Telephone Numbers)	WRNs (Telephone Numbers)	Screening Status (Telephone Numbers)	Eligibility Status (Households)	Subsampling Status (Eligible Persons)	Interview Status (Eligible Persons)
C: complete interview	A = set of all telephone numbers in the released sample	B = set of all resolved telephone numbers	B_1 = set of all resolved WRNs	S = set of WRNs that responded to the screening interview	E = set of households with one or more eligible persons	P_1 = set of selected, eligible persons	C = set of respondents to the main interview
ER: incomplete interview							$P_1 \cap C^c$ = selected, eligible persons that did not complete the main interview
Eligible persons subsampled out						$P \cap P_1^c$ = set of nonselected eligible persons	
J: screened household, no eligible person					$S \cap E^c$ = set of ineligible households		
U1: known household, incomplete screening interview				$B_1 \cap S^c$ = set of WRNs that did not complete the screening interview			
NR: nonresidential, out-of-scope			$B \cap B_1^c$ = set of resolved non-working or non-residential numbers				
D: nonworking, out-of-scope							
I: answering machine (status unresolved)		B^c = set of all unresolved telephone numbers					
NC: noncontact (status unresolved)							
U2: possible household (status unresolved)							

4.6. Forming adjustment cells

A different set of nonresponse adjustment cells may be formed and used at the different steps in the RDD response process. As would be typical of any sample survey, cells in an RDD survey are formed to achieve two ends: (i) units within a cell should be alike with respect to the survey characteristics of interest, and (ii) the survey response rates should vary from cell to cell. Cells are usually, though not necessarily always, nested within the sampling strata. The directory-listed status is often found to be a significant correlate of the nonresponse mechanism at each of the resolution, screening, and interviewing steps and is almost always included in the cell structure within stratum.

At the resolution step, only sampling frame variables are available for cell formation, including such variables as directory-listed status; broad geographic location including census region and state; and environmental variables obtained at the census tract level corresponding to the approximate mapping between the telephone exchange and the tract. The latter include variables related to the distribution of the population with respect to age, sex, race/ethnicity, housing tenure, and income.

At the screening step, both frame variables and variables collected at the resolution step are available for cell formation, and at the interviewing step, frame variables, resolution variables, and variables collected at the screening step are available for cell formation. Usually, the resolution step adds little information, and the cell structure for the screener nonresponse adjustment is forced to rely mainly on frame variables. Depending on the RDD survey, some screening interviews collect substantial new information and some do not. Sometimes the screening interview is minimized to include only essential variables so as not to risk additional nonresponse at this step. The cell structure for the interview nonresponse adjustment may be relatively more or less articulated depending on how many additional useful variables are obtained in the screening interview.

4.7. Adjustment for multiple probabilities of selection

Many households have two or more landlines that may be used for voice communications (excluding the lines used only for fax or computer communications), and the interview respondents in these households have multiple chances of selection into the survey. The increased probabilities of selection were not known at the time the base weights were formed and, therefore, an adjustment for these probabilities is now in order. Without an adjustment, the extant weights would provide an upward biased estimator of the population total. Although we have not emphasized this point until now, at this step in the weighting process, we have an essentially unbiased estimator of the total population of eligible person/WRN pairs. An eligible person with two landlines appears twice in this total. The real parameter of interest, however, is the total (and functions thereof) of the population of eligible persons. A weight adjustment is needed to convert the weighted estimator to be an estimator of the real parameter of interest.

The number of voice landlines is collected during the interview, and the appropriate adjusted weight is given by

$$W_{6i} = W_{5i} / \min(t_i, t_o), \tag{23}$$

where t_i is the number of telephone lines in the household of the ith eligible respondent. The value t_o is used to cap the number of telephone lines used in the weight adjustment, both to control variability and to guard against reporting bias. For example, some surveys take $t_o = 3$.

It is possible that a household with multiple landlines is selected more than once in the sample. With time and expense, one can undertake efforts to identify the lines. To identify the landlines in the sample that are linked to the same household, all landlines of the responding household can be collected and the sample file of responding landlines can be checked to establish the link. However, to avoid the response burden and the costs of collecting and processing the information, and considering the negligible chance of selecting the household more than once in the sample, it is usually assumed in the weighting process that only one landline has been selected per household.

4.8. Adjustment for noncoverage of nonlandline households

Since the frame of a RDD survey only includes landline telephone numbers, the eligible persons living in households without landline service (including both nontelephone households and households with only cell-telephone service) are not covered by traditional RDD surveys. To compensate for this type of undercoverage, a certain poststratification adjustment can be used (Keeter, 1995). The basic idea is to use the households with an interruption in landline telephone service of one week or more during the past year to represent not only themselves but also the households without a landline telephone. Keeter (1995), Brick et al. (1996), and Srinath et al. (2002) showed that the socioeconomic characteristics of persons who live in households with interruptions of one week or more in landline telephone service within the past 12 months are similar to those who live in nontelephone households. This finding resonates with common sense because if a survey is conducted at a point in time when the household service is interrupted, then the household is necessarily considered as a part of the population of nontelephone households. Therefore, interviewed persons living in households with a recent interruption in landline telephone service can be used to represent persons living in nontelephone households.

Given this method, two cells are formed within a stratum by the telephone interruption status of the household as follows: 1) interruption of more than one week during the past 12 months; and 2) no interruption of more than one week during the past 12 months. Let T_{h1} denote the total number of eligible persons in the population either without a landline telephone or with a landline telephone with an interruption in service, and let T_{h2} denote the total number of eligible persons in the population with a landline telephone without an interruption in service, all in stratum h. Then, the adjusted weight for the ith person in C is defined by

$$W_{7i} = \left(T_{hg} \Big/ \sum_{i \in C \cap U_{hg}} W_{6i} \right) W_{6i}, \quad \text{if } i \in C \cap U_{hg}, \tag{24}$$

where g indexes the cell within stratum h, and U_{hg} denotes the population of eligible persons in households in cell (h, g). The population totals T_{h1} and T_{h2} are typically unknown and must be estimated from a census or a reliable reference survey.

4.9. Calibration to population control totals

The aim of the previous step is to reduce bias in the survey estimator due to the noncoverage of the population without landline telephone service. The RDD surveys may also be subject to differential coverage of the population by race/ethnicity and other factors. Like almost any census or survey, some categories of persons are under-reported at a higher rate than others. The penultimate step in RDD surveys is to correct for the differential undercoverage by calibrating the weights to independent population control totals. A poststratification (Holt and Smith, 1979) or raking-ratio type calibration (Brackstone and Rao, 1979) is commonly applied.

The requisite population control totals are obtained from a census or a reference survey and are typically available for socioeconomic characteristics such as age, sex, race/ethnicity, income or poverty status, education attainment, and telephone interruption status. The population control totals may be obtained for the population as a whole or within each stratum.

To illustrate, we give the adjustment corresponding to a raking-ratio calibration. Introduce the following additional notation:

a = iteration within the raking procedure

b = dimension (or classification variable) of the raking structure within the iteration

c = category (or value or the classification variable) within the dimension within the iteration

L^b = number of categories within the bth dimension

T_c^b = population control total for the cth category within the bth dimension of the raking structure

φ_{ci}^b = 1, if the ith eligible person is in the cth category of the bth dimension of the raking structure

= 0, otherwise

Then, the adjusted weights at the bth dimension of the ath iteration are given by

$$
W_{8i}^{a,b} = \left(\sum_{c=1}^{L^b} \varphi_{ci}^b \frac{T_c^b}{\sum_{i \in C} \varphi_{ci}^b W_{8i}^{a,b-1}} \right) W_{8i}^{a,b-1}, \tag{25}
$$

for eligible persons with a completed interview $i \in C$. The raking procedure cycles through each of the dimensions of one iteration before moving forward to the next iteration. The entry weights for the first dimension of an iteration are the exit weights following the last dimension of the previous iteration. The entire process opens with the weights from the prior step in weighting, $\{W_{7i}\}$ for $i \in C$. The process iterates until a user-specified convergence criteria is achieved.

For example, if population control totals are available for five age groups, two sex groups and four race/thnicity groups, then we have the raking structure

Dimension (b)	Classification Variable	Number of Categories (L^b)
1	Age	5
2	Sex	2
3	Race/ethnicity	4

4.10. Trimming of extreme weights

Since the weights in an RDD survey are derived through the foregoing series of adjustments, a few weights in some strata may end up being very large in comparison to other weights in the stratum. To avoid any undue influence of large weights on survey estimates and to control the sampling variance of the estimator, the extreme weights may be trimmed by using a suitable truncation or Winsorization procedure (see Fuller, 1991; Kish, 1992; Potter, 1990). After trimming, the calibration adjustment is reapplied to ensure the consistency of the resulting weights with the population control totals. In some cases, the process may have to be iterated a few times to achieve final weights that are at once void of extreme values and also consistent with the population control totals. Let $\{W_{9i}\}$ or simply $\{W_i\}$ denote the resulting, final weights.

4.11. Estimation and variance estimation

For a RDD survey, the survey weights are used to produce survey statistics and variance estimates in the usual way. To see this, expand the notation to let $\mathbf{Y_{hij}} = (Y_{1hij}, \ldots Y_{phij})'$ be a p-variate characteristic of interest reported by the jth eligible person selected in the ith household in the hth stratum, and W_{hij} be the corresponding final weight. Suppose the goal is to estimate a parameter $\theta = g(\mathbf{Y})$ of the eligible population, where $\mathbf{Y}$ is the vector of population totals and $g(\cdot)$ is a well-behaved differentiable function. Then, the estimator of the parameter of interest is $\hat{\theta} = g(\hat{\mathbf{Y}})$, where

$$\hat{Y}_r = \sum_{(h,i,j) \in C} W_{hij} Y_{rhij} \tag{26}$$

is a typical element of $\hat{\mathbf{Y}}$, $r = 1, \ldots, p$, and C is the set of eligible respondents. The usual Taylor series estimator of the variance is

$$v(\hat{\theta}) = \sum_h \frac{n_h}{n_h - 1} \sum_i \left(\sum_j W_{hij} \hat{V}_{hij} - \frac{1}{n_h} \sum_{i'} \sum_{j'} W_{hi'j'} \hat{V}_{hi'j'} \right)^2 \tag{27}$$

$$\hat{V}_{hij} = \sum_{r=1}^{p} \frac{\partial g(\hat{\mathbf{Y}})}{\partial y_r} Y_{rhij}, \tag{28}$$

where $\sum_h$ denotes a sum over strata, $\sum_i$ is a sum over the respondents' households within stratum, n_h is the number of such households, and $\sum_j$ is a sum over respondents within household. For a superior estimator of variance that takes into account the calibration done in the 9th step in weighting, see Sections 6.12 and 6.13 of Wolter (2007). Notice that the households act as the primary sampling units in an RDD survey.

Alternatively, a replication-type estimator of variance may be used, such as the jackknife estimator

$$v_J(\hat{\theta}) = \sum_h \frac{n_h - 1}{n_h} \sum_i \left(\hat{\theta}_{(hi)} - \hat{\theta}_{(h+)} \right)^2, \tag{29}$$

where

$$\hat{\theta}_{(hi)} = g(\hat{\mathbf{Y}}_{(hi)}) \tag{30}$$

is the estimator derived from the sample after omitting the (h, i)th household,

$$\hat{\theta}_{(h+)} = \frac{1}{n_h} \sum_i \hat{\theta}_{(hi)}, \tag{31}$$

$$\hat{Y}_{r(hi)} = \sum_{h'} \sum_{i'} \sum_{j'} W_{(hi)h'i'j'} Y_{rh'i'j'} \tag{32}$$

is a typical element of $\hat{\mathbf{Y}}_{(hi)}$ for $r = 1, \ldots, p$, and the replicate weights are defined by

$$
\begin{aligned}
W_{(hi)h'i'j'} &= W_{h'i'j'}, &&\text{if } h' \neq h \\
&= W_{h'i'j'} \frac{n_h}{n_h - 1}, &&\text{if } h' = h \text{ and } i' \neq i \\
&= 0, &&\text{if } h' = h \text{ and } i' = i.
\end{aligned}
\tag{33}
$$

For large RDD surveys, it will be prohibitive to compute replicate weights for the drop-out-one-completed-interview version of the jackknife. To better manage survey costs, one may use the survey replicates defined in Section 3 and compute the drop-one-replicate (or random group) version of the jackknife (see Wolter, 2007, Chapter 4).

One may use the estimated variances and covariances together with ordinary normal theory to make inferences regarding relationships and parameters of interest.

4.12. Difference in frame and actual location

In some cases, a telephone number is selected from one stratum, but the interview reveals that the respondent (or the respondent's housing unit) is actually in a different stratum. Actual location may be collected in the interview and used to define the estimation domains for analysis. (Actual location could also be used earlier in the implementation of the last several weight adjustments.)

Since the sample selection probability may vary substantially from one stratum to another, the movement of one or more respondents to a new estimation domain may add variation in the sampling weights within the domain. To protect against any such extra variation in weights, the weight of the reclassified respondent may be truncated if it is very large compared with all other weights in the domain. A multiple of the average weight in the domain is sometimes used as a cap for the weights.

Let δ_i^d be an indicator variable for the dth estimation domain based on actual location collected in the interview. Then the estimator of the domain total is defined by

$$\hat{Y}^d = \sum_{i \in C} W_i \delta_i^d Y_i. \tag{34}$$

Variance estimation occurs as defined in Section 4.10, with the variable $Y_i^d = \delta_i^d Y_i$ replacing the original variable Y_i.

Essential Methods for Design Based Sample Surveys
ISSN: 0169-7161

4

DOI: 10.1016/B978-0-444-53734-8.00004-2

Statistical Disclosure Control for Survey Data

Chris Skinner

1. Introduction

1.1. The problem of statistical disclosure control

Survey respondents are usually provided with an assurance that their responses will be treated confidentially. These assurances may relate to the way their responses will be handled within the agency conducting the survey or they may relate to the nature of the statistical outputs of the survey as, for example, in the "confidentiality guarantee" in the United Kingdom (U.K.) National Statistics Code of Practice (National Statistics, 2004, p. 7) that "no statistics will be produced that are likely to identify an individual." This chapter is concerned with methods for ensuring that the latter kinds of assurances are met. Thus, in the context of this chapter, *statistical disclosure control* (SDC) refers to the methodology used, in the design of the statistical outputs from the survey, for protecting the confidentiality of respondents' answers. Methods relating to the first kind of assurance, for example, computer security and staff protocols for the management of data within the survey agency, fall outside the scope of this chapter.

There are various kinds of *statistical outputs* from surveys. The most traditional are tables of descriptive estimates, such as totals, means, and proportions. The release of such estimates from surveys of households and individuals have typically not been considered to represent a major threat to confidentiality, in particular because of the protection provided by sampling. Tabular outputs from the kinds of establishment surveys conducted by government have, however, long been deemed risky, especially because of the threat of disclosure of information about large businesses in cells of tables which are sampled with a 100% sampling fraction. SDC methods for such tables have a long history and will be outlined in Section 2.

Although the traditional model of delivering all the estimates from a survey in a single report continues to meet certain needs, there has been increasing demand for more flexible survey outputs, often for multiple users, where the set of population parameters of interest is not prespecified. There are several reasons why it may not be possible to prespecify all the parameters. Data analysis is an iterative process, and what analyses are of most interest may only become clear after initial exploratory analyses of the data. Moreover, given the considerable expense of running surveys, it is natural for many commissioners of surveys to seek to facilitate the use of the data by multiple users. But it is usually impossible to prespecify all possible users and their needs in advance. A natural way to provide flexible outputs from a survey to address such needs is to make the survey microdata available, so that users can carry out the statistical analyses that interest them.

However, the release of such microdata raises serious confidentiality protection issues. Of course, statistical analyses of survey data do not require that the identities of the survey units are known. Names, addresses, and contact information for individuals or establishment can be stripped from the data to form an *anonymized* microdata file. The problem, however, is that such basic anonymization is often insufficient to protect confidentiality, and therefore, it is necessary to use one of a range of alternative approaches to SDC and this will be discussed further in Section 3.

1.2. Concepts of confidentiality, disclosure, and disclosure risk

To be precise about what is meant by "protecting confidentiality" requires discussion of definitions. These usually involve the notion of a hypothetical *intruder* who might seek to breach confidentiality. There are thus three key parties: (1) the *respondent* who provides the data, (2) the *agency* which collects the data, releases statistical outputs, and designs the SDC strategy, and (3) the hypothetical *intruder* who has access to these outputs and seeks to use them to disclose information about the respondent. One important notion of disclosure is *identity disclosure* or *identification,* which would occur if the intruder linked a known individual (or other unit) to an individual microdata record or other element of the statistical output. Another important notion is *attribute disclosure,* which would occur if the intruder could determine the value of some survey variable for an identified individual (or other unit) using the statistical output. More generally, *prediction disclosure* would occur if the intruder could predict the value of some survey variable for an identified individual with some uncertainty. When assessing the potential for disclosure for a particular statistical output, it is usual to refer to the *disclosure risk.* This might be defined as the probability of disclosure with respect to specified sources of uncertainty. Or the term might be used loosely to emphasize not only the uncertainty about potential disclosure but also the potential harm that might arise from disclosure (Lambert, 1993). The *confidentiality* of the answers provided by a respondent might be said to be protected if the disclosure risk for this respondent and the respondent's answers is sufficiently low. In this chapter, disclosure risk is discussed in more detail in Sections 2 and 3. For further discussion of definitions of disclosure, see Duncan and Lambert (1986, 1989) and Skinner (1992).

1.3. Approaches to protecting confidentiality

If the disclosure risk is not deemed to be sufficiently low, then it will be necessary to use some method to reduce the risk. There are broadly two approaches, which are referred to here as *safe setting* and *safe data* (Marsh et al., 1994). The safe setting approach imposes restrictions on the set of possible users of the statistical output and/or on the ways that the output can be used. For example, users might be required to sign a licensing agreement or might only be able to access microdata by visiting a secure laboratory or by submitting requests remotely (National Research Council, 2005). The safe data approach, on the other hand, involves some modification to the statistical output. For example, the degree of geographical detail in a microdata file from a national social survey might be limited so that no area containing less than 100,000 households is identified. In this chapter, we focus on the safe data approach and generally refer to methods for modifying the statistical output as SDC methods.

1.4. SDC methods, utility, and data quality

SDC methods vary according to the form of the statistical output. Some simple approaches are as follows:

- *Reduction of detail*, for example, the number of categories of a categorical variable might be reduced in a cross-classified table or in microdata.
- *Suppression*, for example, the entry in a table might be replaced by an asterisk, indicating that the entry has been suppressed for confidentiality reasons.

In each of these cases, the SDC method will lead to some *loss of information* for the user of the statistical output. Thus, the method will reduce the number of population parameters for which a user can obtain survey estimates. Other kinds of SDC methods might not affect the number of parameters which can be estimated but may affect the *quality* of the estimates that can be produced. For example, if *random noise* is added to an income variable to protect confidentiality, then this may induce bias or variance inflation in associated survey estimates. The general term *utility* may be used to cover both the information provided by the statistical outputs, for example, the range of estimates or analyses which can be produced, and the quality of this information, for example, the extent of errors in these estimates. It should, of course, be recognized that survey data are subject to many sources of error, even prior to the application of SDC methods, and the impact of SDC methods on data quality therefore needs to be considered in this context.

Generally, utility needs to be considered from the perspective of a *user* of the statistical outputs, who represents a key fourth party to add to the three parties referred to earlier: the respondent, the agency, and the intruder.

1.5. SDC as an optimization problem: the risk-utility trade-off

The key challenge in SDC is how to deal with the trade-off between disclosure risk and utility. In general, the more the disclosure risk is reduced by an SDC method, the lower will be the expected utility of the output. This trade-off may be formulated as an optimization problem. Let D be the (anonymized) survey data and let $f(D)$ be the statistical output, resulting from the use of an SDC method. Let $Rf(D)$ be a measure of the disclosure risk of the output, and let $Uf(D)$ be a measure of the utility of the output. Then, the basic challenge of SDC might be represented as the constrained optimization problem:

for given D and ε, find an SDC method, $f(.)$, which

maximizes $U[f(D)]$, subject to $R[f(D)] < \varepsilon$.

The elements of this problem need some clarification:

$f(.)$: the *SDC method*—a wide variety of these have been proposed and we shall refer to some of these in this chapter;

$R(.)$: the *disclosure risk function*—we shall discuss ways in which this function may be defined; this is certainly not straightforward, for example, because of its dependence on assumptions about the intruder and because of the challenge of combining the threats of disclosure for multiple respondents into a scalar function;

$U(.)$: the *utility function*—this will also not be straightforward to specify as a scalar function, given the potential multiple uses of the output;

ε: the *maximum acceptable risk*—in principle, one might expect the agency to provide this value in the light of its assurances to respondents. However, in practice, agencies find it very difficult to specify a value of ε, other than zero, that is, no disclosure risk. Unfortunately, for most definitions of disclosure risk, the only way to achieve no disclosure risk is by not releasing any output and this is rarely a solution of interest!

Given these difficulties in specifying $R(.)$ and $U(.)$ as scalar functions and in specifying a value for ε, the above optimization problem serves mainly as conceptual motivation. In practice, different SDC methods can be evaluated and compared by considering the values of alternative measures of risk and utility. For given measures of each, it can sometimes be useful to construct an RU map (Duncan et al., 2001), where a measure of risk is plotted against a measure of utility for a set of candidate SDC methods. The points on this map are expected to display a general positive relationship between risk and utility, but one might still find that, for given values of risk, some methods have greater utility than others and thus are to be preferred. This approach avoids having to assume a single value of ε.

2. Tabular outputs

2.1. Disclosure risk in social surveys and the protection provided by sampling

The main developments in SDC methods for tabular outputs have been motivated by the potential risks of disclosure arising when 100% sampling has been used, such as in censuses or in administrative data. Frequency tables based upon such data sources may often include small counts, as low as zero or one, for example, in tables of numbers of deaths by area by cause of death. Such tables might lead to identity disclosure, for example, if it is public knowledge that someone has died, then it might be possible to identify that person as a count of one in a table of deaths using some known characteristics of that person. Attribute disclosure might also occur. For example, it might be possible to find out the cause of the person's death if the table cross-classifies this cause by other variables potentially known to an intruder.

In social surveys, however, the use of sampling greatly reduces the risks of such kinds of disclosure for two reasons. First, the presence of sampling requires different kinds of statistical outputs. Thus, the entries in tables for categorical variables tend to be weighted proportions (possibly within domains defined by rows or columns) and not unweighted sample counts. Even if a user of the table could work out the cell counts (e.g., because the survey uses equal weights and the sample base has been provided), the survey agency will often ensure that the published cells do not contain very small counts, where the estimates would be deemed too unreliable due to sampling error. For example, the agency might suppress cell entries where the sample count in the cell falls below some threshold, for example, 50 persons in a national social survey. This should prevent the kinds of situations of most concern with 100% data. Sometimes, agencies use techniques of small area estimation in domains with small sample counts and these techniques may also act to reduce disclosure risk.

Second, the presence of sampling should reduce the precision with which an intruder could achieve predictive disclosure. For example, suppose that an intruder could find out from a survey table that, among 100 respondents falling into a certain domain, 99 of them have a certain attribute and suppose that the intruder knows someone in the population who falls into this domain. Then, the intruder cannot predict that this person has the attribute with probability 0.99, since this person need not be a respondent and prediction is subject to sampling uncertainty. This conclusion depends, however, on the identities of the survey respondents being kept confidential by the agency, preventing the intruder knowing whether the known person is a respondent, referred to as *response knowledge* by Bethlehem et al. (1990). In general, it seems very important that agencies do adopt this practice since it greatly reduces disclosure risk while not affecting the statistical utility of the outputs. In some exceptional cases, it may be difficult to achieve this completely. For example, in a survey of children it will usually be necessary to obtain the consent of a child's parent (or other adult) in order for the child to take part in the survey. The child might be assured that their responses will be kept confidential from their parent. However, when examining the outputs of the survey, the parent (as intruder) would know that their child was a respondent.

For the reasons given above, disclosure will not generally be of concern in the release of tables of estimates from social surveys, where the sample inclusion probabilities are small (say never exceeding 0.1). See also Federal Committee on Statistical Methodology (2005, pp. 12–14).

2.2. Disclosure risk in establishment surveys

A common form of output from an establishment survey consists of a table of estimated totals, cross-classified by characteristics of the establishment. Each estimate takes the form $\hat{Y}_c = \sum_s w_i I_{ci} y_i$, where w_i is the survey weight, I_{ci} is a 0–1 indicator for cell c in the cross-classification, and y_i is the survey variable for the ith establishment in the sample s. For example, y_i might be a measure of output and the cells might be formed by cross-classifying industrial activity and a measure of size.

The relevant definition of disclosure in such a setting will often be a form of prediction disclosure. Prediction disclosure for a specific cell c might be defined under the following set-up and assumptions:

- the intruder is one of the establishments in the cell which has the aim of predicting the value y_i for one of the other establishments in the cell or, more generally, the

intruder consists of a *coalition* of m of the N_c establishments in the cell with the same predictive aim;
- the intruder knows the identities of all establishments within the cell (since, e.g., they might represent businesses competing in a similar market).

Given such assumptions, prediction disclosure might be said to occur if the intruder is able to predict the value y_i with a specified degree of precision. To clarify the notion of precision, we focus in the next subsection on the important case where the units in the cell all fall within completely enumerated strata. Thus, $w_i = 1$ when $I_{ci} = 1$ so that $\hat{Y}_c = \sum_{U_c} y_i$, where U_c is the set of all establishments in cell c and N_c is the size of U_c. In this case, the intruder faces no uncertainty due to sampling and this might, therefore, be treated as the worst case.

2.2.1. Prediction disclosure in the absence of sampling

In the absence of sampling, prediction is normally considered from a deterministic perspective and is represented by an interval (between an upper and lower bound) within which the intruder knows that a value y_i must lie. The precision of prediction is represented by the difference between the true value and one of the bounds. It is supposed that the intruder undertakes prediction by combining prior information with the reported value $\hat{Y}_c$.

One approach to specifying the prior information is used in the *prior-posterior rule* (Willenborg and de Waal, 2001), also called the pq rule, which depends upon two constants, p and q, set by the agency. The constant q is used to specify the precision of prediction based upon the prior information alone. Under the pq rule, it is assumed that intruder can infer the y_i value for each establishment in the cell to within $q\%$. Thus, the agency assumes that, prior to the table being published, the intruder could know that a value y_i falls within the interval $[(1 - q/100)y_i, (1 + q/100)y_i]$. The combination of this prior information with the output $\hat{Y}_c = \sum_{U_c} y_i$ can then be used by the intruder to obtain sharper bounds on a true value. For example, let $y_{(1)} \le y_{(2)} \le \cdots \le y_{(N_c)}$ be the order statistics and suppose that the intruder is the establishment with the second largest value, $y_{(N_c-1)}$. Then, this intruder can determine an upper bound for the largest value $y_{(N_c)}$ by subtracting its own value $y_{(N_c-1)}$ together with the sum of the lower bounds for $y_{(1)}, \ldots, y_{(N_c-2)}$ from $\hat{Y}_c$. The precision of prediction using this upper bound is given by the difference between this upper bound and the true value $y_{(N_c)}$, which is $(q/100) \sum_{i=1}^{N_c-2} y_{(i)}$. This cell would be called *sensitive* under the pq rule, that is, judged *disclosive*, if this difference was less than $p\%$ of the true value, that is, if

$$(p/100)y_{(N_c)} - (q/100) \sum_{i=1}^{N_c-2} y_{(i)} > 0. \tag{1}$$

The expression on the left-hand side of (1) is a special case of a *linear sensitivity measure*, which more generally takes the form $R_c = \sum_{i=1}^{N_c} a_i y_{(i)}$, where the a_i are specified weights. The cell is said to be sensitive if $R_c > 0$. In this case, prediction disclosure would be deemed to occur. A widely used special case of the pq rule is the $p\%$ rule, which arises from setting $q = 100$, that is, no prior information is assumed. Another commonly used linear sensitivity measure arises with the (n, k) or *dominance rule*. See Willenborg and de Waal (2001), Cox (2001), Giessing (2001), and Federal Committee on Statistical Methodology (2005) for further discussion.

2.2.2. Prediction disclosure in the presence of sampling

More generally, all cell units may not be completely enumerated. In this case, $\hat{Y}_c$ will be subject to sampling error and, in general, this will lead to additional disclosure protection, provided that the intruder does not know whether other establishments (other than those in the coalition) are sampled or not. The definition of risk in this setting appears to need further research. Willenborg and de Waal (2001, Section 6.2.5) presented some ideas. An alternative model-based stochastic approach might assume that before the release of the table, the prior information about the y_i can be represented by a linear regression model depending upon publicly available covariate values x_i with a specified residual variance. The predictive distribution of y_i given x_i could then be updated using the known value(s) of y_i for the intruder and the reported $\hat{Y}_c$, which might be assumed to follow the distribution $\hat{Y}_c \sim N[Y_c, v(\hat{Y}_c)]$, where $v(\hat{Y}_c)$ is the reported variance estimate of $\hat{Y}_c$. Prediction disclosure could then be measured in terms of the resulting residual variance in the prediction of y_i.

2.3. SDC methods for tabular outputs

If a cell in a table is deemed sensitive, that is, the cell value represents an unacceptably high disclosure risk, a number of SDC approaches may be used.

2.3.1. Redefinition of cells

The cells are redefined to remove sensitive cells, for example, by combining sensitive cells with other cells or by combining categories of the cross-classified variables. This is also called *table redesign* (Willenborg and de Waal, 2001).

2.3.2. Cell suppression

The value of a sensitive cell is suppressed. Depending upon the nature of the table and its published margins, it may also be necessary to suppress the values of "complementary" cells to prevent an intruder being able to deduce the value of the cell from other values in the table. There is a large literature on approaches to choosing complementary cells which ensure disclosure protection. See, for example, Willenborg and de Waal (2001), Cox (2001), and Giessing (2001) and references therein.

2.3.3. Cell modification

The cell values may be modified in some way. It will generally be necessary to modify not only the values in the sensitive cells but also values in some complementary nonsensitive cells, for the same reason as in cell suppression. Modification may be deterministic, for example, Cox et al. (2004), or stochastic, for example, Willenborg and de Waal (2001, Section 9.2). A simple method is *rounding*, where the modified cell values are multiples of a given base integer (Willenborg and de Waal, 2001, Chapter 9). This method is more commonly applied to frequency tables derived from 100% data but can also be applied to tables of estimated totals from surveys, where the base integer may be chosen according to the magnitudes of the estimated totals. Instead of replacing the cell values by single safe values, it is also possible to replace the values by intervals, defined by lower and upper bounds (Giessing and Dittrich, 2006; Salazar, 2003). The method of *controlled tabular adjustment* (Cox et al., 2004) determines modified cell values within such bounds so that the table remains additive and certain safety and statistical properties are met.

2.3.4. Pretabular microdata modification

Instead of modifying the cell values, the underlying microdata may be perturbed, for example, by adding noise, and then the table formed from the perturbed microdata (Evans et al., 1998; Massell et al., 2006).

The statistical output from a survey will typically include many tables. Although the above methods may be applied separately to each table, such an approach takes no account of the possible additional disclosure risks arising from the combination of information from different tables, in particular, from common margins. To protect against such additional risks raise new considerations for SDC. Moreover, the set of tables constituting the statistical output is not necessarily fixed, as in a traditional survey report. With developments in online dissemination, there is increasing demand for the generation of tables which can respond in a more flexible way to the needs of users. This implies the need to consider SDC methods which not only protect each table separately as above but also protect against the risk arising from alternative possible sequences of released tables (see, e.g., Dobra et al., 2003).

3. Microdata

3.1. Assessing disclosure risk

We suppose the agency is considering releasing to researchers an anonymized microdata file, where the records of the file correspond to the basic analysis units and each record contains a series of survey variables. The record may also include identifiers for higher level analysis units, for example, household identifiers where the basic units are individuals, as well as information required for survey analysis such as survey weights and primary sampling unit (PSU) identifiers.

We suppose that the threat of concern is that an intruder may link a record in the file to some external data source of known units using some variables, which are included in both the microdata file and the external source. These variables are often called *key variables* or identifying variables. There are various ways of defining disclosure risk in this setting. See, for example, Paass (1988) and Duncan and Lambert (1989). A common approach, often motivated by the nature of the confidentiality pledge, is to consider a form of *identification risk* (Bethlehem et al., 1990; Reiter, 2005), concerned with the possibility that the intruder will be able to determine a correct link between a microdata record and a known unit. This definition of risk will only be appropriate if the records in the microdata can meaningfully be said to be associated with units in the population. When microdata is subject to some forms of SDC, this may not be the case (e.g., if the released records are obtained by combining original records) and in this case, it may be more appropriate to consider some definition of predictive disclosure (e.g., Fuller, 1993) although we do not pursue this further here.

A number of approaches to the assessment of identification risk are possible, but all depend importantly upon assumptions about the nature of the key variables. One approach is to conduct an empirical experiment, matching the proposed microdata against another data source, which is treated as a surrogate for the data source held by the intruder. Having made assumptions about the key variables, the agency can use record linkage methods, which it is plausible would be available to an intruder, to match units between the two data sets. Risk might then be measured in terms of the number of units for which matches are achieved together with a measure of the match quality (in terms of the proportions of false positives and negatives). Such an experiment, therefore, requires that the agency has information which enables it to establish precisely which units are in common between the two sources and which are not.

The key challenge in this approach is how to construct a realistic surrogate intruder data set, for which there is some overlap of units with the microdata and the nature of this overlap is known. On some occasions a suitable alternative data source may be available. Blien et al. (1992) provide one example of a data source listing people in certain occupations. Another possibility might be a different survey undertaken by the agency, although agencies often control samples to avoid such overlap. Even if there is overlap, say with a census, determining precisely which units are in common and which are not may be resource intensive. Thus, this approach is unlikely to be suitable for routine use.

In the absence of another data set, the agency may consider a reidentification experiment, in which the microdata file is matched against itself in a similar way, possibly after the application of some SDC method (Winkler, 2004). This approach has the advantage that it is not model-dependent, but it is possible that the reidentification risk is overestimated if the disclosure protection effects of sampling and measurement error are not allowed for in a realistic way.

In the remainder of Section 3, we consider a third approach, which again only requires data from the microdata file, but makes theoretical assumptions, especially of a modeling kind, to estimate identification risk. As for the reidentification experiment, this approach must make assumptions about how the key variables are measured in the microdata and by the intruder on known units using external information. A simplifying but "worst case" assumption is that the key variables are recorded in identical ways in the microdata and externally. We refer to this as the *no measurement error assumption*, since measurement error in either of the data sources may be expected to invalidate this assumption. If at least one of the key variables is continuous and the no measurement error assumption is made, then an intruder who observes an exact match between the values of the key variables in the microdata and on the known units could conclude with probability one that the match is correct, in other words, the identification risk would be one. If at least one of the key variables is continuous and it is supposed that measurement error may occur, then the risk will generally be below one. Moreover, an exact matching approach is not obviously sensible and a broader class of methods of record linkage might be considered. See Fuller (1993) for the assessment of disclosure risk under some measurement error model assumptions.

In practice, variables are rarely recorded in a continuous way in social survey microdata. For example, age would rarely be coded with more detail than 1 year bands. And from now on, we restrict attention to the case of categorical key variables. For simplicity, we restrict attention to the case of exact matching, although more general record linkage methods could be used. We focus on a microdata file, where the only SDC methods which have been applied are recoding of key variables or random (sub)sampling. We comment briefly on the impact of other SDC methods on risk in Section 3.4.

3.2. File-level measures of identification risk

We consider a finite population U of N units (which will typically be individuals) and suppose the microdata file consists of records for a sample $s \subset U$ of size $n \leq N$. We assume that the possibility of statistical disclosure arises if an intruder gains access to the microdata and attempts to match a microdata record to external information on a known unit using the values of m categorical key variables $X_1, \ldots, X_m$. (Note that s and $X_1, \ldots, X_m$ are defined after the application of (sub)sampling or recoding, respectively, as SDC methods to the original microdata file.)

Let the variable formed by cross-classifying $X_1, \ldots, X_m$ be denoted by X, with values denoted $k = 1, \ldots, K$, where K is the number of categories or key values of X. Each of these key values corresponds to a possible combination of categories of the key variables. Under the no measurement error assumption, identity disclosure is of particular concern if a record is unique in the population with respect to the key variables. A record with key value k is said to be *population unique* if $F_k = 1$, where F_k denotes the number of units in U with key value k. If an intruder observes a match with a record with key value k, knows that the record is population unique and can make the no measurement error assumption then the intruder can infer that the match is correct.

As a simple measure of disclosure risk, we might therefore consider taking some summary of the extent of population uniqueness. In survey sampling, it is usual to define parameters of interest at the population level and this might lead us to define our measure as the population proportion N_1/N, where $N_r = \sum_k I(F_k = r)$ is the population frequencies of frequencies, $r = 1, 2, \ldots$. From a disclosure risk perspective, however, we are interested in the risk for a specific microdata file it is natural to allow the risk measure to be sample dependent. Thus, we might expect the risk to be higher if a sample is selected with a high proportion of unusual identifiable units than for a sample where this proportion is lower. Thus, a more natural file-level measure is the proportion of population uniques in the sample. Let the sample counterpart of F_k be denoted by f_k, then this measure can be expressed as follows:

$$\Pr(PU) = \sum_k I(f_k = 1, F_k = 1)/n. \tag{2}$$

It could be argued, however, that the denominator of this proportion should be made even smaller, since the only records which might possibly be population unique are ones that are sample unique (since $f_k \leq F_k$), that is, have a key value k such that $f_k = 1$. Thus, a more conservative measure would be to take

$$\Pr(PU|SU) = \sum_k I(f_k = 1, F_k = 1)/n_1, \tag{3}$$

where n_1 is the number of sample uniques and, more generally, $n_r = \sum_k I(f_k = r)$ is the sample frequencies of frequencies. For further consideration of the proportion of sample uniques that are population unique, see Fienberg and Makov (1998) and Samuels (1998).

It may be argued (e.g., Skinner and Elliot, 2002) that these measures may be overoptimistic, since they only capture the risk arising from population uniques and not from other records with $F_k \geq 2$. If an intruder observes a match on a key value with frequency F_k, then (subject to the no measurement error assumption) the probability that the match is correct is $1/F_k$ under the exchangeability assumption that the intruder is equally likely to have selected any of the F_k units in the population. An alternative measure of risk is then obtained by extending this notion of probability of correct match across different key values. Again, on worst case grounds, it is natural to restrict attention to sample uniques. One measure arises from supposing that the intruder starts with the microdata, is equally likely to select any sample unique and then matches this sample unique to the population. The probability that the resulting match is correct is then the simple average of $1/F_k$ across sample uniques:

$$\theta_s = \left[\sum_k I(f_k = 1)/F_k/n_1 \right]. \tag{4}$$

Another measure is

$$\theta_U = \sum_k I(f_k = 1) \Big/ \sum_k F_k I(f_k = 1), \tag{5}$$

which is the probability of a correct match under a scenario where the intruder searches at random across the population and finds a match with a sample unique.

All the above four measures are functions of both the f_k and the F_k. The agency conducting the survey will be able to determine the sample quantities f_k from the microdata but the population quantities F_k will generally be unknown. It is, therefore, of interest to be able to make inference about the measures from sample data.

Skinner and Elliot (2002) showed that, under Bernoulli sampling with inclusion probability π, a simple design-unbiased estimator of θ_U is $\hat{\theta}_U = n_1/[n_1 + 2(\pi^{-1} - 1)n_2]$. They also provided a design consistent estimator for the asymptotic variance of $\hat{\theta}_U - \theta_U$. Skinner and Carter (2003) showed that a design-consistent estimator of θ_U for an arbitrary complex design is $\hat{\theta}_U = n_1/[n_1 + 2(\overline{\pi}_2^{-1} - 1)n_2]$, where $\overline{\pi}_2^{-1}$ is the mean of the inverse inclusion probabilities π_i^{-1} for units i with key values for which $f_k = 2$. They also provided a design-consistent estimator of the asymptotic variance of $\hat{\theta}_U - \theta_U$ under Poisson sampling.

Such simple design-based inference does not seem to be possible for the other three measures in (2)–(4). Assuming a symmetric design, such as Bernoulli sampling, we might suppose that $n_1, n_2, \ldots$ represent sufficient statistics and seek design-based moment-based estimators of the measures by solving the equations:

$$E(n_r) = \sum_t N_t P_{rt}, \quad r = 1, 2, \ldots,$$

where the coefficients P_{rt} are known for sampling schemes, such as simple random sampling or Bernoulli sampling (Goodman, 1949). The solution of these equations for N_t with $E(n_r)$ replaced by n_r gives unbiased estimators of K and N_1 under apparently weak conditions (Goodman, 1949). Unfortunately, Goodman found that the estimator of K can be "very unreasonable" and the same appears to be so for the corresponding estimator of N_1. Bunge and Fitzpatrick (1993) reviewed approaches to estimating

K and discussed these difficulties. Zayatz (1991) and Greenberg and Zayatz (1992) proposed an alternative "nonparametric" estimator of N_1 but this appears to be subject to serious upward bias for small sampling fractions (Chen and Keller-McNulty, 1998).

One way of addressing these estimation difficulties is by making stronger modeling assumptions, in particular by assuming that the F_k are independently distributed as follows:

$$F_k|\lambda_k \sim \text{Po}(\lambda_k) \tag{6}$$

where the λ_k are independently and identically distributed, that is, that the F_k follow a compound Poisson distribution. A tractable choice for the distribution of λ_k is the gamma distribution (Bethlehem et al., 1990) although it does not appear to fit well in some real data applications (e.g., Skinner et al., 1994; Chen and Keller-McNulty, 1998). A much better fit is provided by the log-normal (Skinner and Holmes, 1993). Samuels (1998) discussed estimation of $\text{Pr}(PU|SU)$ based on a Poisson-Dirichlet model. A general conclusion seems to be that results can be somewhat sensitive to the choice of model, especially as the sampling fraction decreases, and that θ_U can be more robustly estimated than the other three measures.

3.3. Record-level measures of identification risk

A concern with file-level measures is that the principles governing confidentiality protection often seek to avoid the identification of *any* individual, that is require the risk to be below a threshold for each record, and such aims may not adequately be addressed by aggregate measures of the form (2)–(5). To address this concern, it is more natural to consider record level measures, that is, measures which may take different values for each microdata record. Such measures may help identify those parts of the sample where risk is high and more protection is needed and may be aggregated to a file level measure in different ways if desired (Lambert, 1993). Although record level measures may provide greater flexibility and insight when assessing whether specified forms of microdata output are "disclosive," they are potentially more difficult to estimate than file level measures.

A number of approaches have been proposed for the estimation of record level measures. For continuous key variables, Fuller (1993) showed how to assess the record level probability of identification in the presence of added noise, under normality assumptions. See also Paass (1988) and Duncan and Lambert (1989). We now consider related methods for categorical variables, following Skinner and Holmes (1998) and Elamir and Skinner (2006).

Consider a microdata record with key value X. Suppose the record is sample unique, that is, with a key value k for which $f_k = 1$, since such records may be expected to be most risky. Suppose the intruder observes an exact match between this record and a known unit in the population. We make the no measurement error assumption so that there will be F_k units in the population which potentially match the record. We also assume no response knowledge (see Section 2.1). The probability that this observed match is correct is

$$\text{Pr}(\text{correct match} \mid \text{exact match}, X = k, F_k) = 1/F_k, \tag{7}$$

where the probability distribution is with respect to the design under a symmetric sampling scheme, such as simple random sampling or Bernoulli sampling. (Alternatively, it could be with respect to a stochastic mechanism used by the intruder, which selects any of the F_k units with equal probability). This probability is conditional on the key value k and on F_k.

In practice, we only observe the sample frequencies f_k and not the F_k. We, therefore, integrate out over the uncertainty about F_k and write the measure as

$$\Pr(\text{correct match} \mid \text{exact match}, X = k, f_k) = E(1/F_k | k, f_k = 1). \tag{8}$$

This expectation is with respect to both the sampling scheme and a model generating the F_k, such as the compound Poisson model in (6). An alternative measure, focusing on the risk from population uniqueness, is

$$\Pr(F_k = 1 | k, f_k = 1). \tag{9}$$

The expressions in (8) and (9) may be generalized for any record in the microdata with $f_k > 1$. A difference between the probabilities in (8) and (9) and those in the previous section is that here we condition on the record's key value $X = k$. Thus, although we might assume $F_k | \lambda_k \sim \text{Po}(\lambda_k)$, as in (6), we should like to condition on the particular key value k when considering the distribution of λ_k. Otherwise, if the λ_k is identically distributed as in the previous section, then we would obtain the same measure of risk for all (sample unique) records. A natural model is a log-linear model:

$$\log(\lambda_k) = z_k \beta, \tag{10}$$

where z_k is a vector of indicator variables representing the main effects and the interactions between the key variables $X_1, \ldots, X_m$, and β is a vector of unknown parameters.

Expressions for the risk measures in (8) and (9) in terms of β are provided by Skinner and Holmes (1998) and Elamir and Skinner (2006). Assumptions about the sampling scheme are required to estimate β. Under Bernoulli sampling with inclusion probability π, it follows from (6) that $f_k | \lambda_k \sim \text{Po}(\pi \lambda_k)$. Assuming also (10), β may be estimated by standard maximum likelihood methods. A simple extension of this argument also applies under Poisson sampling where the inclusion probability π_k may vary with respect to the key variables, for example, if a stratifying variable is included among the key variables. In this case, we have $f_k | \lambda_k \sim \text{Po}(\pi_k \lambda_k)$. Skinner and Shlomo (2008) discussed methods for the specification of the model in (10). Skinner (2007) discussed the possible dependence of the measure on the search method used by the intruder.

3.4. SDC methods

In this section, we summarize a number of SDC methods for survey microdata.

3.4.1. Transformation of variables to reduce detail

Categorical key variables may be transformed, in particular, by combining categories. For example, the variable household size might be *top coded* by creating a single maximum category, such as 8+. Continuous key variables may be *banded* to form ordinal categorical variables by specifying a series of cut-points between which the intervals define categories. The protection provided by combining categories of key variables can be assessed following the methods in Sections 3.2 and 3.3. See also Reiter (2005). Provided the transformation is clear and explicit, this SDC method has the advantage that the reduction of utility is clear to the data user, who may suffer loss of information but the validity of analyses is not damaged.

3.4.2. Stochastic perturbation of variables

The values of potential key variables are perturbed in a stochastic way. In the case of continuous variables, perturbation might involve the *addition of noise*, analogous to the addition of measurement error (Fuller, 1993; Sullivan and Fuller, 1989). In the case of categorical variables, perturbation may consist of misclassification, termed the *Postrandomization Method* (PRAM) by Gouweleeuw et al. (1998). Perturbation may be undertaken in a way to preserve specified features of the microdata, for example, the means and standard deviations of variables in the perturbed microdata may be the same as in the original microdata, but in practice there will inevitably be unspecified features of the microdata which are not reproduced. For example, the estimated correlation between a perturbed variable and an unperturbed variable will often be downwardly biased if an analyst uses the perturbed data but ignores the fact that perturbation has taken place. An alternative is to provide users with the precise details of the perturbation method, including parameter values, such as the standard deviation of the noise or the entries in the misclassification matrix, so that they may "undo" the impact of perturbation when undertaking their analyses. See, for example, Van den Hout and Van der Heijden (2002) in the case of PRAM or Fuller (1993) in the case of added noise. In principle, this may permit valid analyses although there will usually be a loss of precision and the practical disadvantages are significant.

3.4.3. Synthetic microdata

This approach is similar to the previous approach, except that the aim is to avoid requiring special methods of analysis. Instead, the values of variables in the file are replaced by values generated from a model in a way that is designed for the analysis of the synthetic data, as if it were the true data, to generate consistent point estimates (under the assumption that the model is valid). The model is obtained from fitting to the original microdata. To enable valid standard errors as well as consistent point estimators, Raghunathan et al. (2003) proposed that multiple copies of the synthetic microdata are generated in such a way that multiple imputation methodology can be used. See Reiter (2002) for discussion of complex designs. Abowd and Lane (2004) discussed release strategies combining remote access to one or more such synthetic microdata files with much more restricted access to the original microdata in a safe setting.

3.4.4. Selective perturbation

Often concern focuses only on records deemed to be risky and it may be expected that utility will be greater if only a subset of risk records is perturbed. In addition to creating stochastically perturbed or synthetic values for only targeted records, it is also possible just to create missing values in these records, called *local suppression* by Willenborg and de Waal (2001), or both to create missing values and to replace these by imputed values, called *blank and impute* by Federal Committee on Statistical Methodology (2005). A major problem with such methods is that they are likely to create biases if the targeted values are unusual. The data user will typically not be able to quantify these biases, especially when the records selected for blanking depend on the values of the variable(s) which are to made missing. Reiter (2003) discussed how valid inference may be conducted if multiple imputed values are generated in a specified way for the selected records. He referred to the resulting data as *partially synthetic microdata*.

3.4.5. Record swapping

The previous methods focus on the perturbation of the values of the variables for all or a subset of records. The method of record swapping involves, instead, the values of one or more key variables being swapped between records. The choice of records between which values are swapped may be controlled so that certain bivariate or multivariate frequencies are maintained (Dalenius and Reiss, 1982) in particular by only swapping records sharing certain characteristics (Willenborg and de Waal, 2001, Section 5.6). In general, however, it will not be possible to control all multivariate relationships and record swapping may damage utility in an analogous way to misclassification (Skinner and Shlomo, 2007). Reiter (2005) discussed the impact of swapping on identification risk.

3.4.6. Microaggregation

This method (Defays and Anwar, 1998) is relevant for continuous variables, such as in business survey microdata, and in its basic form consists of ordering the values of each variable and forming groups of a specified size k (the first group contains the k smallest values, the second group the next k smallest values, and so on). The method replaces the values by their group means, separately for each variable. An advantage of the method is that the modification to the data will usually be greatest for outlying values, which might also be deemed the most risky. It is difficult, however, for the user to assess the biasing impact of the method on analyses.

SDC methods will generally be applied after the editing phase of the survey, during which data may be modified to meet certain edit constraints. The application of some SDC methods may, however, lead to failure of some of these constraints. Shlomo and de Waal (2006) discussed how SDC methods may be adapted to take account of editing considerations.

3.5. SDC for survey weights and other design information

Survey weights and other complex design information are often released with survey microdata in order that valid analyses can be undertaken. It is possible, however, that such design information may contribute to disclosure risk. For example, suppose a survey is stratified by a categorical variable X with different sampling fractions in different categories of X. Then, if the nature of the sampling design is published (as is common), it may be possible for the intruder to determine the categories of X from the survey weight. Thus, the survey design variable may effectively become a key variable. See de Waal and Willenborg (1997) and Willenborg and de Waal (2001, Section 5.7) for further discussion of how survey weights may lead to design variables becoming key variables. Note that this does not imply that survey weights should not be released; it just means that disclosure risk assessments should take account of what information survey weights may convey. Willenborg and de Waal (2001, Section 5.7.3) and Mitra and Reiter (2006) proposed some approaches to adjusting weights to reduce risk.

In addition to the release of survey weights, it is common to release either stratum or PSU labels or replicate labels, to enable variances to be estimated. These labels will generally be arbitrary and will not, in themselves, convey any identifying information. Nevertheless, as for survey weights, the possibility that they could be used to convey information indirectly needs to be considered. For example, if the PSUs are defined by areas for which public information is available, for example, a property tax rate, and the microdata file includes area-level variables, then it is possible that these variables may enable a PSU to be linked to a known area. As another example, suppose that a PSU is an institution, such as a school, then school level variables on the microdata file, such as the school enrolment size, might enable the PSU to be linked to a known institution. Even for individual level microdata variables, it is possible that sample-based estimates of the total or mean of such variables for a stratum, say, could be matched to published values, allowing for sampling uncertainty.

A standard simple approach to avoiding releasing PSU or replicate identifiers is to provide information on design effects or generalized variance functions instead. Such methods are often inadequate, however, for the full range of uses of survey microdata (Yung, 1997). Some possible more sophisticated approaches include the use of adjusted bootstrap replicate weights (Yung, 1997), adjusted pseudoreplicates or pseudo PSU identifiers (Dohrmann et al., 2002), or combined stratum variance estimators (Lu et al., 2006).

4. Conclusion

The development of SDC methodology continues to be stimulated by a wide range of practical challenges and by ongoing innovations in the ways that survey data are used, with no signs of diminishing concerns about confidentiality. There has been a tendency for some SDC methods to be developed in somewhat ad hoc way to address specific problems, and one aim of this chapter has been to draw out some principles and general approaches which can guide a more unified methodological development. Statistical modeling has provided one important framework for this purpose. Other fields with the potential to influence the systematic development of SDC methodology in the future include data mining, in particular methods related to record linkage and approaches to privacy protection in computer science and database technology.

Acknowledgments

I am grateful to Natalie Shlomo and a reviewer for comments on an earlier draft.

Essential Methods for Design Based Sample Surveys
ISSN: 0169-7161
DOI: 10.1016/B978-0-444-53734-8.00005-4

5

Sampling and Estimation in Household Surveys

Jack G. Gambino and Pedro Luis do Nascimento Silva

1. Introduction

A household survey is a particular type of social survey. In a household survey, we are interested in the characteristics of all or some members of the household. These characteristics typically include a subset of variables such as health, education, income, expenditure, employment status, use of various types of services, etc. Since they became common in the 1940s, a number of major trends in household surveys have been evident. Many of these trends are closely linked to technological advances both in statistical agencies and in society, and have accelerated following the spread of personal computers in the early 1980s. These trends include, but are not limited to, the following.

1.1. Simplification of sample designs

A good example of simplification is the Canadian Labour Force Survey (LFS), which went from as many as four stages of sampling to two stages, and for which the feasibility of using a single-stage design, with an address register as a frame, is currently being studied. In the United Kingdom, the LFS already uses, and the new Integrated Household Survey will use, an unclustered design (see Office for National Statistics, 2004). In the United States, the American Community Survey (ACS) also adopted a stratified unclustered design (see U.S. Census Bureau, 2006b).

1.2. Increasingly complex estimation methods

The increasing power and availability of computers has made it possible to use increasingly complex estimation procedures.

1.3. Increased use of telephone interviewing

As the proportion of households with a telephone has increased, the proportion of interviews conducted in person (across surveys) has decreased. This trend accelerated with the introduction of computer-assisted interviewing. A related trend is the increased use of multiple modes of collection for the same survey. The latter trend is likely to continue as use of the internet as a medium for survey response becomes more popular.

1.4. Increasingly complex questionnaires

The introduction of computer-assisted interviewing has made it possible to have questionnaires with complex skip patterns, built-in edits, and questions tailored to the respondent.

1.5. Increased availability of data on data collection (paradata)

The use of computers for interviewing makes it possible to save data on various aspects of the interviewing process. In addition, interviews can be monitored, providing another source of data.

We discuss some of these trends throughout the chapter. There are a number of other important trends we do not discuss since they are covered in other chapters in this volume. They include increasingly elaborate editing and imputation procedures, the rising importance of confidentiality and privacy, questionnaire design and related research, and the advent of more sophisticated data analysis methods, particularly for data from complex surveys.

2. Survey designs

Much of the theory and practice of survey design was developed from the 1930s to the 1960s. In fact, many methods currently in use, particularly for area-based sampling, are already included in the classic text by Hansen et al. (1953). We will often refer to dwellings, which are sometimes referred to as dwelling units or housing units in other publications. A dwelling may be vacant (unoccupied) or occupied by a household.

2.1. Frames

The traditional frames used for household surveys are area frames and list frames. Alternatives to these include the use of random-digit dialling (RDD) and the internet.

2.1.1. List frames
If a list of population units is available, then it can be used to sample directly, possibly after stratification of the units into more homogeneous groups. Examples include lists of dwellings, lists of households or families, lists of people, and lists of telephone numbers. A major advantage of list frames for surveys is that they are easy to use for sampling and usually lead to relatively straightforward weighting and estimation procedures. On the other hand, it is often difficult and expensive to keep a list up-to-date in light of individual changes such as moves, marriages and divorces, and births and deaths. Certain types of unit pose difficulties. For example, students and workers living temporarily at a work site may be listed twice or missed completely. There are many other situations that can lead to missed units (or undercoverage) or double-counted units (or overcoverage), as well as to units that do not belong to the list (also overcoverage). Nevertheless, the benefits of having an up-to-date list of units are sufficiently great that several countries have invested in the creation and maintenance of permanent lists, including population registers and address registers. For example, Scandinavian

countries have made increasing use of population registers in their censuses, including complete replacement of traditional censuses in some cases (see Statistics Finland, 2004; Statistics Norway, 2005).

2.1.2. Area frames

An area frame is obtained by dividing a country (or province, state, etc.) into many mutually exclusive and exhaustive smaller areas. In principle, therefore, an area frame has complete coverage. In practice, the area frame approach pushes the coverage problems of list frames down to a smaller geographical level since, at some point in the sampling process, a list of ultimate sampling units (dwellings, households, or people) will be needed. Obtaining such lists for many small areas can be very time-consuming and expensive.

The use of area frames has led naturally to multistage designs where clusters form the penultimate stage: the country may be divided into provinces or states, which are divided into counties, say, and so on, until we come to the smallest area units, which we will refer to as (geographical) clusters. We may then start the sample selection process at one of these geographical levels. The simplest case would be direct selection of clusters (possibly after stratification), followed by selection of units (typically dwellings) within the clusters that were selected at the first stage. This has the great advantage that we need a complete list of ultimate (or elementary) units only for clusters that are selected in the sample—in effect, we "localize" the list frame creation problem.

In practice, getting a complete list of units (dwellings and/or people) in a cluster can be difficult. The cluster may contain easy to miss dwellings (e.g., they may not be obvious from the street). There may be problems identifying the cluster boundary, especially if a part of the boundary is an imaginary line or if new construction has occurred in the area. As a result, in the field, it may not be clear who belongs to the cluster.

The use of an area frame with multistage sampling is very common in both developing and developed countries. One important benefit is the reduced travel costs for personal interviewing when dwellings are selected in compact geographical areas such as clusters and higher level sampling units (e.g., a village can be a sampling unit) since the interviewer drives to a cluster and contacts several dwellings in close proximity.

2.1.3. Apartment frames

A list of apartment buildings (typically useful in metropolitan areas) is, in a sense, at the intersection of list frames and area frames: a survey may use an area frame, but whenever apartment buildings are found (because they are new or were missed earlier), they are "removed" from the area frame and put in a separately maintained list frame of apartment buildings. Each apartment building may then be treated as a cluster.

2.1.4. Telephone-based frames

A list of telephone numbers is simply a list frame, as discussed above. An alternative way of using telephone numbers is via RDD. In RDD, telephone numbers are generated at random, avoiding the need for a list of numbers. In practice, the process is more sophisticated than simply generating a string of digits and expecting that the result will be a valid telephone number. Efforts are made to eliminate invalid or business numbers in advance. One can also vary the probability for certain sets of numbers. There is a vast literature on RDD, which we do not cover here. Nathan (2001) includes an extensive list of references on RDD and other telephone-based methods of data collection.

Until recently, the use of mobile (or cell) phones as either a frame or as a mode for conducting interviews has been avoided, but this may be changing in light of the increasing number of households with mobile phones (and without a traditional landline telephone). Problems related to the use of mobile phones for surveys have generated a great deal of interest recently; see, for example, the 2005 Cell Phone Sampling Summit and several sessions at the 2007 AAPOR conference. Blumberg and Luke (2007) present recent results on the rapid increase in the number of cell phone-only households in the United States using data from the most recent National Health Interview Survey. They also look at the demographic and health-related characteristics of these households. For example, they find that homeowners are much less likely to be in a cell phone-only household than renters.

In this section, we have noted that survey statisticians face a variety of problems in constructing and maintaining frames. Yansaneh (2005) discusses these further and presents possible solutions to some of the problems.

2.2. Units and stages

We have already had occasion to mention sampling units and sampling stages in this chapter. We now discuss these more formally in the following.

2.2.1. Persons and households

In the surveys under consideration, interest is usually in persons, families, or households. We usually get to these units via the dwelling (for our purposes, we define a dwelling as a set of living quarters; a formal definition can be quite involved—e.g., the formal definition of private dwelling on the Statistics Canada web site is more than 500 words long). As discussed above, we get to the dwelling via an area frame, a dwelling frame, or a telephone number. Although formal definitions of units may be quite involved, smooth implementation in the field may require simplifications to be practicable.

In some surveys, the ultimate unit of interest is not the household but one person within the household. This introduces a problem of representativity of the sample if persons within households are selected by a naive method: certain groups (e.g., age groups) may be over- or under-represented. For example, Beland et al. (2005) cite a Canadian example where a naive approach (an individual in the household is selected with equal probability) yields 8.2% of the sample in the 12–19 age range, whereas the percentage in the population is 12.4. For people aged 65 and older, the corresponding percentages were 21.5 and 14.5. Thus, young people would be under-represented and old people over-represented. Solutions to this problem are discussed in the study by Beland et al. (2005) and by Tambay and Mohl (1995). Clark and Steel (2007) discuss optimal choice of the number of persons to select from each household.

2.2.2. Clusters

We already mentioned clusters when we discussed frames. A cluster is usually a compact geographical area containing a few dozen to a few hundred households. In urban areas, clusters are typically formed by combining contiguous block faces or blocks. In rural areas, many countries use census enumeration areas as clusters, but there may also be natural clusters such as villages that can be used. Sometimes these clusters vary widely in size, which is undesirable, and may require using sampling with probabilities proportional to size. Apartment buildings are sometimes used as clusters, especially in large metropolitan areas where a significant proportion of the population lives in apartment complexes.

2.2.3. The role of census units

In most countries, the process of conducting a population census entails the formation of a hierarchy of geographical units. It is natural to consider these "ready-made" units when designing household surveys. We have mentioned the use of a census unit, namely an enumeration area, as a sampling unit in surveys. More generally, census units are used in surveys for a variety of purposes: the biggest census units may form geographical strata (examples are provinces and states), and the smallest census units may be useful building blocks for the formation of primary sampling units (PSUs) and (optimal) strata. In countries with big populations, PSUs can be very large (e.g., a whole city can be a PSU). In the United States, national household surveys often use counties or groups of counties as PSUs. Typically, the largest PSUs are self-representing, i.e., they are selected with probability 1, which means they are really strata rather than PSUs.

2.3. Stratification

Subnational geographical areas such as provinces, states, and regions form the highest level of stratification both because they have well-defined, stable boundaries and because they are often of interest for policy-making. In most cases, these subnational areas are too big and need to be divided into finer geographical strata. Once the lowest level of geographical stratification is reached, there may be enough PSUs (e.g., census enumeration areas) in some geographical strata to form optimal strata within the latter. Optimality is defined by some measure of homogeneity and the PSUs are grouped into final strata that are as homogeneous as possible.

In some cases, it is more convenient to use *implicit* stratification via ordering of the units—similar units are placed near each other. Then, some form of systematic sampling is used to select the sample to ensure that no major subpopulations are left out of the sample. A special case is geographical ordering to ensure that no major areas are left out of the sample and also to achieve approximately proportional allocation of the sample between areas.

Textbooks on survey sampling tend to devote little space to stratum formation and even then, they emphasize the case of a single variable. In practice, several variables are usually of interest, and a compromise stratification is needed. A common tool for stratum formation based on several variables is cluster analysis. For a recent approach to stratification using a spatial cluster analysis algorithm that minimizes distances between PSUs in a stratum on selected variables considering the spatial location of PSUs, see Palmieri Lage et al. (2001).

2.4. Sample size

The determination of sample size for household surveys is complicated by the fact that most surveys are interested in several variables, so the standard textbook formulas based on a single variable are not adequate. In addition, one must decide on a criterion: standard error versus coefficient of variation (CV), that is, to aim to control either absolute or relative error. Finally, if a clustered design will be used, the "IID" (independent and identically distributed) or "SRS" sample size formulas are inadequate—they will likely understate the required sample size. The design effect (deff) is a measure of this phenomenon. The deff is defined as the ratio of the variance of an estimator under the actual design to the variance of the estimator under simple random sampling (SRS), assuming that the sample size is the same for both designs. Thus,

$$\text{deff}(\hat{\theta}; p) = V_{p,n}(\hat{\theta})/V_{\text{SRS},n}(\hat{\theta}), \tag{1}$$

where $\hat{\theta}$ denotes an estimator of a parameter θ, p denotes a complex survey design, $V_{p,n}(\hat{\theta})$ and $V_{\text{SRS},n}(\hat{\theta})$ denote the variances of $\hat{\theta}$ under the designs p and SRS, respectively, with n defined as the number of sampled households.

A common approach to sample size determination in complex surveys is to use information from similar surveys or census data to obtain (or assume) design effects and population variances for key variables, and use these *deffs* and variances to determine n. This follows from using (1) in the following manner. Suppose a sample size n_0 can be determined using the standard SRS formulas so that a specified variance v is achieved for the estimator of a key parameter, that is, n_0 solves $V_{\text{SRS},n_0}(\hat{\theta}) = v$. Then, if the same sample size n_0 was used with the complex design p, (1) implies that $V_{p,n_0}(\hat{\theta})/\text{deff}(\hat{\theta}; p) = v \Leftrightarrow V_{p,n_0}(\hat{\theta}) = v \times \text{deff}(\hat{\theta}; p)$. Hence, to obtain the same variance v using a complex design p, we need to solve $V_{\text{SRS},n}(\hat{\theta}) = v/\text{deff}(\hat{\theta}; p)$, which leads to the simple solution corresponding to multiplying the initial sample size n_0 by deff, that is,

$$n = n_0 \times \text{deff}(\hat{\theta}; p) \tag{2}$$

For surveys where proportions are the target parameters, the above solution is simple, since sample sizes under SRS can be determined easily using the fact that the variances of sample proportions are maximized when the population proportion is ½ (see, e.g., Cochran, 1977, Chapter 3). This is a conservative solution which is feasible even if little or no information is available about the possible range of the population proportions. In cases where the target proportions are far from ½, especially near 0 (rare subpopulations), it may be useful to consider using sample sizes that aim to provide specified levels of relative variance or CV of the estimated proportions. In either case, the theory for sample size determination under SRS is quite simple, and the adjustment (2) may be applied to determine sample sizes for a complex design.

In practice, one must make sure that the "right" design effects are used. For example, if regression estimators will be used for the actual survey, one should use the corresponding *deffs* and not *deffs* of simple means or totals; otherwise, the formulas may give incorrect sample size requirements. The numerator and denominator in the *deff* formula should agree both in terms of key design features (e.g., stratification) and choice of estimator. It is misleading to have, say, a ratio estimator in the numerator and a total in the denominator, or stratification in the numerator but not in the denominator. Finally, if $V_{\text{SRS},n}(\hat{\theta})$ is to be estimated using data from a complex survey, the estimate of V is not the usual SRS one since the actual complex design (e.g., clustering) needs to be taken into account. Two useful references on design effects are Lê and Verma (1997) and Park and Lee (2004).

An area where household survey practice needs to improve is in making estimates of design effects widely available. Such estimates are not regularly provided, especially in less developed survey organizations. Survey designers are then left with the task of having to estimate *deff*s from the survey microdata (when these are available and carry sufficient information about the design that variances can be correctly estimated), or alternatively, from published estimates of standard errors for certain parameter estimates (again, if available). In either case, this can be time-consuming for those unfamiliar with the survey or having limited access to detailed information about its design.

All surveys have nonresponse, which must be taken into account at the design stage. In addition to taking an anticipated rate of nonresponse into account, designers of household surveys that use the dwelling as a sampling unit also need to take the proportions of vacant and ineligible dwellings into account. These can be quite stable for large areas but may vary for smaller ones, even over short time periods. Attention must be paid to dwellings of temporary residence, such as those commonly found in beach and mountain resorts, where the resident population is sometimes smaller than the temporary population. Similar care is needed to account for addresses that are not residential if using an address frame where it is not possible to determine beforehand which ones are occupied by households. Two options for addressing these issues include: a) increasing sample size by dividing the initial sample size by the expected proportion of eligible and responding dwellings, in which case the selected sample is fixed but the effective sample is random; b) using a form of inverse sampling, where the required number of responding eligible dwellings is fixed, but the total number of selected dwellings is random. In both cases, weighting is required to compensate for the unequal observed eligibility and response rates. Such weighting requires precise tracking of eligibility and response indicators during fieldwork.

Allocation of the sample to strata, both geographical and optimal, requires a compromise among the important variables that the survey will measure. In addition, the designer must make compromises between different geographical levels (national, subnational, and so on). For many variables, simply allocating the sample proportional to population size is nearly optimal for national estimates. However, this allocation will likely be poor for subnational estimates. For example, if the country is divided into R regions and the national sample size is n, then a good allocation for regional estimates is likely to be to give each region about n/R units. Unless the regions have approximately equal populations, the two allocations (proportional and equal) are likely to be very different, and a compromise must be found. A common approach is to allocate the sample proportional to the square root (or some other power) of population size (see, e.g., Kish, 1988). Singh et al. (1994) describe a pragmatic solution to the specific problem of producing good estimates for both the nation and relatively small areas within it, which has been used for the Canadian LFS since the late 1980s. In this approach, most of the sample, say two-thirds, is allocated to produce the best possible national estimates. The remaining sample is then allocated disproportionately to some smaller areas to ensure a minimal level of quality in each area. As a result, large metropolitan areas get little or no sample in the second allocation round and, conversely, sparsely populated small areas get much of their sample in that round.

2.5. Sample selection

One aspect where household and business surveys tend to differ most is sample selection. The methods used by business surveys are discussed in Chapter 6. Household surveys that sample from a list can use methods similar to those. For example, the ACS and the U.K. Integrated Household Sample Survey have unclustered samples of addresses selected from address lists. One advantage of this type of frame is the ability to coordinate samples over time, either to ensure that adjacent survey waves have overlap or to avoid such overlap, when it is important to get a fresh sample of addresses in each wave.

However, for multistage household surveys, methods where the probability of selection of a PSU is proportional to its size are more common. These are referred to as probability proportional to size or PPS methods. They are discussed in most standard textbooks on sampling (see, e.g., Cochran, 1977, Chapter 9A). Under multistage PPS sampling, even though PSUs are selected with unequal probabilities, we can have a *self-weighting* design, in which all ultimate sampling units in a stratum have the same final design weight (see Section 5.1). To illustrate this, consider a two-stage design and suppose PSUs are selected using PPS sampling, with size defined as the number of second-stage units in a PSU. Thus, if M_i is the size of the ith PSU and M is the total size of all the PSUs in the stratum, then the probability that PSU i is selected is $p_i = nM_i/M$, where n is the number of PSUs selected in the stratum. Now, select the same number m of second-stage units from each sampled PSU using SRS. Then the second-stage inclusion probability is $p_{2ij} = m/M_i$ for all j in PSU i. Hence, the overall inclusion probability for each second-stage unit is $p_i p_{2ij} = nM_i/M \times m/M_i = nm/M = 1/d$. The design is then *self-weighting* because all units in the sample have the same design weight d.

In practice, this textbook procedure is often not useful since unit sizes M_i become out of date quickly. In fact, since the sizes are typically based on a recent census or an administrative source, they are likely to be out of date as soon as they become available. To preserve self-weighting in this more realistic situation, an alternative is to use systematic sampling at the second stage instead of SRS and fix the sampling interval over time. For example, if we should select every Kth unit according to the census counts, then we continue to select every Kth unit thereafter. One undesirable consequence of this procedure is that the sample size in each PSU is no longer constant. If a PSU has grown by 10%, then its sample will also grow by 10% and conversely for decreases in size. Since populations tend to grow over time, the former is the more serious problem. At the national level, this implies that the total sample size, and therefore costs, will gradually increase. One way to deal with this growth is to randomly drop enough units from the sample to keep the total sample size stable. This is the approach used by the Canadian LFS and the U.S. Current Population Survey (CPS).

An alternative is to design a sample which is self-weighting, but to allow the weights to vary over time for households selected from different PSUs. This weight variation would happen anyway if nonresponse varies between PSUs and if simple weight adjustments are applied at the PSU level. This design will be slightly less efficient than the corresponding self-weighting design but will not suffer from the cost-increase problem described above. Its main disadvantage is that varying household weights lead to more complex estimation procedures but this disadvantage is less important with the increased availability of modern computer facilities and software.

Regardless of the approach used, in practice, having a pure self-weighting design may not be attainable. For example, even if we implement the fixed sampling interval method described above, there will almost certainly be PSUs in the sample that have grown to such an extent that it will be necessary to subsample from them to control costs and balance interviewer workloads.

3. Repeated household surveys

3.1. Repeated versus longitudinal surveys

Many household surveys are repeated over time with the same or very similar content and methodology to produce repeated measurements of key indicators that are used to assess how demographic, economic, and social conditions evolve. In fact, many descriptive analyses reported in household survey publications discuss how major indicators changed in comparison to previous survey rounds. The idea that a household survey is to be repeated introduces a number of interesting aspects of survey design and estimation, which we consider in this section. We start by establishing a distinction between repeated and longitudinal surveys.

Longitudinal surveys require that a sample of elementary survey units (say households or individuals) is followed over time, with the same units observed in at least two survey data collection rounds or waves. Observation of the selected units continues for a specified length of time, a number of waves, or until a well-specified event takes place (e.g., the person reaches a certain age). Longitudinal designs are essential if the survey must provide estimates of parameters that involve measures of change at the individual level.

Repeated surveys collect data from a specified target population at certain (regular) intervals using the same (or at least comparable) methodology. They do not require that the same elementary units should be followed over time but are often designed such that there is some overlap of units in successive survey waves. They also include surveys for which the samples on different occasions are deliberately non-overlapping or even completely independent.

When a repeated survey uses samples that are at least partially overlapping at the elementary unit level, it includes a longitudinal component, which may or may not be exploited for analysis. In such cases, the distinction between longitudinal and repeated surveys becomes blurred and the key to separating them is the main set of outcomes required from the survey. If the main parameters to be estimated require pairing measurements on the same elementary units from at least two survey waves, we classify the survey as longitudinal. Otherwise, we call it a repeated survey.

Longitudinal surveys are discussed in Chapter 2. In this section, we focus on some design and estimation issues regarding *repeated household surveys*. A more detailed classification of surveys in terms of how their samples evolve in time is provided by Duncan and Kalton (1987) and Kalton and Citro (1993).

The traditional design of household surveys requires specifying a sample selection procedure coupled with an estimation procedure that provides adequate precision for key parameters. Sample sizes are determined by taking account of the survey budget, cost functions describing the relative costs of including additional primary and elementary sampling units, and design effects if available. If the survey is to be repeated, the process by which the sample evolves in time is an additional element that must be designed for. We refer to this process as the *rotation scheme* or *rotation design* of the survey.

3.2. Objectives, rotation design, and frequency for repeated household surveys

The key to efficient design for repeated household surveys is to match the sample selection mechanism, survey frequency, rotation scheme, and estimation procedures to satisfy the survey objectives at minimum cost for a fixed precision or maximum precision for a fixed cost. Key references on this topic are Binder and Hidiroglou (1988) and Duncan and Kalton (1987). For a good example of an in-depth discussion of how the survey objectives affect the rotation design in the case of a household labor force survey, see Steel (1997).

We first consider the problem of specifying what the key objectives of inference are for a repeated household survey. They may include the following:

(a) Estimating level: estimating specified population parameters at each time point;
(b) Estimating change: estimating (net) change in parameters between survey waves;
(c) Estimating averages: estimating the average value over several survey waves;
(d) Cumulating samples of rare populations or for small domains over time.

Kalton and Citro (1993) list several other objectives that require a survey to be repeated over time, but these require the longitudinal component of the survey to be of primary interest. Here, we focus only on repeated surveys, where the main objectives do not require the longitudinal component.

Considering objective (d), the best possible rotation design is to have completely non-overlapping samples in the various survey waves. This design maximizes the speed with which new observations from the rare target population can be found. For example, consider a survey which needs to cumulate observations of people who migrated from a foreign country to their current place of residence during the five years preceding interview time. To be able to have a sufficiently large sample of this subpopulation, it may be necessary to use either a cross-sectional screening survey with a very large sample size, or alternatively, a repeated survey with smaller samples at each wave, screening for this subpopulation, thus providing a sample of the intended size after a number of waves has been completed.

Note, however, that this would often be a secondary objective for a repeated survey because the alternative of using a large cross-sectional survey would probably be more cost-effective than the non-overlapping repeated survey option described above. Nevertheless, given an existing repeated survey (overlapping or not), it may be cost-effective to include the screening questions and additional survey modules as required for the measurement of this "rare" target subpopulation. Advantages and limitations of these two competing approaches must be carefully considered before choosing the survey design for any particular application.

It follows that the main objectives leading to a repeated survey design are likely to be (a), (b), (c), or a combination of these. For these objectives, alternative rotation designs affect the precision of estimators for each type of target parameter. Let U_t

denote the target population at time t and let θ_t denote the value of a target parameter at time t, where t could refer to years, quarters, months, weeks, etc. We assume that the definition of the target population is fixed over time, for example, all adults living in private dwellings. However, the size and composition of this target population may change over time, because people die, migrate, or reach the age limit to be included in the survey from a given time point onwards. Changes in the parameter θ_t over time can thus be caused by changes in both the composition of the population and in the values of the underlying characteristics of members of the population.

A repeated survey may have as key objective the estimation of the series of values of $\theta_1, \theta_2, \ldots, \theta_t, \ldots$, that is, the main goal is to get the best possible estimates for the *level* θ_t at each point in time (objective (a)). Alternatively, the target parameter may be the *change* between times t and $t - 1$, defined as $\theta_t - \theta_{t-1}$ (objective (b)). In some situations, the target parameter may be an average of the values of the parameters at different time points (objective (c)), and a simple example of this is $\bar{\theta}_{t,2} = (\theta_{t-1} + \theta_t)/2$, the (moving) average of the parameter values at two successive time points.

Denoting by $\hat{\theta}_t$, an (approximately) unbiased estimator of θ_t, we have the following results on the precision of estimators of these types of target parameters. For the estimation of *change*, the variance of the simple (approximately) unbiased estimator $\hat{\theta}_t - \hat{\theta}_{t-1}$ is given by

$$V(\hat{\theta}_t - \hat{\theta}_{t-1}) = V(\hat{\theta}_t) + V(\hat{\theta}_{t-1}) - 2\text{COV}(\hat{\theta}_{t-1}; \hat{\theta}_t). \tag{3}$$

If successive measurements on the same unit for the survey variable defining the parameters θ_t are positively correlated over time, then with some degree of overlap between the samples at times $t - 1$ and t, the covariance term in the right-hand side of (3) would be positive. In this case, the estimation of the change would be more efficient with overlapping samples than with completely independent samples at times t and $t - 1$. Independent samples lead to complete independence between $\hat{\theta}_t$ and $\hat{\theta}_{t-1}$, in which case the variance of the difference is

$$V(\hat{\theta}_t - \hat{\theta}_{t-1}) = V(\hat{\theta}_t) + V(\hat{\theta}_{t-1}). \tag{4}$$

So for the estimation of change, some overlap of samples in successive waves increases the precision of the estimator when the underlying characteristic is positively correlated for measurements on successive occasions.

For the estimation of averages over time, the variance of the simple (approximately) unbiased estimator $\bar{\hat{\theta}}_{t,2} = (\hat{\theta}_t + \hat{\theta}_{t-1})/2$ is given by

$$V\left[\frac{1}{2}(\hat{\theta}_t + \hat{\theta}_{t-1})\right] = \frac{1}{4}\left[V(\hat{\theta}_t) + V(\hat{\theta}_{t-1}) + 2\text{COV}(\hat{\theta}_{t-1}; \hat{\theta}_t)\right]. \tag{5}$$

Here, the positive correlation of successive measurements of the underlying survey characteristic would lead to reduced precision with overlapping surveys, and having independent or non-overlapping samples would be more efficient. This discussion illustrates the importance of regularly publishing estimates of the correlations over time between key measurements such that these are available to inform survey design or redesign.

These two examples illustrate the need to specify clearly what the inferential objectives are for the survey; otherwise, one may end up with an inefficient rotation design. In addition, they also indicate that if in a given situation, both changes and averages over time are required, overlapping samples increase efficiency for change but reduce it for averaging. This poses a problem to the survey designer and calls for an explicit assessment of the relative importance of the different survey objectives so that decisions regarding the rotation design are not misguided. The estimation of change is usually the harder of these two objectives, often requiring larger sample sizes than would be needed for the estimation of level itself. Hence, the survey could be designed to achieve the required level of precision for estimating change both in terms of sample design, sample size, and rotation design, and still be able to provide estimates of acceptable accuracy for averages over time.

Another important design parameter of a repeated survey is the frequency of the survey, which again is closely linked with the survey objectives. Surveys having a short interval between waves (say monthly or quarterly) are better for tracing respondents over time, provide better recall of information between surveys because successive interviews constitute useful benchmarking events, and are generally capable of providing more frequently updated estimates for the target parameters. Also, shorter survey intervals are better for monitoring more volatile target parameters. On the other hand, the shorter the interval between surveys, the larger the burden on respondents with overlapping surveys, which may increase nonresponse and lead to response conditioning, a well-known source of bias in repeated surveys. Longer survey intervals may suffice for less volatile parameters.

Labor force surveys provide an example where the key outputs are required monthly or quarterly, given the need to assess how employment and unemployment totals and rates evolve in the short term. In the European Union, member states are required to carry out continuous labor force surveys, that is, surveys which measure the labor force status of the people every week during the year, to report the results of such surveys at least quarterly (see the Council of the European Union, 1998; the European Parliament and Council of the European Union, 2002). It is interesting to note that the key concept in an LFS requires establishing each person's economic activity status in a specified reference week. For this reason, several countries conduct their LFS in a fixed or prespecified week every month, with the reference week being the week before the interview (see Table 1 below). This however places a heavy burden on the statistical agency, which then must have a workforce capable of handling all the required data collection of a country's LFS within a single (or sometimes two) week(s). Other countries, while still aiming to measure the same concept, use moving reference weeks to be able to spread the data collection activities over a longer period of time. In the United Kingdom, the sample for a quarter is split into 13 weekly assignments to create an efficient fieldwork design, and hence the sample size for every week is about 7.7% of the total sample size for a quarter. However, this choice has implications for both the estimation and analysis of the resulting indicators, which we do not discuss here (for further details, see Steel, 1997).

At the other end of the spectrum, demographic and health surveys are carried out with intervals of up to five years between successive survey waves because the main parameters of interest in these surveys are expected to vary slowly. The same is true for the case of household income and expenditure surveys in many countries—see, for example, the Seventeenth International Conference of Labour Statisticians (2002)— although there is a trend to increasing the frequency of such surveys in other countries.

A quick note on retrospective versus prospective data collection designs: for most repeated surveys, prospective data collection designs are adopted. We call a design

Table 1
Summary of characteristics of selected LFS rotation designs

Item	Survey					
	U.S. CPS	Canadian LFS	Australian LFS	U.K. LFS	Japan LFS	Brazilian LFS
Frequency	Monthly	Monthly	Monthly	Quarterly	Monthly	Monthly
Reference week	Fixed	Fixed	Fixed	Moving	Fixed	Moving
Collection period	1 week	1 week	2 weeks	13 weeks	2 weeks	4 weeks
Rotation design	4-8(2)	6-0(1)	8-0(1)	1-2(5)	2-10(2)	4-8(2)
Monthly overlap (%)	75	83	88	0	50	75
Quarterly overlap (%)	25	50	63	80	0	25
Yearly overlap (%)	50	0	0	20	50	50
Sample size (HH) per month	60,000	53,000	29,000	20,000	40,000	41,600

prospective when the data will be collected for a period similar to the current reference period every time a household is sampled. A design is called retrospective if in any given survey wave (say the first one), the household is asked to provide data for several past reference periods (say if this is a month, data are required for at least two previous months, which may or may not be the latest ones). Retrospective designs can be useful if there is a need to limit the total number of visits to a household, and yet information needs to be available on an individual basis for different time periods so that some form of longitudinal analysis is possible. However, caution is required given the well-known adverse effects of respondent recall of information for periods not too close to the time of interview or for information regarding events that may not be easily remembered.

3.3. Estimation strategies for some basic objectives and rotation schemes

Cochran (1977) considered the case when $\theta_t = \overline{Y}_t$ is the population mean of a survey variable y, that is, the target parameter is the *level at each time point.* Under SRS from the population at each time period, there are gains to be made from using samples with some overlap in adjacent survey waves, but these gains are modest unless the correlation ρ of the measurements of the survey variable y in two successive time periods is high (say bigger than 0.7). The gains in efficiency are made by using an estimator that combines, in an optimal way, the mean of the unmatched portion of the sample at time t with a regression estimator of the mean based on the portion of the sample at time t, which is matched to units in the sample at time $t - 1$. In this case, the optimal proportion of sample overlap between two successive survey waves would not exceed 50%, and this would be the limiting proportion of overlap required to maximize efficiency gains.

If the target parameter is the *change* between times t and $t - 1$, then the best rotation design requires matching larger proportions of sampling units in successive survey waves. Using more than 50% overlap would not be optimal for level estimation but would not result in substantial losses in efficiency compared with the estimators of level under the optimal overlap for level. Hence, in a survey where both estimation of level and change are important, sample overlap more than 50% may be used as a compromise to obtain large efficiency gains for the estimates of change and modest efficiency gains for estimates of level.

If cost considerations are added, in household surveys it is often the case that the cost of the first interview of a selected household is higher than in subsequent occasions. For example, in many countries, labor force surveys have most or all of their first interviews conducted in person, and most subsequent interviews are carried out over the telephone (see the U.S. CPS, the Canadian LFS, the U.K. LFS, the Australian LFS, etc.). In such cases, cost considerations would suggest retaining the largest possible portion of the sample in two successive survey waves. However, one has to consider the added burden of keeping the same respondents in the sample for several waves and the impact this may have on nonsampling errors, such as potential increases to nonresponse and attrition rates, as well as other more subtle forms of errors such as "panel conditioning"—see Chapter 2 for definitions of technical terms.

Many commonly used rotation designs have the number of times in sample equal to at most 8, which means a maximum overlap of samples in successive waves of 87.5% (as is the case in the Australian LFS, which is a monthly survey using a rotation scheme called in-for-8, where a selected household is in the sample for eight consecutive months).

Now consider a repeated survey where the target parameter is an average level over three waves, represented here by $\bar{\theta}_{t,3} = (\theta_{t-1} + \theta_t + \theta_{t+1})/3$. Here, the best rotation design is selecting independent samples every time, because with this design, the variance of the average is simply 1/3 of the average of the variances of the estimates for the individual survey periods. An example of a survey where the prime target is estimating averages is the ACS, designed to replace the "long form" sample in the decennial census in the United States. The idea is that survey data for periods of five years should provide equivalent data to those formerly obtained using the decennial census sample.

The above discussion reveals that precise knowledge of the key survey inference objectives is required for an efficient rotation design to be selected. Estimating averages over time (objective (c)) requires independent or non-overlapping samples at each survey wave, whereas estimating change (objective (b)) requires samples with high overlap in the survey comparison periods (base and current). Estimating level (objective (a)) suggests that a moderate amount of overlap is required but it retains some efficiency gains compared with independent samples even if the overlap is somewhat bigger than the optimal.

The advice above is based on variance considerations only. Repeated large-scale household surveys must often satisfy several of these objectives and for different survey characteristics. Hence, design choices are more complex and less likely to be based only on variance efficiencies. There are several reasons for not using completely independent repetition of cross-sectional designs across time. First, sample preparation costs and time are likely to be substantially bigger. For example, in many household surveys, there are substantial costs associated with selection of new PSUs, such as listing or frame updating costs, staff hiring or relocating, infrastructure, etc. Second, if a survey is repeated over time, it is not unlikely that users will use the survey results for comparisons over time and completely independent surveys would be very inefficient for this purpose. In addition, nonsampling error considerations must also play an important part in specifying rotation designs such that respondent burden, attrition and measurement error are kept under control.

3.4. Examples of non-overlapping repeated surveys

In this section, some alternative rotation designs used by some major household surveys around the world are highlighted to illustrate how different objectives lead to different design options. We start by describing perhaps the largest repeated cross-sectional survey (i.e., no overlap of samples in adjacent survey waves) in existence: the ACS, see U.S. Census Bureau, 2006b.

The ACS selects a stratified systematic sample of addresses. The survey has a five-year cycle, and each address sampled in a given year is deliberately excluded from samples selected in the four subsequent years (negative coordination of samples in successive years). This is achieved by randomly allocating each address in the Master Address File used as the frame for the survey into one of five subframes, each containing 20% of the addresses in the frame. Addresses from only one of the subframes are eligible to be in the ACS sample in each given year, and a subframe can be used only once in every five years. New addresses are randomly allocated to one of the subframes.

The main objective of the ACS is to provide estimates for small areas, replacing the previous approach of using a long form questionnaire for a large sample of households collected during the Decennial Censuses in the United States. It was designed to provide for sample accumulation over periods of up to five years, after which the sample for each small area would be of similar size to what would have been obtained in the Decennial Census. The survey data are then used to estimate parameters that can be seen as moving averages of five years, with five years of survey data being used to provide estimates for the smallest areas for which results are published, and fewer years being used to provide estimates for broader geographies. Currently, single-year estimates are published annually for areas with a population of 65,000 or more. Multi-year estimates based on three successive years of ACS samples are published for areas with populations of 20,000 or more. Multi-year estimates based on five successive years of ACS samples will be published for all legal, administrative, and statistical areas down to the block-group level regardless of population size.

Another very large survey using repeated non-overlapping cross-sectional samples is the French Population Census. From 2004 onwards, the "census" of France's resident population started using a new approach, which replaced the traditional enumeration previously conducted every eight or nine years. The 1999 general population census of France was the last one to provide simultaneous and exhaustive coverage of the entire population. The new "census" in fact uses a large sample stratified by area size and requires cumulating data over five years to provide national coverage. Data are collected in two months every year (during January–February). All small municipalities (those with fewer than 10,000 inhabitants) are allocated to one of five "balanced" groups. For each group of small municipalities, a comprehensive census (no subsampling of dwellings or households) is carried out once every five years. Large municipalities (those with 10,000 or more inhabitants) carry out a sample survey of about 8% of their population every year. So at the end of a five-year cycle, every small municipality has carried out a census and, in the large municipalities, a sample of around 40% of the households will be available. This comprehensive sample is then used to replace the previous census for all purposes. Once the system is fully in place, rolling periods of five years may be used to provide census-like results, which were previously updated only once every eight or nine years.

Household income and expenditure surveys in many countries provide another important example of repeated cross-sectional surveys that use non-overlapping designs. The current recommendations issued by the International Labour Organization (ILO) on this type of survey (see the Seventeenth International Conference of Labour Statisticians, 2002) specify that such surveys should be conducted with intervals not exceeding five years. The recommendations do not specify that non-overlapping designs are required, but in many countries, this is the preferred method, given the considerable burden that such surveys place on participating households.

Another type of repeated survey design implemented around the world is the use of a panel sample of PSUs, with complete refreshment of the list of households sampled in the selected PSUs. The series of Demographic and Health Surveys (DHS) adopts this design whenever possible. Here, the gains from retaining the same set of PSUs are not as important in terms of variance reduction for estimates of change as if the same households were retained, but there are potential advantages in terms of costs of survey taking and also perhaps less volatile estimates of change between successive waves of this survey in a given country or region. For a more detailed discussion, see Macro International (1996, p. 29).

3.5. Rotation designs in labor force surveys

Labor force surveys conducted in most countries provide the most prominent application of overlapping repeated survey designs. For such surveys, intervals between survey waves are usually very short (months or quarters in most countries). In some countries, data collection is continuous throughout the year and publication periods may again be monthly or quarterly. In the United States and Canada, monthly surveys are used, with a single reference week every month. Rotation designs for these surveys are 6-0(1)[1] for the Canadian LFS, and 4-8(2) for the U.S. CPS. In both surveys the estimation of change in labor force indicators between adjacent months is a prominent survey objective. In both countries, some form of composite estimation (see section 5.3) is used to estimate the indicators of interest. Seasonally adjusted estimates derived from the time series of the composite estimates are also published, and are prominent in the analysis of survey results contained in monthly press releases issued by the corresponding statistical agencies.

In Australia, the LFS uses the 8-0(1) rotation design, and the key estimates highlighted in the publications are the estimates of the trend derived from the time series of the sample estimates. This is a unique example of a survey where the major indicators are based on time series modeling of the basic survey estimates. Because the targets for inference here are not simply the values of the unknown parameters, but of rather complex functions of these (the trend of the corresponding time series), this brings in some interesting design issues. McLaren and Steel (2001) studied options for designs for surveys where the key objective is trend estimation and concluded that monthly surveys using rotation designs 1-2(m) are the best. This study illustrates quite clearly the impact that the choice of objectives has on rotation designs: if the trend is the main

---

[1] We use the convention *in-out* (*times*) to denote the number of waves that a household is included in the sample, then the number of waves that it is left *out* of the sample, and the number of *times* that this pattern is repeated. A similar convention was proposed by McLaren and Steel (2001).

target, monthly surveys need not have monthly overlap. The same is not true, though, if the target parameter is the simple difference of the relevant indicators, without any reference to the underlying trend. The U.K. LFS is a quarterly survey, originally motivated by the aim of measuring quarter-on-quarter change, using a rotation design which may be described as equivalent to a monthly 1-2(5), whereby the sampled households enter the survey with an interview on a single month of the quarter, rest for two months, then return for another four successive quarters. Interestingly, this same rotation pattern (more precisely, a 1-2(8) pattern) is used by the Canadian LFS in Canada's three northern territories, where three-month estimates are published.

Table 1 displays some information on key aspects of rotation designs used in labor force surveys around the world. The U.K. LFS is a model close to the LFS design adopted throughout the European Union. The use of this model for LFSs has spread beyond the European Union. In 2005–2006, Statistics South Africa started an ambitious project to replace its semiannual LFS with a quarterly survey with a rotation design in 2008. A similar project is under way in Brazil, where an integrated household survey using a 1-2(5) rotation design will replace the current annual national household survey and the monthly LFS in 2009.

3.6. Some guidance on efficiency gains from overlapping repeated surveys

Once a decision has been taken that the survey has to have some sample overlap over time, it becomes important to decide how much and which methods to use to control how the sample evolves over successive survey occasions. Cochran (1977, Section 12.11) provides some useful guidance on the choice of how much overlap to have. His results are all based on an assumed SRS design. Most household surveys use more complex sample designs. However, the sample design structure (stratification, clustering, sample sizes, selection probabilities, and estimator) is often held fixed over time. We can express the variance of survey estimates at each time point in terms of the product

$$V_p(\hat{\theta}_t) = V_{\text{SRS}}(\hat{\theta}_t) \times \text{deff}_t, \tag{6}$$

where $V_p(\hat{\theta}_t)$ is the variance of the survey estimator under the complex survey design adopted to carry out the survey, $V_{\text{SRS}}(\hat{\theta}_t)$ is the variance that the survey estimator would have under a simple random sample design with the same sample size, and deff_t is the corresponding design effect. If we assume that the design effect is approximately constant over time (i.e., $\text{deff}_t = \text{deff} \ \forall t$), then the advice provided for SRS is relevant to compare the relative merits of complex surveys for alternative rotation designs.

Suppose that a SRS of size n is used on two successive occasions (t and $t + 1$) and that the population is assumed fixed (no changes due to births or deaths), but the measurements may change. On the second occasion, a SRS of $m < n$ units sampled at t are retained (overlap part) and $n - m$ units are replaced by newly selected ones, also sampled using SRS, from the units not sampled at t. Under this scenario and assuming that the finite population correction can be ignored, the variance of the sample mean $\overline{y}_t$, the simplest estimator of the population mean $\overline{Y}_t$ (the level) of a survey variable y at each time point t, is given by

$$V_{\text{SRS}}(\overline{y}_t) = S_t^2/n, \tag{7}$$

where S_t^2 is the population variance of the survey variable y at time t. Clearly, the variances of the simple estimates of level do not depend on m, the size of the matched or overlapping portion of the sample at time $t+1$. However, assuming that the variance of the survey variable is constant over time, that is, $S_t^2 = S_{t+1}^2 = S^2$, it follows that the variance of the estimate of change, namely, the difference in the population means, is given by

$$V_{\mathrm{SRS}}(\bar{y}_{t+1} - \bar{y}_t) = 2\frac{S^2}{n}\left(1 - \frac{m}{n}\rho\right), \tag{8}$$

where ρ is the correlation of observations of the survey variable in two adjacent time periods.

As ρ is often positive, it becomes clear that no overlap ($m = 0$) is the least efficient strategy for estimation of change and that a panel survey (complete overlap or $m = n$) is the most efficient, with a reduction factor of $1 - \rho$. The overlap fraction m/n provides an attenuation of the variance reduction when a rotation design with less than 100% overlap is adopted.

Now with some sample overlap, there are alternative estimators of the level on the second (and subsequent) occasion(s). If an optimal estimator (see Cochran, 1977, eq. 12.73) is used with optimal weight and optimum matching proportion of $m/n = \sqrt{1-\rho^2}/(1+\sqrt{1-\rho^2})$, its variance would be given approximately by

$$V_{\mathrm{SRS}}\left(\bar{y}_{t+1}^{\mathrm{opt}}\right) = \frac{S^2}{n}\frac{\left(1+\sqrt{1-\rho^2}\right)}{2}. \tag{9}$$

For values of ρ above 0.7, the gains are noticeable, and the optimum matching proportion is never bigger than 50%.

The above discussion demonstrates a clear link between the selection of an estimator and the choice of a rotation design, particularly in terms of the proportion of sample overlap, given a specified survey objective (in the above, estimation of the current population mean). This discussion is at the heart of substantial developments in the literature on estimation from repeated surveys, reviewed in Binder and Hidiroglou (1988), and subsequently, in Silva and Cruz (2002). Some large-scale repeated surveys make use of composite estimators (see Section 5.3), and the prime examples are again the U.S. CPS (see U.S. Census Bureau and U.S. Bureau of Labor Statistics, 2002), and the Canadian LFS (see Gambino et al., 2001).

After reviewing efficiency gains for estimators of both level and change under the simplified SRS scenario discussed above, Cochran (1977, p. 354) suggests that "retention of 2/3, 3/4, or 4/5 from one occasion to the next may be a good practical policy if current estimates and estimates of change are both important." But this large overlap of successive surveys will only be advantageous if the estimators utilized are capable of exploiting the survey overlap as would the "optimal" estimator discussed above.

We conclude this section by pointing out that in addition to considerations of sampling error, it is essential that designers of repeated surveys consider the implications of alternative rotation designs in terms of nonsampling errors. The longer the households are retained in the sample, the more likely they are to drop out (attrition/nonresponse), as well as to start providing conditioned responses (measurement error). After a certain point, the combined adverse effects of nonsampling errors are more likely to overshadow any marginal gains in efficiency, so it is vital not to extend the length of survey participation beyond this point.

4. Data collection

The traditional modes of data collection for household surveys are personal interview, telephone interview, and questionnaire mail out followed usually by self-completion by the respondent. Recently, a variety of new methods have started to be used. These include use of the internet (html questionnaires), the telephone (where the respondent enters his replies using the telephone keypad), and self-completion using a computer (the respondent either enters his responses using the keyboard or gives them orally, and the computer records them).

Gradually, the use of paper questionnaires is diminishing and being replaced by computer-based questionnaires. The latter have several advantages, including the possibility of having built-in edits that are processed during each interview and the elimination of the data capture step needed for paper questionnaires. An important effect of these changes is that the file of survey responses is relatively clean from the outset. Computer-based questionnaires also make it possible to have very complex questionnaires with elaborate skip patterns. Even in some very large-scale household surveys, such as national population censuses, computers are now being used instead of traditional paper questionnaires: this has been the case in Colombia and Brazil, where handheld computers were used for population censuses in 2006 and 2007. The Brazilian case illustrates the potential for such devices to affect data collection because for the first time, the population census and an agricultural census have been integrated into a single field operation, with households in the rural area providing both the population and agricultural census information in a single interview (for details see the web site of the Brazilian official statistics agency IBGE).

One benefit of using computer-assisted interviewing that is receiving increased attention is the wealth of information about the data collection process that it makes available. Such data about the data collection process, and more generally about other aspects of the survey process, are referred to as paradata (see Scheuren, 2005). This information can be invaluable to improving the collection process. It provides answers to questions such as: which parts of the questionnaire are taking the most time? Which questions are being corrected most often? Which are triggering the most edit failures? This information can then be used to review concepts and definitions, improve the questionnaire, improve interviewer training, and so on. Granquist and Kovar (1997) advocate the use of such information as one of the primary objectives of survey data editing, but one that is often not so vigorously pursued in practice. The ultimate goal is to improve the survey process in subsequent surveys or survey waves.

A major change in field operations in developed countries has been the increase in the proportion of interviews conducted by telephone rather than in person. This has had an impact on both the types of people who conduct interviews and the way their workload is organized. For example, when most interviews were conducted in person, often in the daytime, the interviewer needed a car. With the introduction of computer-assisted interviewing from a central facility, the interviewer no longer needs a car and the number of evening interviews can increase since it is possible to have an evening shift that conducts interviews in different time zones. As a result of factors such as these, the demographic characteristics of interviewers have changed (e.g., the number of university students working part-time as interviewers has increased in Canada).

The introduction of computer-assisted interviewing from a central facility also makes it possible to monitor interviews as they happen. In some statistical agencies, elaborate quality assurance programs based on monitoring have been introduced to improve the data collection process (identify problems, target interviewer training needs, etc.).

Monitoring interviews conducted in a central facility has benefits, but until recently, the computers used for personal interviewing, namely, laptop computers and handheld devices, were not powerful enough to implement something similar to monitoring for personal interviews. However, recording of personal interviews (on the same computer used to enter responses to survey questions) has now become feasible. Biemer et al. (2000) discuss the application of computer audio-recorded interviewing (CARI) to the National (United States) Survey of Child and Adolescent Well-being. Another form of interview monitoring that has recently become feasible is to use computers equipped with a Global Positioning System device, which can be used to record the coordinates of dwellings visited for interview at the time of arrival or at the start of the interview. This enables survey organizations to supervise work in ways that were not previously feasible with paper-and-pencil type interviews. Devices like these were used to carry out a mid-decade population census in Brazil and are being considered for the redesigned South African LFS.

The use of computers in survey sampling extends well beyond computer-assisted interviewing. Since the whole survey process can be monitored using software tools, this opens new possibilities for improving data collection. In addition to the tools already mentioned in this section, such as live monitoring of interviews, computers make it easier to keep track of progress on many fronts, such as response rates by various categories (geographical, age group, etc.). Hunter and Carbonneau (2005) provide a high-level overview of what they refer to as active management.

The increased use of telephone interviewing is motivated by cost considerations, but it also introduces problems. We have already mentioned problems associated with the use of mobile telephones in Section 2.1 in the context of frame coverage. There are other aspects of telephone interviewing that make it more difficult to get a response. These include the use of answering machines and call display (caller ID) to screen calls and the apparently greater difficulty for the interviewer to establish a rapport with the respondent by telephone than in person. The recent Second International Conference on Telephone Survey Methodology, held in 2006, was devoted to the subject. A monograph containing selected papers from the conference will be published.

In recent years, there has been increased interest in finding better ways to survey rare populations (or, more accurately, small groups within larger populations) and groups that are difficult to survey, such as the homeless and nomadic populations. Statistics Canada (2004) devoted a methodology symposium to the topic and some of the latest research in this area is covered in the proceedings of that conference.

4.1. Combining data from different sources

Another way to reduce survey costs is to try to make use of existing data and, in particular, data from administrative sources. We focus on the combined use of administrative and survey data, although there are cases where administrative data can be used on their own. In addition to its low cost (since it was collected for some other purpose and therefore already "paid for"), a great benefit of using administrative data is the reduction in burden on survey respondents. In addition, if the concepts used by the survey (e.g., to define income) are close to those on which the administrative data are based, then the administrative data may be more accurate than the same data obtained via a survey. We give two examples of the integration of survey and administrative data. The Canadian Survey of Labour and Income Dynamics (SLID) asks survey respondents whether they prefer to answer several questions on income or, alternatively, to give permission to Statistics Canada to access the information from their income tax records. In recent surveys, 85% of respondents have chosen the latter option. Because of this success, the approach is being extended to other surveys such as the Survey of Household Spending and even the census of population.

A second example involves the long-standing longitudinal Survey of Income and Program Participation (SIPP) conducted by the U.S. Census Bureau, which is being replaced by the Dynamics of Economic Wellbeing System (DEWS). The DEWS will make extensive use of administrative data files to augment survey data (see U.S. Census Bureau, 2006a). A key motivating factor behind this change is the reduction of both costs and response burden. By using administrative data files over time, the new approach also avoids the problem of attrition common to all longitudinal surveys. More generally, data can be combined across multiple surveys and administrative sources. This is often the only way to obtain adequate estimates for small domains. Two sessions at Statistics Canada's 2006 methodology symposium included papers on this topic (see Statistics Canada, 2006).

Merkouris (2004) presented a regression-based method for combining information from several surveys. This method is essentially an extended calibration procedure whereby comparable estimates from various surveys are calibrated to each other, accounting for differences in effective sample sizes. The method has been applied successfully: data from the Canadian SLID was used to improve estimates for the much smaller Survey of Financial Security. Merkouris (2006) adapted the procedure to small domain estimation. A related approach was adopted in the Netherlands (Statistics Netherlands, 2004) to compile a whole "virtual census" using data from several sources, such as a population register, some other administrative records, and selected household surveys. The methodology, called *repeated weighting*, is described in detail in Houbiers et al. (2003).

5. Weighting and estimation

5.1. Simple estimation of totals, means, ratios, and proportions

Estimation in household sample surveys is often started using "standard" weighting procedures. Assuming that

- a two stage stratified sampling design was used to select a sample of households,
- every member in each selected household was included in the survey, and
- the sample response was complete,

then the standard design-weighted estimator for the population total $Y = \sum_h \sum_{i \in U_h} \sum_{j \in U_{hi}} y_{hij}$ of a survey variable y has the general form

$$\hat{Y} = \sum_h \sum_{i \in s_h} \sum_{j \in s_{hi}} d_{hij} y_{hij}, \tag{10}$$

where d_{hij} is the design weight for household j of PSU i in stratum h, y_{hij} is the corresponding value for the survey variable y, U_h and s_h are the population and sample sets of PSUs in stratum h, respectively, of sizes N_h and n_h, U_{hi} and s_{hi} are the population and sample sets of households in PSU i of stratum h, having sizes M_{hi} and m_{hi}, respectively.

The design weight d_{hij} is the reciprocal of the inclusion probability of household j of PSU i in stratum h, which can be calculated as the product of the inclusion probability π_{hi} for PSU hi and the conditional probability $\pi_{j|hi}$ of selecting household j given that PSU hi is selected. Design weights for multistage designs having more than two stages of selection can be computed using similar recursion algorithms where each additional stage requires computing an additional set of inclusion probabilities conditional on selection in preceding stages.

Although the design weight is simply the reciprocal of a unit's inclusion probability, in practice, its computation can be quite involved. For example, in the relatively simple case of the Canadian LFS, the design weight is the product of the following: the first-stage (PSU) inclusion probability, the second-stage (dwelling) inclusion probability, the cluster weight (a factor, usually equal to 1, that accounts for subsampling in PSUs whose population has increased significantly since the last redesign), and the stabilization weight (a factor to account for the high-level subsampling that the LFS uses to keep the national sample size stable over time; see Section 2.5 in this chapter). For some surveys, the computation of design weights can be much more complex than this, particularly for longitudinal surveys. In addition, there are further adjustments to the design weight needed to account for nonresponse and, in some cases, for coverage errors, unknown eligibility (in RDD surveys, for example), and so on.

Estimators of totals similar to (10) are available for designs having any number of stages of selection. The estimator of the population mean $\bar{Y} = \sum_h \sum_{i \in U_h} \sum_{j \in U_{hi}} y_{hij} / \sum_h \sum_{i \in U_h} M_{hi}$ would be obtained simply by substituting the design-weighted estimators of the totals in the numerator and denominator leading to

$$\bar{y}_d = \sum_h \sum_{i \in s_h} \sum_{j \in s_{hi}} d_{hij} y_{hij} \Big/ \sum_h \sum_{i \in s_h} \sum_{j \in s_{hi}} d_{hij}. \tag{11}$$

For most household surveys, the overall population size $M_0 = \sum_h \sum_{i \in U_h} M_{hi}$ is not known and the estimator that could be obtained from (14) by replacing the estimated population size in the denominator by M_0 is not available. However, even if this alternative estimator was available, (11) would still be the usual choice because for many survey situations encountered in practice, it would have smaller variance. Note that (11) is a special case of the estimator

$$\hat{R}_d = \sum_h \sum_{i \in s_h} \sum_{j \in s_{hi}} d_{hij} y_{hij} \Big/ \sum_h \sum_{i \in s_h} \sum_{j \in s_{hi}} d_{hij} x_{hij} \tag{12}$$

for the ratio of population totals $R = Y/X = \sum_h \sum_{i \in U_h} \sum_{j \in U_{hi}} y_{hij} \Big/ \sum_h \sum_{i \in U_h} \sum_{j \in U_{hi}} x_{hij}$, where the variable x in the denominator is equal to 1 for every household. Another special case of interest occurs when the survey variable is simply an indicator variable. In this case, its population mean is simply a population proportion, but the estimator (11) is still the estimator applied to the sample observations of the corresponding indicator variable.

The above estimators would be used also to obtain estimates on characteristics of persons simply by making the corresponding y and x variables represent the sum of the values observed for all members of sampled households. This assumes that survey measurements are taken for every member of sampled households, which is a common situation in practice. However, if subsampling of household members takes place, there would be an additional level of weighting involved, but the above general approach to estimation would still apply.

5.2. Calibration estimation in household surveys

Despite their simplicity, such design-weighted estimators are not the ones most commonly used in the practice of household surveys. Instead, various forms of calibration estimators (Deville and Särndal, 1992) are now commonly used. Calibration estimators of totals are defined as

$$\hat{Y}_C = \sum_h \sum_{i \in s_h} \sum_{j \in s_{hi}} w_{hij} y_{hij}, \tag{13}$$

where the weights w_{hij} are such that they minimize a distance function

$$F = \sum_h \sum_{i \in s_h} \sum_{j \in s_{hi}} G(w_{hij}, d_{hij}) \tag{14}$$

and satisfy the calibration equations

$$\sum_h \sum_{i \in s_h} \sum_{j \in s_{hi}} w_{hij} \mathbf{x}_{hij} = \mathbf{X}, \tag{15}$$

where $\mathbf{x}_{hij}$ is a vector of auxiliary variables observed for each sampled household, $\mathbf{X}$ is a vector of population totals for these auxiliary variables, assumed known, and $G(w, d)$ is a distance function satisfying some specified regularity conditions. A popular choice of distance function is the standard "chi-square" type distance defined as

$$G(w_{hij}, d_{hij}) = (w_{hij} - d_{hij})^2 / q_{hij} d_{hij}, \tag{16}$$

where the q_{hij} are known constants to be specified.

Calibration estimators for ratios and means (as well as proportions) follow directly from using the weights w_{hij} in place of the design weights d_{hij} in the expressions (11) and (12).

Calibration estimators have some desirable properties. First, weights satisfying (15) provide sample "estimates" for the totals of the auxiliary variables in $\mathbf{x}$ that match exactly the known population totals for these variables. If the population totals of the auxiliary variables have been published before the survey results are produced, then using calibration estimators for the survey would guarantee that the survey estimates are coherent with those already in the public domain. This property, although not essential from an estimation point of view, is one of the dominant reasons why calibration is so often used in household surveys. It appeals to survey practitioners in many instances as a way of enforcing agreement between their survey and publicly available totals for key demographic variables. Särndal calls this the cosmetic property of calibration estimators.

The second desirable property is *simplicity*, namely the fact that given the weights w_{hij} calibration estimates are linear in y. This means that each survey record can carry a single weight to estimate all survey variables. Calculation of the estimates for totals, means, ratios, and many other parameters is straightforward using standard statistical software, after the calibration weights have been obtained and stored with each household survey record. In the case of some commonly used distance functions, the calibrated weights are given in a closed form expression and are easy to compute using a range of available software (e.g., CALMAR, GES, BASCULA, G-CALIB-S, R survey package, etc.).

The third property of such calibration estimators is their *flexibility* to incorporate auxiliary information that can include continuous, discrete, or both types of benchmark variables at the same time. If the auxiliary totals represent counts of the numbers of population units in certain classes of categorical (discrete) variables, then the values of the corresponding **x** variables are simply indicators of the units being members of the corresponding classes. Cross-classification of two or more categorical variables can also be easily accommodated by defining indicator variables for the corresponding combinations of categories.

Calibration estimators also yield some degree of *integration* in the sense that some widely used estimators are special cases, for example, ratio, regression, and poststratification estimators (Särndal et al., 1992, Chapter 7) as well as incomplete multiway poststratification (Bethlehem and Keller, 1987).

In addition, if the calibration is performed at the level of the household, all members of the same household will have a common calibration weight w_{hij}, which is a "natural" property since this is the case for the original design weights d_{hij}. If there are auxiliary variables referring to persons, such as age and sex, the calibration at the household level is still possible, provided the auxiliary variables **x** include the counts of household members in the specified age-sex groups for which population auxiliary information is available. This is the approach called "integrated household (family) weighting" by Lemaitre and Dufour (1987).

These are powerful arguments for using calibration estimators. However, when doing so, users must be aware of some difficulties that may be encountered as well. Some of the issues that should be of concern when performing calibration estimation in practice include as follows:

- Samples are often small in certain weighting classes;
- Large numbers of "model groups" and/or survey variables;
- Negative, small (less than 1) or extreme (large) weights;
- Large number of auxiliary variables;
- Nonresponse;
- Measurement error.

The last issue in this list (measurement errors and their effect on calibration) is discussed in Skinner (1999). All the other issues are considered in Silva (2003).

Calibration estimators may offer some protection against nonresponse bias. Poststratification and regression estimation, both special cases of calibration estimators, are widely used techniques to attempt to reduce nonresponse bias in sample surveys. Särndal and Lundström (2005) even suggest "calibration as a standard method for treatment of nonresponse." Calibration estimators are approximately design unbiased if there is complete response for any fixed choice of auxiliary variables. Under nonresponse bias, however, calibration estimators may be biased even in large samples. Skinner (1999) examined the impact of nonresponse on calibration estimators. His conclusions are as follows:

- "the presence of nonresponse may be expected to lead to negative weights much more frequently";
- "the calibration weights will not converge to the original design weights as the sample size increases";
- "the variance of the calibration estimator will be dependent on the distance functions $G(w, d)$ and revised methods of variance estimation need to be considered."

The intended bias reduction by calibration will only be achieved, however, if the combined nonresponse and sampling mechanisms are ignorable given the $\mathbf{x}$ variables considered for calibration. This suggests that the choice of $\mathbf{x}$ variables has to take account of the likely effects of nonresponse, and in particular, should aim to incorporate all $\mathbf{x}$ variables for which auxiliary population data is available that carry information about the unknown probabilities of responding to the survey.

The bias of the calibration estimator will be approximately zero if $y_{hij} = \boldsymbol{\beta}'\mathbf{x}_{hij}$ for every unit in the population, with a nonrandom vector $\boldsymbol{\beta}$ not dependent on the units. (e.g., see Bethlehem, 1988; Särndal and Lundström, 2005, and also Chapter 15 of Särndal et al., 1992). In household surveys, this is an unlikely scenario, and even under models of the form $y_{hij} = \boldsymbol{\beta}'\mathbf{x}_{hij} + \varepsilon_{hij}$, the residuals may not be sufficiently small to guarantee absence of bias due to nonresponse. Särndal and Lundström (2005, Section 9.5) examine additional conditions under which calibration estimators are nearly unbiased and show that if the reciprocals of the response probabilities are linearly related to the auxiliary variables used for calibration, then the calibration estimators will have zero "near bias."

Hence, the key to successfully reduce nonresponse bias in estimating for household surveys is to apply calibration estimation using auxiliary variables that are good linear predictors of the reciprocals of the response probabilities.

Gambino (1999) warns that "nonresponse adjustment can, in fact, increase bias rather than decreasing it," and consequently, that "the choice of variables to use for nonresponse adjustment should be studied even more carefully in the calibration approach than in the traditional approach" for nonresponse compensation.

5.3. Composite estimation for repeated household surveys

For repeated surveys with partial overlap of sample over time, we can use information for the common (matching) sample between periods to improve estimates for the current period t, as we saw in Section 3.6. The common units can be used to obtain a good estimate of change $\hat{\Delta}_{t-1,t}$ between periods $t-1$ and t, which can then be added to the estimate $\hat{\theta}_{t-1}$ to produce an alternative estimate to $\hat{\theta}_t$. An optimal linear combination of these two estimates of θ_t is referred to as a composite estimate. The U.S. CPS has used such estimates since the 1950s. Initially, the CPS used the K-composite estimator

$$\hat{\theta}_t' = (1-K)\hat{\theta}_t + K(\hat{\theta}_{t-1}' + \hat{\Delta}_{t-1,t})$$

with $K = 1/2$. This was later replaced by the AK-composite estimator

$$\hat{\theta}_t' = (1-K)\hat{\theta}_t + K(\hat{\theta}_{t-1}' + \hat{\Delta}_{t-1,t}) + A(\hat{\theta}_{u,t} - \hat{\theta}_{m,t})$$

with $A = 0.2$ and $K = 0.4$, where m and u denote the matched and unmatched portions of the sample (see Cantwell and Ernst, 1992). Note that the term on the far right involves the difference between estimates for the current time point based on the current unmatched and matched samples, respectively. One drawback to using the K- and AK-composite estimators is that the optimal values of A and K depend on the variable of interest. Using different values for different variables will lead to inconsistencies in the sense that parts will not add up to totals (e.g., labor force $\neq$ employed + unemployed). One solution to this problem, called composite weighting, was introduced into the CPS in 1998. Coefficients of $A = 0.4$ and $K = 0.7$ are used for employed and $A = 0.3$ and $K = 0.4$ are used for unemployed, with Not-in-Labour Force being used as a residual category to ensure additivity. Then, a final stage of raking is used to rake to control totals based on composited estimates (see Lent et al., 1999).

The Canadian LFS introduced a regression (GREG) approach, called regression composite estimation, that does not have the consistency problem and has other benefits as well (see Fuller and Rao, 2001; Gambino et al., 2001; Singh et al., 2001).

To implement regression composite estimation, the X matrix used in regression is augmented by columns associated with *last* month's composite estimates for key variables, that is, some of last month's composite estimates are used as control totals. Thus, the elements of the added columns are defined in such a way that, when the final weights of this month are applied to each new column, the total is a composite estimate from the previous month. Therefore, the final calibration weights will respect both these new control totals and the ones corresponding to the original columns of X (typically, age-sex and geographical area population totals).

There are several ways to define the new columns, depending on one's objectives. In the Canadian LFS, a typical new column corresponds to employment in some industry, such as agriculture. If one is primarily interested in estimates of level, the following way of forming columns produces good results. For person i and times $t-1$ and t, let $y_{i,t-1}$ and $y_{i,t}$ be indicator variables that equal 1 whenever the person was employed in agriculture, and 0 otherwise. Then let

$$x_i^{(L)} = \begin{cases} \bar{y}_{t-1}' & \text{if } i \in u \\ y_{i,t-1} & \text{if } i \in m, \end{cases}$$

where $\bar{y}_{t-1}'$ is last month's composite estimate of the proportion of people employed in agriculture. The corresponding control total is last month's estimate of the number of people employed in agriculture, that is, $\hat{Y}_{t-1}'$. Thus, applying the final (regression) weights to the elements of the new column and summing will produce last month's estimate. The superscript L is used as a reminder that the goal here is to improve estimates of level.

If estimates of change are of primary interest, the following produces good results:

$$
x_i^{(C)} = \begin{cases} y_{i,t} & \text{if } i \in u \\ y_{i,t} + R(y_{i,t-1} - y_{i,t}) & \text{if } i \in m, \end{cases}
$$

where $R = \sum w_i / \sum_m w_i$ and $1/R$ is (approximately) the fraction of the sample that is common between successive occasions.

Using the L controls produces better estimates of level for the variables added to the X matrix as controls. Similarly, adding C controls produces better estimates of change for the variables that are added. Singh et al. (2001) present efficiency gains for C-based estimates of level and change and refer to earlier results on L-based estimates.

Although we can add both L and C controls to the regression, this would result in a large number of columns in the X matrix, which can have undesirable consequences. Fuller and Rao (2001) proposed an alternative that allows the inclusion of the industries of greatest interest while allowing a compromise between improving estimates of level and improving estimates of change. They proposed taking a linear combination of the L column and the C column for an industry and using it as the new column in the X matrix, that is, use

$$
x_i = (1 - \alpha)x_i^{(L)} + \alpha x_i^{(C)}.
$$

This is the method currently used by the Canadian LFS. For a discussion on the choice of α, see Gambino et al. (2001). They also discuss some of the subtleties involved in implementing the above approach that we have not considered here. Note the importance of having good tracking or matching information for the survey units and also the need to apply composite estimation in line with the periods defining rotation of the sample (i.e., monthly rotation leads to monthly composite estimators, etc.).

5.4. Variance estimation

Variance estimation for multistage household surveys is often done using approximate methods. This happens because sampling fractions are very small, exact design unbiased variance estimators are complex or unavailable (e.g., when systematic sampling is used at some stage), or estimators are not linear. There are two main alternative approaches, which are as follows:

- Approximate the variance and then estimate the approximation;
- Use some kind of resampling or replication methodology.

Wolter (2007) provides a detailed discussion on variance estimation in complex surveys, and many of the examples discussed in the book come from multistage household surveys. Skinner et al. (1989) also discuss in detail variance estimation under complex sampling designs, not only for standard estimators of totals, means, ratios, and proportions but also for parameters in models commonly fitted to survey data.

In the first approach, which we refer to as the "approximation approach," we approximate the variance of the estimator under the complex design assuming that the selection of PSUs (within strata) had taken place with replacement, even though this was not actually the case. If the estimator of the target parameter is linear, this is the only approximation required to obtain a simpler variance expression and then use the sample to estimate this approximate variance. This is the so-called *ultimate cluster* approach introduced by Hansen et al. (1953). If, in addition, the estimators are nonlinear, but may be written as smooth functions of linear estimators (such as estimators of totals), Taylor series methods are used to approximate their variance using functions of variances of these linear estimators obtained under the assumption of with-replacement sampling of PSUs. Obtaining design-unbiased variance estimators for these variance approximations simplifies considerably, and for a large set of designs and estimators, the corresponding variance estimators are available in explicit form and have been incorporated in statistical software. Such software includes special modules in general statistical packages like SAS, SPSS, STATA, and R (see the "survey" package—Lumley, 2004). It also includes specialized packages such as SUDAAN, PC-CARP, and EPI-INFO.

In contrast, resampling methods start from a completely different perspective. They rely on repeatedly sampling from the observed sample to generate "pseudoestimates" of the target parameter, which are subsequently used to estimate the variance of the original estimator. Let $\hat{\theta}$ denote the estimator of a vector target parameter θ, obtained using the "original" survey weights w_i. Then a resampling estimator of the variance of $\hat{\theta}$ is of the form

$$\hat{V}_R(\hat{\theta}) = \sum_{r=1}^{R} K_r(\hat{\theta}^{(r)} - \hat{\theta})(\hat{\theta}^{(r)} - \hat{\theta})' \qquad (17)$$

for some specified coefficients K_r, where r denotes a particular replicate sample selected from s, R denotes the total number of replicates used, and $\hat{\theta}^{(r)}$ denotes the pseudoestimate of θ based on the rth replicate sample. These replicate samples may be identified in the main sample data set by adding a single column containing revised weights corresponding to each sample replicate, and for each of these columns, having zero weights for units excluded from each particular replicate sample. The constants K_r vary according to the method used to obtain the replicate samples. Three alternative approaches are popular: *jackknife*, *bootstrap*, and *balanced repeated replication*. For details, see Wolter (2007).

Approximation-based methods are relatively cheap in terms of computation time, provided the survey design and the target parameters are amenable to the approximations, and one has the required software to do the calculations. Their main disadvantages are the need to develop new approximations and variance estimators whenever new estimators are employed and the somewhat complex expressions required for some cases, especially for nonlinear estimators.

Resampling methods are reasonably simple to compute, provided the survey data contains the necessary replication weights. They are, however, more costly in terms of computation time, a disadvantage which is becoming less important with the increase in computer power. In addition, the methods are quite general and may apply to novel situations without much effort from a secondary analyst. The burden here lies mostly on the survey organization to compute and store replicate survey weights with each data record.

Modern computer software is available for survey data analysis that is capable of computing variance estimates without much effort, provided the user has access to the required information on the survey design (Lumley, 2004).

For domain estimates, when the sample within a domain is sufficiently large to warrant direct inference from the observed sample, the general approaches discussed above can be applied directly as well. However, many emerging applications require more sophisticated methods to estimate for small domains (small areas). This topic is covered in a large and growing literature and will not be treated here. We note only that when small area estimation methods are used, the variance estimation becomes more complex. The reader is referred to Rao (2003) for a comprehensive review of this topic.

6. Nonsampling errors in household surveys

This section considers briefly some issues regarding nonsampling errors in household surveys, a topic which requires, and has started to receive, more attention from survey statisticians. We identify some factors that make it difficult to pay greater attention to the measurement and control of nonsampling errors in household surveys, in comparison to the measurement and control of sampling errors, and point to some recent initiatives that might help to improve the situation.

Data quality issues in sample surveys have received increased attention in recent years, with a number of initiatives and publications addressing the topic, including several international conferences (see the list at the end of the chapter). Unfortunately, the discussion is still predominantly restricted to developed countries, with little participation and contribution of experiences coming from developing countries. We reach this conclusion after examining the proceedings and publications issued after these various conferences and initiatives.

After over 50 years of widespread dissemination of (sample) surveys as a key observation instrument in social science, the concept of sampling errors and their control, measurement and interpretation has reached a certain level of maturity. Treatment of nonsampling errors in household surveys is not as well developed, especially in developing and transition countries. Lack of a widely accepted unifying theory (see Lyberg et al., 1997, p. xiii; Platek and Särndal, 2001; and subsequent discussion), lack of standard methods for compiling information about and estimating parameters of the nonsampling error components, and lack of a culture that recognizes these errors as important to measure, assess, and report on imply that nonsampling errors, their measurement and assessment receive less attention in many household surveys carried out in developing or transition countries. This is not to say that these surveys are of low quality but rather to stress that little is known about their quality levels.

This has not happened by chance. The problem of nonsampling errors is a difficult one. Such errors come from many sources in a survey. Efforts to counter one type of error often result in increased errors of another kind. Prevention methods depend not only on technology and methodology but also on culture and environment, making it harder to generalize and propagate successful experiences. Compensation methods are usually complex and expensive to implement properly. Measurement and assessment are hard to perform. For example, how does one measure the degree to which a respondent misunderstands or misinterprets the questions asked in a survey (or, more precisely, the impact of such problems on survey estimates)? In addition, surveys are often carried out with very limited budgets, with publication deadlines that are becoming tighter in order to satisfy the increasing demands of information-hungry societies. In this context, it is correct for priority to be given to prevention rather than measurement and compensation, but this leaves little room for assessing how successful prevention efforts were, thereby reducing the prospects for future improvement.

Even if the situation is not good, some new developments are encouraging. The recent attention given to the subject of data quality by several leading statistical agencies, statistical and survey academic associations, and even multilateral government organizations, is a welcome development. The main initiatives that we shall refer to here are the General Data Dissemination System (GDDS) and the Special Data Dissemination Standard (SDDS) of the International Monetary Fund (IMF, 2001), which are trying to promote standardization of reporting about the quality of statistical data by means of voluntary adherence of countries to either of these two initiatives. According to the IMF, "particular attention is paid to the needs of users, which are addressed through guidelines relating to the quality and integrity of the data and access by the public to the data." These initiatives provide countries with a framework for data quality (see http://dsbb.imf.org/dqrsindex.htm) that helps to identify key problem areas and targets for quality improvement. Over 60 countries have now subscribed to the SDDS, having satisfied a set of tighter controls and criteria for the assessment of the quality of their statistical output.

A detailed discussion of the data quality standards promoted by the IMF or other organizations is beyond the scope of this chapter, but readers are encouraged to pursue the matter with the references cited here. Statistical agencies or other survey agencies in developing countries can use the available standards as starting points (if nothing similar is available locally) to promote greater quality awareness both among their members and staff, and perhaps also within their user communities.

Initiatives like these are essential to support statistical agencies in developing countries to improve their position: their statistics may be of good quality but they often do not know how good they are. International cooperation from developed towards developing countries and also among the latter is essential for progress towards better measurement and reporting about nonsampling survey errors and other aspects of survey data quality. A good example of such cooperation was the production of the volumes *Household Sample Surveys in Developing and Transition Countries* and *Designing Household Survey Samples: Practical Guidelines* by the United Nations Statistics Division (see United Nations, 2005a,b).

7. Integration of household surveys

The integration of household surveys can mean a variety of things, including

- Content harmonization, that is, the use of common concepts, definitions, and questions across surveys;
- Integration of fieldwork, including the ability to move cases among interviewers using different modes of collection, both for the same survey and possibly across surveys;

– Master sample, that is, the selection of a common sample that is divided among surveys, possibly using more than one phase of sampling;
– The use of common systems (collection, processing, estimation, and so on).

In the past twenty years or so, there have been efforts in several national statistical agencies to create general systems for data collection, sampling, estimation, etc. that would be sufficiently flexible that most surveys conducted by a given agency could use these systems. These efforts have met with mixed success—it appears inevitable that some surveys will make the claim that their requirements are unique.

The United Kingdom's Office for National Statistics (ONS) is currently in the process of moving its major household surveys into an Integrated Household Survey. This endeavor includes the integration of fieldwork: interviewers, and the systems they use, will be able to work on several surveys during the same time period. The integration extends to all steps in the survey process. All respondents will be asked the same core questions and different subsets of respondents will be asked additional questions from modules on a variety of topics. Details are available at the ONS web site.

At Statistics Canada, different approaches to creating a master sample for household surveys have been studied. One option, to have a distinct first-phase sample in which all respondents get a small, core set of questions, was rejected due partly to the substantial additional cost of a separate first phase. In addition, the benefits of using first-phase information to select more efficient second-phase samples only accrue to surveys that target subpopulations (e.g., travelers are important in the Canadian context), whereas most major surveys, such as the LFS and health surveys, are interested in the population as a whole. To make the first phase more useful, the core content would have to be increased to the point where it affects response burden and jeopardizes response rates. The preferred option at Statistics Canada is to make the "front end" of major surveys such as the LFS the same (corresponding to the core content of phase one of the two-phase approach) and then to pool the samples of all these surveys to create a master sample for subsampling.

In parallel with the study of design options, Statistics Canada is working on content harmonization for key variables. The objective is not only to harmonize questions on important variables such as income and education but also to create well-tested software modules that new surveys can use without needing to develop them themselves. The goal is to have different versions of certain modules, and each survey would choose a version depending on its requirements. For example, for income, there would be a short set of questions and a long set. A survey where income is of primary interest would select the long set and, conversely, most other surveys would select the short set.

Several developing countries, such as South Africa and Vietnam, have developed master samples. Pettersson (2005) discusses the issues and challenges faced by developing countries in the creation of a master sample. These include the availability of maps for PSUs, the accuracy of information, such as population counts, about such units and how to deal with regions that are difficult to access. Of course, many of these challenges are also faced by developed countries, but usually not to the same degree. The development of integrated household survey programs in developing countries has been a United Nations priority for some time. A discussion of efforts in this area, as well as further references, can be found in United Nations (2005b).

8. Survey redesign

Major ongoing surveys such as labor force surveys need to be redesigned periodically. Redesigns are necessary for several reasons.

- Changes in geography, such as municipal boundary changes, may result in the need for domain estimation and these changes accumulate over time. A redesign provides an opportunity to align survey strata with the latest geographical boundaries.
- The needs of users of the survey's outputs change over time, in terms of geography, frequency, and level of detail. These changing needs can be taken into account during a redesign of the survey.
- As the population changes, the sample may no longer be "in the right place" because of uneven growth and migration. A redesign is an opportunity to reallocate the sample.
- Related to the previous point, inclusion probabilities (and therefore weights) become increasingly inaccurate. This is not a concern for the bias of survey estimates but it is for their variance (efficiency).
- A redesign provides an opportunity to introduce improvements (new methods, new technology).
- For surveys with a clustered design, if all sampled clusters are carefully relisted as part of the redesign process, this will reduce undercoverage (missed dwellings) and put all clusters on the same footing (until cluster rotations start occurring).

Because of these benefits, surveys invest in periodic redesigns even though they can be very expensive. Typically, redesigns take place shortly after a population census since data from the census and census geography are key inputs for the design of household surveys.

9. Conclusions

In the introduction, we mentioned some major trends in household surveys since the 1940s. We conclude this chapter by taking a nonexhaustive look at current and future challenges. We have already noted that the theory of sample design is well-developed for traditional household surveys. A traditional area where there is scope for further development is the coordination of surveys. The U.S. Census Bureau recently conducted a study comparing four methods based on either systematic sampling or permanent random numbers for their household surveys (see Flanagan and Lewis, 2006). The goal of the study was to find the best method for avoiding selection of a given household in more than one survey over a certain time period. Studies of this type are needed in other contexts as well.

Most new developments are likely to stem from technological changes, particularly the internet. Currently, the internet is a useful medium for data collection, but it is not as useful as a basis for selecting representative samples of people or households. Perhaps this will change in the future: will there come a time when each individual will have a unique and persistent internet address? We have already mentioned the challenges (in developed countries) and opportunities (in developing countries) posed by the increased use of mobile phones. The future of telephone surveys depends on the development of the mobile phone industry and its impact on landline telephone usage.

Like the theory of sample design, estimation theory for sample surveys is mature, especially for relatively simple parameters such as totals, means, ratios, and regression coefficients. However, there is still a great deal to do for analytical problems, especially those associated with longitudinal surveys. Another area of active research is small area estimation, which we mentioned only briefly in this chapter.

Perhaps the biggest estimation-related challenges in household surveys are associated with nonsampling errors: how to measure them and how to fix them or take them into account. Despite their importance, space considerations prevented us from addressing this topic here, and Section 6 of this chapter barely skimmed the surface. We expect that there will continue to be a great deal of research on topics such as nonresponse and imputation, errors and biases due to reporting problems (including work on questionnaire design and cognitive research), and variance estimates that reflect more than simply sampling variability.

A common element underlying the challenges mentioned in the previous two paragraphs is the need for statistical models. Traditionally, national statistical agencies have favored purely design-based methods where possible, minimizing the use of explicit models. To deal with the problems now facing them, survey statisticians in these agencies recognize the need to use models explicitly in many areas, such as imputation and small area estimation.

Finally, we mention the influence of cost considerations on household survey methodology. In most countries, there is constant pressure to reduce survey costs. In countries with a high penetration of landline telephones, this has led to increased use of telephone interviewing, but we have noted that there is a reversal under way and that there is scope to use the internet as a response medium to counteract this reversal. We expect that efforts to improve the survey collection process using paradata and other technology-based tools such as interviewer monitoring will continue (see Groves and Heeringa, 2006). Sharing of experiences in this area among national statistical agencies (e.g., what works, what are the savings) would be beneficial.

Acknowledgments

The authors are grateful to Ibrahim Yansaneh, Charles Lound, and Stephen Miller, whose comments helped to improve the chapter.

Essential Methods for Design Based Sample Surveys
ISSN: 0169-7161
© 2009 Elsevier B.V. All rights reserved
DOI: 10.1016/B978-0-444-53734-8.00006-6

6

Sampling and Estimation in Business Surveys

Michael A. Hidiroglou and Pierre Lavallée

1. Introduction

A *business survey* collects data from businesses or parts thereof. These data are collected by organizations for various purposes. For instance, the System of National Accounts within National Statistical Offices of several countries uses them to compile annual (and sometimes quarterly) data on gross product, investment, capital transactions, government expenditure, and foreign trade. Business surveys produce a number of economic statistics such as: *production* (outputs, inputs, transportation, movement of goods, pollution, etc.); *sales* (wholesale and retail services, etc.); *commodities* (inputs, outputs, types of goods moved, shipments, inventories, and orders); *financial statements* (revenues, expenses, assets, liabilities, etc.); *labor* (employment, payroll, hours, benefits, employee characteristics); and *prices* (current price index, industrial price index).

Business surveys differ in a number of ways from social surveys, throughout the survey design. The frame of businesses is highly heterogeneous in terms of size and industrial classification of its units, whereas the one associated with social surveys is more homogenous. Business surveys usually sample from business registers (or equivalent list frames) that contain contact information, such as name, address, contact points, from administrative files. Social surveys, on the other hand, often use area frames to select households, and eventually individuals from within these households.

The literature on the conduct of business surveys is relatively sparse. Deming's (1960) book is the only sampling book that specifically focuses on business surveys. The two recent International Conferences on Establishment Surveys (1993 and 2000) resulted in two books specially dedicated to establishment surveys: Cox et al. (1995) and ICES-II (2001).

This chapter is structured as follows. In Section 2, we discuss sampling frames for business surveys. In Section 3, we will discuss how administrative data form an important component of business surveys. In Section 4, commonly used procedures for stratifying a business register and allocating samples will be introduced. In Section 5, methods for sample selection and rotation will be discussed, highlighting procedures that minimize response burden. The remaining Sections 6 and 7 will include brief coverage of data editing, outlier detection, imputation, and estimation, as they are covered in more depth in other chapters of this book.

2. Sampling frames for business surveys

2.1. Basic concepts

A *business* is an economic unit (establishment, farm, etc.) engaged in the production of goods or the provision of services that uses resources (labor, capital, raw materials) to produce these goods or services. Businesses operate in economic sectors that include retail trade, wholesale trade, services, manufacturing, energy, construction, transportation, agriculture, and international trade. A *business survey* is one that collects data used for statistical purposes from a sample of businesses or firms.

Businesses are characterized by a set of attributes that include *identification data*, *classification data*, *contact data*, and *activity status*. *Identification data* uniquely identify each unit with name, address, and alphanumeric identifiers. *Classification data* (size, industrial and regional classifications) are required to stratify the population and select a representative sample. *Contact data* are required to locate units in the sample, including the contact person, mailing address, telephone number, and previous survey response history. *Activity status* indicates whether a business is active (live, in-season) or inactive (dead, out-of-season). *Maintenance and linkage data* are needed to monitor and follow businesses through time. They include dates of additions and changes to the businesses and linkages between them. Collectively, the identification, classification, contact, maintenance, and linkage data items are referred to as *frame data*.

A business is also characterized by its legal structure, or its operating structure. Administrative files usually reflect how businesses are structured with respect to their *legal* arrangements, but do not reflect associated *operating structures*. The *legal* structure provides the basis for ownership, entering into contracts, employing labor, and so forth. It is via the legal structure that a business is registered with the government, and subsequently submits tax returns and/or payroll deductions and value-added taxes. The *operating* structure reflects the way the business makes and enacts decisions about its use of resources, production of goods and services, and how its accounting systems keep track of production inventories and personnel (salaries and wages, number and types of employees). These structures are reflected, and maintained, on a business register by representing their linkages with the associated business. The linkages are maintained by regular profiling of the businesses or signals triggered by survey feedback, or from updates from administrative files.

The sampling of businesses takes place by usually transforming operating structures into standardized units known as *statistical units*. The transformation takes into account decision-making autonomy, homogeneity of industrial activity, and the data available from each operating unit. Statistical units are usually represented as a hierarchy, or series of levels that allow subsequent integration of the various data items available at different levels within the organization. The number of levels within the hierarchy differs between statistical agencies. For example, the Canadian Business Register has four such levels: enterprise, company, establishment, and location (see Colledge, 1995 for definitions). In the United Kingdom, the business register has two levels: establishment and local unit (see Smith et al., 2003, for definitions). Statistical units are characterized by size (e.g., number of employees, income), geography, and industry. Statistical units are used for sampling purposes. Such units are called *sampling units*, and the level of the hierarchy that is sampled depends on the data requirements of the specific business survey.

Businesses either have a *simple* or *complex* structure. A simple business engages in a single type of activity at a single location. The vast majority of businesses have a simple structure that consists of a single legal unit that owns and controls a single operating unit. A complex business engages in a range of economic activities taking place in many locations, and can be linked to several legal units that in turn control several operating units.

The *target population* is the set of units about which data are required for a specific business survey. *Target units* within that population can be any of: legal units, operating structures, administrative units linked to businesses, or statistical units. For example, the target population could be the set of all locations that have industrial activity in the industries associated with that survey. The sampling units are at a level equal or higher than the target units.

Data collection arrangements between the statistical agency and a sampled business (defined at the statistical unit level) are established via *collection units*. Three attributes associated with a collection unit are:

- Coverage—defining the relationship between the business from which the data are being acquired and the level within the business (i.e., enterprise, location) for which the data are required;
- Collection mode—the means of obtaining the data (e.g., questionnaire, telephone interview, administrative record, etc.);
- Contact—the respondent name, address, and telephone number within the business operating structure.

Figures 1 and 2 illustrate how statistical units and collection units are related for a simple or complex business, respectively.

Collection units provide one of several means for updating the business frame in terms of frame data: others include administrative data updates and profiling. Collection

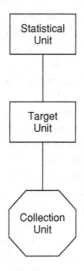

Fig. 1. Simple business.

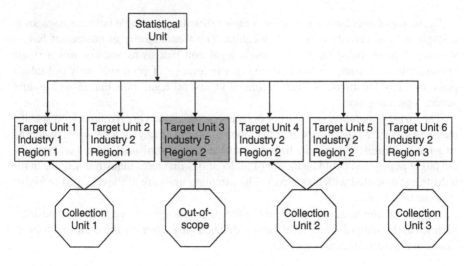

Fig. 2. Complex business.

units also represent the vehicle for monitoring respondent contacts and assessing respondent burden. Collection units are created only for statistical units in a survey sample, and are survey specific. Collection units are automatically generated with rules that depend on statistical-operating links and data availability. They can be modified manually, as need be, to take into account information related to nonstandard reporting arrangements requested by respondents.

2.2. Types of sampling frames

Sampling frames for business populations, such as retail stores, factories, or farms, are constructed so that a sample of units can be selected from them. There are two main types of sampling frames used for business surveys: *list frames* and *area frames*.

A list frame is list of businesses with their associated frame data (such as administrative identification, name, address, and contact information). This list, also known as a *business register*, should represent as closely as possible the real-life universe of businesses. Business surveys are carried out in most countries by sampling businesses from the business register. For National Statistical Agencies, administrative files are by far the preferred way to maintain the business register, as they are relatively inexpensive to acquire from government tax collecting agencies by the surveying agency. Examples of administrative files provided by tax collecting agencies to National Agencies include the Unemployment Insurance system in the Bureau of Labor Statistics in the United States and the Value Added Tax files in Britain. In Canada, a wide range of administrative files maintained by the Canada Revenue Agency is available to Statistics Canada's Business Register. These include files on corporate tax, individual tax, employee payroll deductions, goods and services tax, importers.

An area frame is a collection of geographic areas (or area segments) used to select samples using stratified multistage designs. All businesses within the selected areas are enumerated. The use of area frames for business surveys presents both advantages and disadvantages. An advantage is that it ensures the completeness of the business frame. However, their use presents a number of disadvantages. It is expensive to list and maintain a list of businesses within an area frame, as they have to be personally enumerated. The sampling design is inefficient on account of the clustering of the selected segments, and the high skewness of the data associated with businesses.

A business survey may be based on more than one of these frames, and all possible combinations have been used by National Statistical Agencies. For instance, in New Zealand, business surveys were at one time based solely on an area frame. Business surveys have always been based on a list frame in the United Kingdom, whereas Canada and the United States used at one time a combination of list and area frames. Area frames are much more costly to develop than list frames. Consequently, their use as a sampling frame for business surveys is warranted if they represent a significant portion of the estimates, or if they represent the only means to obtain a list of businesses. It is for that reason that Canada abandoned the area sample component of its retail and wholesale businesses in the late 1980s: the lack of an area frame was compensated by adjusting the weights to account for undercoverage. The United States followed suit (see Konschnik et al., 1991) in the early 1990s for their monthly retail trade surveys. The joint use of area frames with list frames results in a multiple frame. Kott and Vogel (1995) provide an excellent discussion of problems and solutions encountered in this context. From hereon, the discussion will focus on the building of business sampling frames using administrative files.

2.3. Maintenance

A business universe is very dynamic. There are five main types of changes: (i) *births* due to brand new business formation, mergers or amalgamations; (ii) *deaths* resulting from either splits or physical disappearances of exiting businesses; (iii) *structural* changes in existing businesses; (iv) *classification* changes of existing businesses in terms of industry, size, and/or geography; and (v) *contact information* changes. In the case of mergers or amalgamations, the statistical units are to be linked prior to this change are inactivated and the resulting statistical units are birthed with a new identifier. Also, if a business splits, the parent statistical units are inactivated and the resulting descendents are birthed. Such changes are tracked by a combination of (a) continuously matching the administrative files to the business register; (b) profiling of existing businesses on the business register; and (c) using feedback from surveys that use the business register as their sampling frame.

The ideal system would keep all such changes up-to-date on the business register. The reality is that this is not always possible, and errors in coverage (missing, extraneous, duplicate units), classification (size, industry, geography), and contact information (name, address, telephone number) do occur. Reasons for coverage errors include improper matching of the business register to administrative files, delays in updating new births and structural changes to existing units, and delayed removal of deaths from the register. Haslinger (2004) describes the problems associated with matching administrative files to a business register in further detail. Hedlin et al. (2006) propose a methodology for predicting undercoverage due to delays in reporting new units.

Survey feedback updates the classification status, structures, and contact information of existing businesses. Although survey feedback from surveys is beneficial for updating a register, it can result in biased estimates if changes in classification stratification and/or activity status (live to dead) are used for the same sampled units in future occasions of the same survey. The problem can be avoided using updating the frame with a source independent of the sampling process. If the frame is simultaneously updated through several sources, indicators on the frame that reflect the source of the update will allow the application of independent updates in an unbiased manner. If an independent update source is not available, then domain estimation is required. This is achieved by maintaining a copy of the original status of the stratification informa- tion of the in-scope sampling units that allows computation of the survey weights as units were first selected. Domain estimation can then take place reflecting any changes in stratification and activity status of the sampled units. Although domain estimation results in unbiased estimates, their variability will eventually become too large.

We illustrate how an independent source can be used to handle dead units. Dead units in a sample are representative of the total number of dead units on the business frame. Such units are initially retained on the frame and treated as zeroes in the sample. Given that the independent administrative source identifies dead units, how do we use it? We restrict ourselves to two occasions, and for a single stratum, to illustrate how this can be handled during estimation.

On survey occasion 1, a sample s_1 of n_1 units is selected using simple random sam- pling without replacement (srswor) from a population U_1 of size N_1. On the second occasion, the universe U_2 consists of all the original units in U_1 as well as a set U_b of N_b universe births that have occurred between the two occasions. Suppose that a subset of U_1 has died, between the creation of U_1 and data collection for s_1. This sub- set denoted as U_d, consists of N_d unknown dead units. Suppose that the independent administrative source identifies A_d of the N_d unknown dead units, where $A_d < N_d$. Dur- ing data collection of s_1, n_d deaths are also observed in sample s_1, and $a_d (a_d < n_d)$ of these deaths are also identified by the administrative source. The sample s_1 is enlarged with a representative sample s_b of size $n_b = f_1 N_b$, where $f_1 = n_1/N_1$, selected using srswor from U_b. If all the deaths are retained in the sample, then the resulting sample consists of $n_2 = n_1 + n_b$ units of which n_d are known to be dead.

Suppose that the parameter of interest is the population total $Y_2 = \sum_{k \in U_2} y_k$. An unbiased estimator of Y_2 is given by $\hat{Y}_{2,\text{HT}} = \frac{N_1}{n_1} \sum_{k \in s_2} y_k$ where $s_2 = s_1 \bigcup s_b$. Note that at least n_d units are dead in that sample (because some more deaths have occurred during the collection of data after the second selection). A more efficient estimator of Y_2 is given by the poststratified estimator $\hat{Y}_{2,\text{PS}} = \frac{N_2}{\hat{N}_2} \hat{Y}_{2,\text{HT}}$ where $\hat{N}_2 = \frac{N_1}{n_1} (n_1 - a_d + n_b)$ and $N_2 = N_1 - A_d + N_b$. This estimator is of a ratio form and is therefore approximately unbiased.

3. Administrative data

Administrative data have been increasingly used by many National Statistical Agencies for a number of years. Data are becoming more readily available in computer readable format and because that their potential to replace direct survey data reduces overall sur- vey costs. Brackstone (1987) classified administrative data records into six types, based on their administrative purpose: the regulation of the flow of goods and people across national borders; legal requirements to register particular events; the administration of benefits or obligations; the administration of public institutions; the administration of government regulation of industry; and the provision of utilities (electricity, phone, and water services).

3.1. Uses

Brackstone (1987) divided the use of administrative data into four categories: direct tabulation, indirect estimation, survey frames, and survey evaluation. These categories have been refined into seven types of use (see Lavallée, 2007a).

3.1.1. Survey frames
Administrative files have long been used by National Statistical Agencies to build and maintain their business register. The objective is to use the business register to select samples for all business surveys.

3.1.2. Sample design
Administrative data can be used as auxiliary variables for improving the efficiency of sample designs in terms of sample allocation, for example.

3.1.3. Partial data substitution in place of direct collection
Some of the variables on an administrative file can be used instead of corresponding variables collected by a direct survey. The practice of partial data substitution has been adapted by Statistics Canada for both annual and subannual surveys: annual tax data on incorporated businesses are used to replace direct collection of financial variables for annual surveys; and Goods and Services Tax (GST) data are used for monthly surveys. Erikson and Nordberg (2001) point to similar practices in Sweden's Structural Business Survey: administrative data replace direct data collection for small enterprises that have less than 50 employees.

3.1.4. Edit and imputation
Administrative data can be used to assess the validity of collected variables. For example, we should expect expenses on wages and salaries, collected via a direct survey, to be smaller than the total expenses of the business. As total expenses are also available for the corresponding unit on the administrative file, one is in a better position to decide whether a collected value is valid. In the event that collected data have been declared as incorrect, administrative data may be used to replace them, provided that the concepts and definitions are comparable between the survey and administrative data.

3.1.5. Direct estimation
As administrative data are often available on a census basis, estimates such as totals and means are obtained by summing the corresponding administrative data. Although the resulting estimates are free of sampling errors, they will be subject to all of the nonsampling errors associated with administrative data.

3.1.6. Indirect estimation
Administrative data can be used as auxiliary data to improve the precision of collected data. Calibration procedures such as those given in Deville and Särndal (1992) are used for that purpose.

3.1.7. Survey evaluation

Once the survey process has been completed, administrative data can be used to evaluate the quality of the resulting process. Validation compares the survey-based estimates to corresponding administrative values to ensure that the results make sense. Such survey evaluations can be done at the microlevel (i.e., the record level), and at the macro level (i.e., the estimate level) as well. For example, in the Current Employment Statistics survey conducted by the Bureau of Labor Statistics, administrative data available on a lagged basis are used for that purpose.

3.2. Advantages and disadvantages

The use of administrative data offers both advantages and disadvantages which we discuss briefly. We begin with the advantages. First, administrative data are often the only data source for essential statistics (e.g., births, customs transactions). Second, because most of the administrative data are available in computer form, considerable savings in terms of capture costs are realized. This does not, however, reduce processing costs to edit, impute, and transform them into a usable format for a specific application. Third, they can also contribute to the reduction of response burden. Fourth, as administrative data are often available on a census basis, there are no sampling errors associated with statistics obtained from them. Another consequence of their availability on a census basis is that it is possible to produce statistics for any domain of interest, including those with a very small number of units. The production of domain estimates is, however, constrained by the availability of that describes the domains on the administrative files.

We next note some of the disadvantages of using administrative data: some of them are similar to those associated with direct surveying. First, there is limited control on data timeliness, content, and quality as the administrative data originator's main objective may not be to use these data for statistical purposes. This will have a negative effect on national agencies' statistical programs. For example, a problem may occur if the frequency for compiling administrative data is changed in mid-stream (e.g., changing from monthly to quarterly). Furthermore, even though there are automated procedures for assigning industrial classification codes to administrative records, the resulting codes may be erroneous because of the limited available information describing industrial activity. Administrative data may as well not be checked as thoroughly as possible at source, and this means that the user needs to build edit checks that verify data consistency. Data in error (missing or failing edit checks) are either corrected using logical checks or imputed. Second, because administrative data have a limited number of variables, they need to be supplemented with data collected by direct surveys. Third, there are coverage problems if the population represented by the administrative data differs from the target population of the survey that uses them.

3.3. Calendarization

Administrative data can cover time periods that differ markedly from the reporting periods required by surveys. For example, the ideal reporting period for an annual survey would be a calendar year, while the one for a monthly survey would be a calendar month. These time periods are also known as *reference periods*. When the reference periods of the administrative data and the survey requirements differ, the administrative data are transformed using a method known as *calendarization*. In Canada, reference periods may differ between different records of an administrative file, and may even change within the same record.

We will assume, without loss of generality, that the calendarization of the administrative data is required at a monthly calendar level. The following is a summary of Quenneville et al. (2003). Formally, the objective of calendarization is to generate monthly estimates for a variable of interest y over a selected range of T months, called the *estimation range*, from the set of N transactions of a given unit. The available reporting periods may either partially or fully cover the months in the estimation range. If the set of transactions does not cover all the estimation range, there are *gaps* between some of the transactions and after the last transaction. The generated monthly estimates $\hat{\theta}_t$ for each month t are called interpolations when they are within the span of the transactions. These interpolations provide monthly estimates for all the months associated with the transactions, as well as the gaps. The generated monthly estimates $\hat{\theta}_t$ are called extrapolations when they are outside the span of the transactions. These extrapolations provide monthly estimates for transactions not yet received. Figure 3 illustrates some of the ideas given in this paragraph.

Calendarization benchmarks a monthly indicator series x to the administrative data y. The monthly indicator series is a series obtained from another data source that reflects the seasonal pattern of the series to be calendarized. The indicator series x is in fact used for taking seasonality into account. The benchmarking procedure is based on a regression model with autocorrelated errors. It is a generalization of the method of Denton (1971), which now explicitly recognizes the timing and the duration of the data.

The benchmarking model for calendarization is represented by two linear equations. The first one, given by $y_k = \sum_{t=1}^{T} \gamma_{k,t} \theta_t + \varepsilon_k$, specifies the relationship between the reported value y_k of each of the N transactions and the unknown, but true, interpolations θ_t. This is the key to calendarization. It states that a transaction y_k corresponds to the temporal weighted sums of the true interpolations θ_t over its reporting period. The quantity $\gamma_{k,t}$, called the *coverage fraction*, is the fraction of month t covered by y_k. For example, if y_k covers from July 1 to August 31, the coverage fractions are equal to 31/31 for July, 31/31 for August, and 0 for all the other months. As another example,

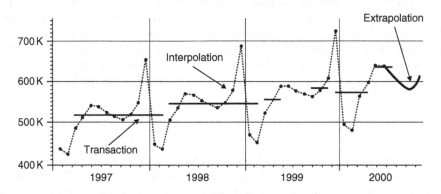

Fig. 3. Calendarization for a monthly series.

if y_k covers from June 16 to August 17, the coverage fractions are equal to 15/30 for June, 31/31 for July, 17/31 for August, and 0 for all the other months. It is assumed that $E(\varepsilon_t) = 0$, $V(\varepsilon_k) = \sigma_k^2$, $\text{Cov}(\varepsilon_k, \varepsilon_{k'}) = 0$ for $k \neq k'$.

The second linear equation, $x_t = \theta_t + c_t e_t$, states that the monthly indicator series x_t is the sum of the true interpolation θ_t and a measurement error $c_t e_t$, $t = 1, \ldots T$. It is assumed that the indicator series x_t is available for all the months $t = 1, \ldots T$ in the estimation range. It is further assumed that $E(e_t) = 0$, and $E(e_t e_{t'}) = \rho(|t - t'|)$ where $\rho(l)$ is the autocorrelation at lag $l = 0, 1, \ldots, T-1$ of a stationary and invertible Auto-Regressive Moving Average (ARMA) process (Box and Jenkins, 1976). We also have $E(\varepsilon_k e_t) = 0$. The quantities c_t are known constants proportional to a power of $|x_t|$. Note that the indicator series needs to be rescaled to the level of the data by multiplying it by the factor $\left(\sum_{k=1}^{N} y_k\right)\Big/\left(\sum_{k=1}^{N} \sum_{t=1}^{T} \gamma_{k,t} x_t\right)$.

These equations can be written in matrix notation as: $\mathbf{y} = \gamma\theta + \varepsilon$, where $E(\varepsilon) = 0$ and $\text{Cov}(\varepsilon) = \mathbf{V}_\varepsilon$; and $\mathbf{x} = \theta + \mathbf{C}e$, $E(e) = \mathbf{0}$, $\text{Cov}(e) = \rho_e$, where $\mathbf{y}$ is the column vector containing the reported values y_k, and so on for $\mathbf{x}$, ε, and e. We define $\gamma = [\gamma_{k,t}]_{N \times T}$, $\mathbf{V}_\varepsilon = \text{diag}(\sigma_k^2)$, $\mathbf{C} = \text{diag}(c_t)$, and $\rho_e = [\rho(|t - t'|)]_{T \times T}$.

Using a Generalized Least Squares procedure such as the one given in Dagum et al. (1998), the estimated monthly interpolations $\hat{\theta}_t$ are obtained from $\hat{\theta} = \mathbf{x} + \mathbf{C}\rho_e \mathbf{C}\gamma'(\gamma\mathbf{C}\rho_e\mathbf{C}\gamma' + \mathbf{V}_\varepsilon)^{-1}(\mathbf{y} - \gamma\mathbf{x})$. The estimated interpolations can be shown to exactly satisfy the benchmarking constraint by setting $\mathbf{V}_\varepsilon = 0$ and premultiplying both sides of the previous equation in $\hat{\theta}$ by the matrix γ. This leads to $\gamma\hat{\theta} = \mathbf{y}$, which shows that the estimated interpolations exactly satisfy the benchmarking constraint $\mathbf{y} = \gamma\theta$, because we set $\mathbf{V}_\varepsilon = 0$.

In Canada, calendarization of the Goods and Service Tax data provided by the Canada Revenue Agency has contributed significantly to reducing survey costs and response burden of conducting monthly business surveys in a number of industrial sectors that include wholesale, retail, manufacturing, and services.

4. Sample size determination and allocation

4.1. Choice of sampling unit

The sampling of a business universe is usually done in two steps. First, the in-scope target universe is defined, and a set of target units are obtained. Second, the sampling unit is defined at some level of the statistical units. The sampling level will be at least at the level of the target units. For example, suppose that locations and establishments are the only two types of statistical units on the business register. Given that the target unit is the location, the sampling unit could either be the location or the establishment.

4.2. Stratification of sampling units

Once the population of businesses has been partitioned into sampling units, the sampling units are stratified. The selection of samples is done independently in each of these strata. The strata are usually based on geography (e.g., Canadian provinces and major metropolitan centers), standard industrial classification (e.g., restaurants, agents and brokers, garages, department stores), and some measure of size (e.g., number of employees, gross business income, net sales). Cochran (1977) gives four main reasons for stratification. First, it reduces the variances of survey estimates, if they are correlated with the stratification variables. Second, stratification may be dictated by administrative convenience if, for example, the statistical agency has field offices. Third, sampling problems may differ markedly in different parts of the population such that each part should be considered independently. Finally, reliable estimates for designated subpopulations that have high overlap with the design strata can be obtained as a byproduct.

The selection of samples in business surveys frequently uses simple random sampling techniques applied to each stratum. A feature of most business populations is the skewed nature of the distribution of characteristics such as sales, employment output, or revenue. A "certainty" or "take-all" stratum of the very largest sampling units is usually created to reduce the variances of estimates: all sampling units within the certainty strata are selected in the sample. Noncertainty strata are then formed and the remaining sampling units are placed in them according to their size.

The optimality of stratification breaks down over time, resulting in a less efficient sample design. Deterioration of stratification of the frame requires that the whole frame be restratified. A new sample that is optimal with respect to the newer stratification is then selected, in general with as much overlap as possible with the previous sample. This overlap ensures continuity of the estimates in a periodic survey, and is less expensive than a complete redraw from a collection perspective.

Factors that affect realized precision include: population size; overall sample size; stratification of the frame in terms of the number of strata and the stratum allocation scheme; the construction of stratum boundaries for continuous stratification variables; the variability of characteristics in the population; the expected nonresponse; cost, time, and operational constraints; and the targeted precision of summary statistics such as means and totals of the target variables.

4.3. Allocation

4.3.1. Notation

We introduce notation to deal with allocation for a single x-variable (univariate allocation). The finite population U of N units is divided into L nonoverlapping subpopulations or strata U_h, $h = 1, 2, \ldots, L$, with N_h units each, and $N = \sum_{h=1}^{L} N_h$. A sample s_h of size n_h, $h = 1, \ldots, L$, is selected independently by simple random sampling without replacement (*srswor*) within each h-th stratum, yielding an overall sample of size $n = \sum_{h=1}^{L} n_h$. Let x_{hk} (known from a previous census or survey) denote the k-th observation within stratum h. An unbiased estimator of the population total $X = \sum_{h=1}^{L} X_h = \sum_{h=1}^{L} \sum_{k \in U_h} x_{hk}$ is given by $\hat{X} = \sum_{h=1}^{L} N_h \bar{x}_h$, where $\bar{x}_h = \sum_{k \in s_h} x_{hk}/n_h$, and its associated population variance is

$$V(\hat{X}) = \sum_{h=1}^{L} N_h^2 \left(\frac{1}{n_h} - \frac{1}{N_h} \right) S_h^2$$

$$= \left(\sum_{h=1}^{L} A_h/n_h \right) - D \tag{1}$$

where $A_h = N_h^2 S_h^2$, $D = \sum_{h=1}^{L} N_h S_h^2$, $\overline{X}_h = X_h/N_h$, and $S_h^2 = \sum_{k \in U_h} \left(x_{hk} - \overline{X}_h\right)^2 / (N_h - 1)$.

The x-values for the current population will not be known. However, an estimate $V'(\hat{X})$ of the population variance $V(\hat{X})$, based on a sample s_h' of size n_h' from a pilot survey, a past survey, or from administrative data, can be used as a substitute. Here, $V'(\hat{X}) = \left(\sum_{h=1}^{L} A_h'/n_h'\right) - D'$ with $A_h' = N_h^2 \hat{S}_h'^2$, $D' = \sum_{h=1}^{L} N_h \hat{S}_h'^2$, $\hat{S}_h'^2 = \sum_{k \in s_h'} (x_{hk} - \overline{x}_h')^2/(n_h' - 1)$, and $\overline{x}_h' = \sum_{k \in s_h'} x_{hk}/n_h'$.

4.3.2. Some allocation schemes

Let a_h denote the proportion allocated to the h-th stratum, where $0 \le a_h \le 1$ and $\sum_{h=1}^{L} a_h = 1$. The number of units of units allocated to a given stratum h is given by $n_h = na_h$ for $h = 1, 2, \ldots, L$.

Assume that the cost of collecting data is the same for all units. The allocation of the sample s of size n to strata s_h can be carried out in two ways:

i Require that the variance of $\hat{X}$ should be minimal given that the overall sample size n is fixed or the overall cost is fixed;

ii Specify a tolerance on the precision of the estimate $\hat{X}$ as a predetermined coefficient of variation c, that is $c^2 = V(\hat{X})/X^2$. In that case, the objective is to minimize the sample size n (or total cost), and is computed using the chosen allocation scheme. Substituting $n_h = n a_h$ and $V(\hat{X}) = c^2 X^2$ into (Eq. 1) and solving for n, we obtain:

$$n = \left(c^2 X^2 + D\right)^{-1} \left(\sum_{h=1}^{L} A_h/a_h\right) \tag{2}$$

A number of allocation schemes for stratified *srswor* are summarized using the above notation.

4.3.2.1. N-proportional allocation ($a_h = N_h/N$).

This scheme is generally superior to simple random sampling of the whole population if the strata averages $\overline{X}_h$ differ considerably from each other. A slight reduction in variance results only if the strata means are similar. It is often used in business surveys to equalize the sampling weights between strata whose units are known to have a high probability to change classification.

4.3.2.2. X-Proportional allocation ($a_h = X_h/X$).

X-proportional allocation is used in business surveys because distribution of data is quite skewed. The largest units are sampled with near certainty and the remaining units are sampled with probability less than one.

4.3.2.3. Optimal allocation $\left(a_h = \left(\sum_{h=1}^{L} (N_h S_h)\right)^{-1} (N_h S_h)\right)$.

More sample units are allocated to the larger strata and/or strata that have the highest variances. This type of allocation is also known as Neyman allocation (see Neyman, 1934). Optimal allocation is similar to X-proportional allocation if $S_h/\overline{X}_h$ is assumed constant across strata.

The difficulty with this allocation is that the population variance S_h^2, or its estimate $\hat{S}_h^{\prime 2}$, may be unstable.

4.3.2.4. $\sqrt{N}$ or $\sqrt{X}$-proportional allocation $\left(a_h = \left(\sum_{h=1}^{L} \sqrt{N_h} \right)^{-1} \sqrt{N_h} \right.$ or $\left. \left(\sum_{h=1}^{L} \sqrt{X_h} \right)^{-1} \sqrt{X_h} \right)$. This scheme results in good reliability of strata estimates $\hat{X}_h$, but it is not as efficient as Neyman allocation for the overall estimate $\hat{X}$. This type of allocation was first proposed by Carroll (1970), and provides fairly similar coefficients of variation for stratum totals $\hat{X}_h$. Bankier (1988) extended the concept by considering a_h as $\left(\sum_{h=1}^{L} (X_h)^q \right)^{-1} (X_h)^q$ where $0 \leq q \leq 1$. Note that setting q to 0.5 results in the Carroll allocation.

4.4. Some special considerations

Nonresponse, out-of-datedness of the frame, initial over-allocation of units to strata, and minimum sample size within strata are additional factors to account for in the computation of the sample size.

4.4.1. Nonresponse
Nonresponse reduces the effective sample size, and hence the reliability of summary statistics. Assume the sample size is $n = \sum_{h=1}^{L} n_h$ units and the nonresponse rates (known from experience) are expected to be $r_h (h = 1, 2, \ldots, L)$, where $0 \leq r_h < 1$ within each stratum. The resulting effective sample size would be $n_{\text{eff}} = \sum_{h=1}^{L} n_h (1 - r_h) < n$ after data collection. The sample size can be increased to $n_h' = n_h / (1 - r_h)$ within each stratum h to compensate for the nonresponse. This increase assumes that the nonrespondents and respondents have similar characteristics. If they differ, a representative sample of the nonrespondents needs to be selected to represent the nonresponding part of the sample.

4.4.2. Out-of-date frame
The impact of an out-of-date frame should be reflected in the sample size determination and allocation method. The out-of-datedness of a frame occurs because the classification (geography, industry, status: live or dead) of the units is not up to date. In our case, we just focus on estimating the total of the live units for a variable y, given that a number of dead units are present on the frame but identified as active. Consequently, a representative portion of them will be included in the sample. The universe of "active" units is labeled as U. The corresponding universe of live units (but unknown) is denoted as U_ℓ, where $U_\ell \subset U$. Let the parameter of interest be the domain total $Y_\ell = \sum_{k \in U} y_{k\ell}$, where $y_{k\ell}$ is equal to y_k if $k \in U_\ell$ and zero otherwise. We need to determine the sample size n, such that: (i) the allocation to the design strata is $n_h = n \, a_h (0 < a_h < 1)$ and (ii) the targeted coefficient of variation c is satisfied, that is, $V(\hat{Y}_\ell) = c^2 Y_\ell^2$. Simple random samples s_h of size n_h are selected from $U_h (h = 1, \ldots, L)$, without replacement. The corresponding estimator is $\hat{Y}_\ell = \sum_{h=1}^{L} \hat{Y}_{h\ell}$ with $\hat{Y}_{h\ell} = \sum_{k \in s_h} (N_h / n_h) \, y_{k\ell}$. We obtain the required sample size n using $V\left(\hat{Y}_\ell \right) = c^2 Y_\ell^2 = \sum_{h=1}^{L} N_h^2 \left(\frac{1}{n_h} - \frac{1}{N_h} \right) S_{h\ell}^2$

where $S_{h\ell}^2 = (N_h - 1)^{-1} \sum_{k \in U_h} \left(y_{k\ell} - \overline{Y}_{h\ell} \right)^2$ with $\overline{Y}_{h\ell} = \sum_{k \in U_h} y_{k\ell}/N_h$. The required sample size is

$$
n = \frac{\sum_{h=1}^{L} N_h^2 \left(\tilde{S}_{h\ell}^2 + (1 - P_{h\ell}) \tilde{\overline{Y}}_{h\ell}^2 \right) P_{h\ell}/a_h}{c^2 \left(\sum_{h=1}^{L} N_h \overline{Y}_{h\ell} P_{h\ell} \right)^2 + \sum_{h=1}^{L} N_h \left(\tilde{S}_{h\ell}^2 + (1 - P_{h\ell}) \tilde{\overline{Y}}_{h\ell}^2 \right) P_{h\ell}},
$$

where $\tilde{S}_{h\ell}^2 = (N_{h\ell} - 1)^{-1} \sum_{k \in U_{h\ell}} \left(y_k - \tilde{\overline{Y}}_{h\ell} \right)^2$, $\tilde{\overline{Y}}_{h\ell} = \sum_{k \in U_{h\ell}} y_{k\ell}/N_{h\ell}$, $N_{h\ell}$ is the number of units belonging to domain $U_{h\ell} = U_\ell \cap U_h$, and $P_{h\ell} = N_{h\ell}/N_h$ is the expected proportion of units that belong to U_ℓ and initially sampled in stratum h. Note that the case of $P_{h\ell} = 1$ yields the usual sample size formula. The mean and variance components can be estimated from previous surveys. The required sample sizes at the stratum level are then simply $n_h = n\, a_h$ for $h = 1, \ldots, L$. It is not recommended to use an approximation of the type $n_h^* = n_h/P_h$ to compensate for unknown dead units, where n_h is computed ignoring the existence of unknown dead units in the universe.

4.4.3. Over-allocation
Optimum allocation (Neyman), X-proportional or $\sqrt{X}$-allocation may result in sample sizes n_h that are larger than the corresponding population sizes N_h for some strata. The resulting overall sample size will be smaller than the required sample size n. Denote the set of strata where over-allocation has taken place as "OVER." Such strata are sampled with certainty, that is, $n_h = N_h$, with total sample size $n_{\text{OVER}} = \sum_{h \in \text{OVER}} n_h$. The remaining set of strata, denoted as "NORM," is allocated the difference $n - n_{\text{OVER}}$ using the chosen allocation scheme. That is, for $h \in \text{NORM}$, $n_h' = (n - n_{\text{OVER}}) a_h'$ where a_h' is computed according to the given allocation scheme, with $\sum_{h \in \text{NORM}} a_h' = 1$. The process is repeated until there is no over-allocation. A similar procedure is used in the case where the overall sample size is chosen to satisfy reliability criteria. The only difference is that $n_h = a_h' \dfrac{\sum_{h \in \text{NORM}} A_h/a_h'}{c^2 X^2 + D'}$.

4.4.4. Minimal sample size
A minimal sample size within each stratum is a requirement to protect against empty strata occurring on account of nonresponse. It also provides some protection against allocations that are poor for characteristics not considered in the sample design. A minimal sample size of three to five units is quite often used in large-scale surveys: at least two units are required to estimate variances unbiasedly. Denote as $m_h (h = 1, 2, \ldots, L)$ the minimal sample size within the h-th stratum: m_h will most likely be the same for all strata. The minimum sample size may be applied before or after the allocation of given sample size n has been established. If it is applied before, a sample size m' is initially set aside for minimum size requirements across all strata, where $m' = \sum_{h=1}^{L} m_h'$ and $m_h' = \min\{N_h, m_h\}$. The remaining sample size $n - m'$ is allocated to the population strata of size $N_h - m_h'$ using the chosen allocation method. If the minimum size is applied after allocation, the sample size for the h-th stratum is $n_h' = \min\{\max[n_h, m_h], N_h\}$. The sum of the overall sample $\sum_{h=1}^{L} n_h'$ may be greater than n.

4.4.5. Equalization of the coefficient of variation among strata

A property of power allocation is that the coefficient of variation for the estimates of the totals will be fairly similar for each stratum. However, it may be required to have exactly equal levels of reliability for estimates of the totals at the stratum level: that is, $V(\hat{X}_h)/X_h^2 = c_1$ for all $h = 1, \dots, L$, where c_1 is not known and is bounded between 0 and 1. If the overall coefficient of variation has been fixed at c, it follows that $c_1 = c\sqrt{X/\sum_{h=1}^L X_h^2}$, and the sample size within each stratum is $n_h = A_h/(c_1^2 X_h^2 + D_h)$. If the overall sample size has been fixed at n, we solve iteratively $f(c_1) = n - \sum_{h=1}^L \dfrac{A_h}{c_1^2 X_h^2 + D_h}$, an increasing function in c_1, using the Newton–Raphson procedure.

4.4.6. Simultaneous level of reliability for two stratification variables

Assume that the population has been stratified at the *geography* ($h = 1, \dots, L$) and *industry* ($\lambda = 1, \dots, M$) levels. Specified coefficients of variation of totals are required at the subnational and industry levels: let these be $c_h.$ and $c_{.\lambda}$, respectively. The sampling takes place within a further size stratification of these LM possible cross-classifications. The required sample size for each of these levels can be computed if we can obtain the corresponding coefficient of variation given the marginal (i.e., *geography* and *industrial* reliability constraints). This coefficient can be obtained using a raking procedure (see Deming and Stephan, 1940). Let $X_{h\lambda}$ be the population total of a given variable of interest (say x), and let $X_h.$ and $X_{.\lambda}$ be the associated marginal totals. The $h\lambda$-th coefficient of variation at the r-th iteration is given by

$$c_{h\lambda}^{(r)} = c_{h\lambda}^{(r-1)} \frac{(c_h. X_h.)(c_{.\lambda} X_{.\lambda})}{\sqrt{\sum_{h=1}^L c_{h\lambda}^{(r-1)} X_{h\lambda}^2} \sqrt{\sum_{\lambda=1}^M c_{h\lambda}^{(r-1)} X_{h\lambda}^2}}$$

The starting point for this algorithm is $c_{h\lambda}^{(0)} = (\dot{c}_h. + \dot{c}_{.\lambda})/2$, where $\dot{c}_h.$ and $\dot{c}_{.\lambda}$ are marginal coefficients of variation given by $\dot{c}_h. = c_h. X_h./\sqrt{\sum_{\lambda=1}^M X_{h\lambda}^2}$ for $h = 1, 2, \dots, L$ and $\dot{c}_{.\lambda} = c_k X_{.\lambda}/\sqrt{\sum_{h=1}^L X_{h\lambda}^2}$ for $\lambda = 1, 2, \dots, M$. In practice, five iterations are sufficient to stabilize the $c_{h\lambda}^{(r)}$ values. The sample size required to achieve the required marginal coefficients of variation for each $h\lambda$-th cell is then $n_{h\lambda} = \left(\left(c_{h\lambda}^{(R)} \right)^2 X_{h\lambda}^2 + D_{h\lambda} \right)^{-1} A_{h\lambda}$, where $A_{h\lambda} = N_{h\lambda}^2 S_{h\lambda}^2$, $D_{h\lambda} = N_{h\lambda} S_{h\lambda}^2$, and $c_{h\lambda}^{(R)}$ is the coefficient of variation at the final iteration R.

4.5. Construction of self-representing strata

4.5.1. Using known auxiliary data x

Stratification of a population into natural strata based on geography and industrial activity usually increases the efficiency of a sample design. Further stratification by size of business (employment, sales) always increases the efficiency of the sample design in business surveys because business populations are typically composed of a few large units (accounting for a good portion of the total for the variable of interest) and many small units. It is therefore desirable to construct stratum boundaries that split the businesses into a take-all stratum containing the largest units (being sampled with certainty) and a number of take-some strata containing the remaining units (sampled with a given probability). The resulting stratification offers two advantages. First, the overall sample size required to satisfy reliability criteria (denoted as c) is dramatically reduced (or alternatively, the variance of the estimated total is minimized for a fixed overall sample size). Second, because the largest units are sampled with certainty, the chance of observing large values for the units selected in the take-some strata is reduced.

Consider a population U of size N where the units have the size measures, $x_1, x_2, \ldots, x_N$. Define order statistics $x_{(1)}, x_{(2)}, \ldots, x_{(N)}$ where $x_{(k)} \leq x_{(k+1)}$, $k = 1, \ldots, N-1$.

We first provide two approximations due to Glasser (1962) for fixed sample size n and Hidiroglou (1986) for fixed coefficient of variation c for splitting the universe into a take-all and a take-some stratum. Glasser's (1962) rule for determining an optimum cut-off point is to declare all units whose x value exceeds $\overline{X}_N + \sqrt{N S_N^2 / n}$ as belonging to the take-all stratum, where $\overline{X}_N = \sum_{k \in U} x_k / N$ and $S_N^2 = \sum_{k \in U} (x_k - \overline{X}_N)^2 / (N-1)$. Hidiroglou's (1986) algorithm is iterative. The take-all boundary $B_r (r = 1, 2, \ldots)$ at the r-th iteration is given by

$$B_r = \overline{X}_{N-T_{r-1}} + \left\{ \frac{(n - T_{r-1} - 1)}{(N - T_{r-1})^2} c^2 X^2 + S_{N-T_{r-1}}^2 \right\}^{1/2}$$

where T_{r-1} is the number of take-all units at the $(r-1)$-th iteration, and $\overline{X}_{N-T_{r-1}}$ and $S_{N-T_{r-1}}^2$ are the corresponding take-some stratum population mean and variance. The process is started by setting T_0 to zero, and the iterative process continues until $0 < (1 - T_r / T_{r-1}) < 0.10$ has been met. Convergence usually occurs after two to five iterations.

Lavallée and Hidiroglou (1988) provided a procedure for stratifying skewed populations into a take-all stratum and a number of take-some strata, such that the sample size is minimized for a given level of precision. They assumed power allocation of the sample for the take-some strata, as this type of allocation tends to equalize coefficients between the strata. Their algorithm uses Dalenius's (1950) representation of a finite population in terms of a continuous population. That is, given a continuous density function g of the auxiliary variable x in the range $(-\infty, \infty)$, the conditional mean and variance of the h-th stratum U_h can be expressed as $\mu_h = \int_{b_{(h-1)}}^{b_{(h)}} y g(y) / W_h$ and $\sigma_h^2 = \int_{b_{(h-1)}}^{b_{(h)}} y^2 g(y) / W_h - \mu_h^2$ where $W_h = \int_{b_{(h-1)}}^{b_{(h)}} g(y) dy$. The overall sample size is given by

$$n = N W_L + \frac{N \sum_{h=1}^{L-1} W_h^2 \sigma_h^2 / a_h}{N \left(c \sum_{h=1}^{L} W_h \mu_h \right)^2 \mu^2 + \sum_{h=1}^{L-1} W_h \sigma_h^2} \tag{3}$$

where $a_h = \dfrac{(W_h \mu_h)^p}{\sum_{h=1}^{L-1} (W_h \mu_h)^p}$ for $h = 1, \ldots, L - 1$. Hidiroglou and Srinath (1993) proposed a more general form of a_h, given by $a_h = \gamma_h / \sum_{h=1}^{L-1} \gamma_h$ where $\gamma_h =$

$W_h^{2q_1}\mu_h^{2q_2}\sigma_h^{2q_3}$, $q_i \geq 0$ ($i = 1, 2, 3$). A number of different allocations are obtained with various choices of the q_i's. For example, Neyman allocation is obtained by setting $q_1 = q_3 = 0.5$ and $q_2 = 0$.

The optimum boundaries $b_1, b_2, \ldots, b_{L-1}$, where $x_{(1)} \leq b_1 < \ldots < b_{L-1} \leq x_{(N)}$, are obtained by taking the partial derivatives of (3) with respect to each b_h, $h = 1, \ldots L-1$, equating them to zero, and solving the resulting quadratic equations iteratively using a procedure suggested by Sethi (1963). The initial values are set by choosing the boundaries with an equal number of elements in each group. Although the Lavallée–Hidiroglou method is optimal, Slanta and Krenzke (1996) and Rivest (2002) noted that it does not always converge, and that convergence depends on providing the algorithm with reasonable initial boundary values.

Gunning and Horgan (2004) recently used the geometric progression approach to stratify skewed populations. They based their algorithm on the following observation stated in Cochran (1977): when the optimum boundaries of Dalenius (1950) are achieved, the coefficients of variation ($CV_h = S_h/\overline{X}_h$) are often found to be approximately the same in all strata. Assuming that the x variable is approximately uniformly distributed within each stratum, their boundaries are $b'_h = a\tau^h$ for $h = 1, 2, \ldots, L-1$ where $a = x_{(1)}$ and $\tau = \left(x_{(n)}/x_{(1)}\right)^{1/L}$. The advantages of Gunning–Horgan's procedure are that it is simple to implement, and that it does not suffer from convergence problems. However, two weaknesses of the procedure are that it does neither stratify a population according to an arbitrary sample allocation rule (represented by a_h), nor does it require the existence of a take-all stratum. The stratification boundaries obtained by the Gunning–Horgan procedure could be used as starting points for the Lavallée–Hidiroglou algorithm to ensure better convergence.

4.5.2. Using models to link auxiliary data x and survey variable y
A number of authors have developed models between the known auxiliary data x and the survey variable y. They include Singh (1971), Sweet and Sigman (1995), and Rivest (2002). The last three authors incorporated the impact of the model in the Lavallée–Hidiroglou algorithm. As Sweet and Sigman (1995) and Rivest (2002) demonstrated, the incorporation of the model could lead to significant improvements in the efficiency of the design.

5. Sample selection and rotation

As mentioned earlier, strata are often cross-classifications of industry and geography by size. These strata are either completely enumerated (take-all) or sampled (take-some). We denote the required sampling fraction within a take-some stratum as f (the subscript h is dropped in this section to ease the notation). It is equal to unity for the take-all strata.

The sampling mechanism of in-scope units in business surveys needs to account for a number of factors. First, the units should be selected using a well-defined probability mechanism that yields workable selection probabilities for both estimation (π_k's, $k = 1, \ldots, N$) and variance estimation ($\pi_{k,k'}$'s). Note that $\pi_k \approx f$ within the strata. Second, the resulting samples should reflect the changing nature of the universe in terms of births, deaths, splits, mergers, amalgamations, and classification changes. Third, the selection should allow for sample rotation of the units to alleviate response burden across time. Fourth, there should be some control of the overlap of the sampled units between various business surveys occurring concurrently. Fifth, if there are significant changes in the stratification of the universe, it should be possible to redraw a sample that reflects the updated stratification and sampling fractions.

Response burden occurs *within surveys* and *across surveys*. Response burden within surveys is minimized if a selected business remains in sample for as few occasions as possible. Response burden across surveys is minimized if a business is selected in as few surveys as possible at the same time. However, these preferences will not normally agree with what is best for a survey in terms of reliability within and between occasions.

Two types of coordination can be distinguished for selecting several samples from the same frame: they are *negative* and *positive coordination*. *Negative coordination* implies that response burden is reduced, by ensuring that a business is not selected in too many surveys within a short time frame. *Positive coordination* implies that the overlap is maximized as much as possible between samples.

5.1. Selection procedures

Poisson sampling and its variants form the basis for sampling business surveys in most national agencies. This method allows for response burden control within and across surveys. Poisson sampling as defined by Hájek (1964) assigns each unit in the population of size N a probability of inclusion in the sample denoted as $\pi_k = np_k, k = 1, \ldots, N$. Here, n is the required sample size and p_k is usually linked to some measure of size of the unit k. Ohlsson (1995) provides the following procedure for selecting a Poisson sample of expected sample size n. A set of N independent uniform random numbers u_k is generated, where $0 \le u_k \le 1$. If these random numbers are fixed and not regenerated for the same units between two survey occasions, they are called permanent random numbers (PRN). A starting point α is chosen in the interval [0, 1]. A population unit k is included in the sample if $\alpha < u_k \le \alpha + np_k$, provided $\alpha + np_k \le 1$. If $\alpha + np_k > 1$, it is included in the sample if $(\alpha < u_k \le 1) \cup (0 < u_k \le \alpha + np_k - 1)$. The value of α is usually set to zero when a survey sample is first selected. Sampling from a stratified universe occurs by assigning the required p_k's and sample sizes within each stratum. Births to a business universe are easily accommodated with Poisson sampling: a PRN is generated for the birth unit, and it is selected using the previously stated algorithm. Rotation of the sample takes place by incrementing α by a constant κ on each survey occasion. The constant κ reflects the required rotation rate.

A special case of Poisson sampling is *Bernoulli* sampling: the p_k's are equal to $1/N$ within each stratum. It should be noted that, conditioning on the realized sample size, Bernoulli sampling is equivalent to *srswor*. Poisson and Bernoulli sampling are often not used in practice, because the realized sample sizes may vary too much around the expected sample sizes. A number of procedures have been developed over the years to control this weakness. These include collocated sampling (Brewer et al., 1972), sequential Poisson sampling (Ohlsson, 1995), and Pareto sampling (Rosén, 2001). Statistics Canada uses Bernoulli sampling to sample tax records from Canada Revenue Agency's administrative tax files. PRNs are created by transforming the unique identifying numbers on the administrative files to pseudorandom numbers using a hashing algorithm. This approach, introduced by Sunter (1986), maximizes sample overlap between sampling occasions.

Bernoulli variants have been used in a number of agencies. Statistics Sweden samples from their business register using sequential *srswor*. Sequential *srswor*, described in Ohlsson (1990a,b), involves the selection of the first n_h units within the ordered list of the PRNs within each stratum of size N_h. The synchronized sampling methodology used by the Australian Bureau of Statistics, developed by Hinde and Young (1984), is quite similar to the one developed at Statistics Sweden. It differs from the Swedish one with respect to its definition of the start and end points of the sampling intervals. The start and end points are equal to the PRNs associated with the units at the time of selection. In-scope population units are selected if they belong to these sampling intervals. The start point is in sample but the end point is not. A desired sample size n is achieved by including the start point, and the remaining $n - 1$ successive PRNs. The incorporation of births and deaths is done by moving the start or end points to the right to prevent units reentering the sample. The procedure allows for rotation, as well as periodic restratification of the frame. Negative or positive coordination is achieved by allowing different surveys to use well defined intervals on the [0, 1) interval. More details of this methodology are available in McKenzie and Gross (2001). Sampling of the business register at Statistics Canada is a blend of collocated sampling described in Brewer et al. (1972), and the panel sampling procedure given by Hidiroglou et al. (1991). Details of the procedure are given in Srinath and Carpenter (1995).

A variant of Poisson sampling, known as Odds Ratio Sequential Poisson sampling, has been used to sample businesses in the petroleum industry. Saavedra and Weir (2003) provide more details of the methodology, which is really Pareto sampling as described in Rosén (2001). This method provides fixed sample sizes in the Poisson context, and the resulting probabilities of selection closely approximate the desired probabilities to be proportional to size.

5.2. Accounting for response burden

The methodology used by the "Central Bureau voor de Statistiek" (CBS) in the Netherlands incorporates PRNs to control sample rotation across and within their business surveys, while accounting for response burden. De Ree (1999) briefly described this methodology: at the time of initial sample selection, sampling units on the business register are assigned a PRN, and ranked accordingly. A PRN remains associated with a given unit on the register throughout its life. However, the manner in which samples are selected may vary. Businesses can be selected several times successively for a specific survey (a subannual survey), or by several different surveys. Each time a business is surveyed, its associated response-burden coefficient is increased. After each selection, the ordering changes so that businesses with a lower cumulated response burden are placed before businesses with a higher cumulated response burden. Earlier versions of this methodology are presented in more detail in Van Huis et al. (1994). The CBS sampling system has useful features: (i) it integrates sampling amongst several surveys, and takes into account the response burden; (ii) it allows user specified parameters for defining the sampling and rotation rates. However, there is some limitation in the choice of stratification.

Rivière (2001) discusses a somewhat different approach, whereby the sample selection procedure does not change, but the initially assigned random numbers are systematically permuted between units for different coordination purposes: smoothing out the burden, minimizing the overlap between two surveys, or updating panels. Permutations of the random numbers are carried out within intersections of strata that are referred to as microstrata. The microstrata method was developed in 1998 in the framework of Eurostat's SUPCOM project on sample coordination. The methodology was initially implemented in a program known as SALOMON in 1999. Improvements to SALOMON resulted in a program known as MICROSTRAT in 2001.

In the microstrata method, the initial procedure is to assign a random number to every unit on the business register that is in-scope for sampling. As with the CBS methodology, every unit is also assigned a response-burden coefficient that cumulates every time the unit is selected for a given survey. The random numbers never change but they can be permuted between the units. The permutation of the random numbers is controlled by the cumulated burden similarly to the CBS procedure. The most important difference from the CBS methodology is that the permutations are done within the microstrata. A microstratum is the largest partition that can be defined so as to sort the units by increasing response burden without introducing bias. Using this technique, the random numbers remain independent and identically distributed with a uniform distribution.

The main drawback to microstratification is the possible creation of microstrata so small that the sample coordination becomes ineffective. However, this can be avoided using a different sorting procedure. On the other hand, microstratification has several benefits. The method has good mathematical properties and gives a general approach for sample coordination in which births, deaths, and strata changes are automatically handled. There is no particular constraint on stratification and rotation rates of panels. It is unbiased, as shown by Rubin–Bleuer (2002).

6. Data editing and imputation

6.1. Data editing

Business survey data are not free of errors and this holds true whether they have been collected by direct surveys or obtained through administrative sources. Errors are detected and corrected by editing data both at the *data capture* and *estimation* stages. A number of data editing procedures are used to detect errors in the data. The associated edits are based on a combination of subject-matter experts' knowledge, as well as data analysis. Edits are either applied to each individual observation or across a number of them. The former is known as *microediting*, whereas the latter as *macroediting*.

Microediting takes place both during data capture and estimation. Microediting can be manual (e.g., a human declaring data in error) or automated (e.g., a computer rejecting data using predetermined editing rules). Edits associated with microediting include validation edits, logical edits, consistency edits, range edits, and variance edits. Validity edits verify the syntax within a questionnaire. For example, characters in numeric fields are checked, or the number of observed digits is ensured to be smaller or equal to the maximum number of positions allowed within the questionnaire. Range edits identify whether a data item value falls within a determined acceptable range. Consistency edits ensure that two or more data items (mainly financial variables) within a record do not have contradictory values. They follow rules of subject-matter experts to verify that relationships between fields are respected. Variance edits isolate cells with suspiciously high variances at the output stage, that is, when the estimates and variances have been produced. Erroneous or questionable data are corrected, identified for follow-up, or flagged as missing to be later imputed. Edits may be differentiated to declare resulting errors as either fatal or as suspicious.

Macroediting is carried out at the estimation stage. Errors in the data set missed by microediting are sought out via the analysis of aggregate data. The objective of the procedure is to detect suspicious data that have a large impact on survey estimates (Granquist, 1997). If this impact is quite large, suspicious data can be considered as outliers. Macroediting offers a number of advantages. First, significant cost savings can be obtained without loss of data quality. Second, data quality can be improved by redirecting resources and concentrating on editing of high impact records. Third, timeliness improvements are achieved by cutting down survey processing time and subject-matter experts' data analysis time. Finally, follow-ups are reduced, thereby relieving respondent burden.

A drawback of microediting is that too many records can be flagged for follow-up without accounting for the relative importance of individual records and the high cost in editing all records. This is remedied by *selective* editing, which cuts down on checking all records declared in error by focusing on a subset. Selective editing (also known as significance editing) selects records in error for follow-up if it is expected that the corrected data will have a large impact on the estimates. Such methods have been developed at Statistics Canada (Latouche and Berthelot, 1992), the Australian Bureau of Statistics (Lawrence and McDavitt, 1994; Lawrence and McKenzie, 2000), and the Office for National Statistics (Hedlin, 2003). Records in error that are not followed-up are imputed.

Latouche and Berthelot (1992) defined a score function to determine which records to follow up. Their score function was based on the magnitude of historical change for that record, the number of variables in error for that record, and the importance of each variable in error. One of the score functions that they suggested is given by

$$\text{Score}_k(t) = \sum_{q=1}^{Q} \frac{w_k(t)E_{k,q}(t)I_q\left(x_{k,q}(t) - x_{k,q}(t-1)\right)}{\sum_s w_k(t)x_{k,q}(t-1)}$$

where $E_{k,q}(t)$ equals 1 if there is an edit failure or partial nonresponse, and 0 otherwise; $w_k(t)$ is the weight for unit k at time t, and I_q reflects the relative importance for variable q. For example, if variable x_q is considered more crucial or important than variable $x_{q'}$, then this is reflected in the score function by assigning a larger value to I_q than $I_{q'}$. Suspicious records are ranked by their associated score. Records with scores above a given threshold are followed up.

Hedlin's (2003) procedure differs from Latouche and Berthelot's (1992) procedure in that his score function minimizes the bias incurred by accepting records in error. For a sample s, let the clean data be denoted as $y_1, y_2, \ldots y_n$ and the raw data as $z_1, z_2, \ldots z_n$. The score for z_k is computed as $\text{Score}(z_k) = w_k \times |z_k - E(z_k)|/\hat{Y}$, where $\hat{Y} = \sum_{k \in s_d} w_k y_k$, and s_d is part of the current sample for a specified domain d of interest (say the d-th industrial sector). The $E(z_k)$ term is usually the previous "clean" value of that record y_k, or the median of the corresponding domain. A record k will be rejected if Score(z_k) exceeds a prespecified threshold based on historical data. The threshold is set so that the coverage probability of the estimate is nearly unchanged.

6.2. Detection of outliers

Outliers are a common feature of almost all surveys. This is especially true for business surveys due to the highly skewed nature of their data. Outliers may result in unrealistically high or low estimates of population parameters, such as totals.

Outliers can come from two sources. First, they can be erroneous values, due to data entry or measurement problems for example. Second, they can be improbable or rarely occurring, but valid values. Erroneous values that are detected as outliers should be corrected, or removed from the dataset. On the other hand, improbable values identified as outliers should be left in the dataset, but special treatments should be applied to them to reduce their effects on estimates.

Chambers (1986) classifies outliers in sample surveys into two groups: *representative* and *nonrepresentative*. Outliers are representative if they have been correctly recorded and represent other population units similar in value to the observed outliers. Nonrepresentative outliers are those that are either incorrectly recorded or unique in the sense that there is no other unit like them. Errors that lead to outliers should be detected and corrected at the editing stage. In what follows, we focus on outliers that are free of error, and such outliers may either be representative or nonrepresentative.

Suppose that we have observed a sample s of size n with values y_k, and associated weights $w_k, k = 1, \ldots, n$. Outliers will be *influential* if the joint effect of the data and associated weight is significant. This is so whether they are representative or nonrepresentative. A typical example is a frame that is out-of-date in terms of size classification of its units. Suppose that a unit classified as small or medium size should have been classified as a large unit. The joint effect of the sampling weight w_k and large observed value y_k may result in declaring unit k as an influential observation.

In this section, we focus on a number of procedures to detect outliers. We present a number of those used in practice for business surveys. The treatment of outliers is discussed by Beaumont and Rivest.

6.2.1. Top-down method
This simple procedure sorts the largest entered values (top 10 or 20) and starts the manual review from the top or the bottom of the list. Units that have an abnormally large contribution to an estimator of interest such as the sample total are flagged and followed up.

Let $y_{(1)} \leq \ldots \leq y_{(n)}$ denote the ordered values of the observed y-values in the sample s. The cumulative percent distribution of the top j units to the y total of all the sampled units is computed. Unweighted or weighted versions of the cumulative percent distribution are computed. The unweighted version identifies units that may be in error. Once the unweighted top-down method is performed, the weighted version provides us with an idea of units that will be influential on account of their very large $w_k y_k$ product.

We can illustrate how the unweighted cumulative percent distribution is computed. The computations for the weighed version are identical with the exception of incorporating the weights w_k into the computations. The cumulative percent contribution $P_{(j)}$

to the total for each of the j top units is given by $P_{(j)} = 100 \times \sum_{k=j}^{n} y_{(k)}/Y_s$, where. For $j = n$, we have $P_{(n)} = 100 \times y_{(n)}/Y_s$. For $j = n - 1$, we have $P_{(j)} = 100 \times (y_{(n-1)} + y_{(n)})/Y_s$, and so on. More details on the top-down method are available in Granquist (1987).

6.2.2. Standardized distance

Let $z_k = w_k y_k$ be the product of the sampling weight w_k and observed value y_k. Let m_z and σ_z be the estimates of the location and scale of z_k. A typical measure used to detect outliers is the standardized distance $\delta_{z,k} = (z_k - m_z)/\sigma_z$. A unit k is identified as an outlier if the absolute value of $\delta_{z,k}$ is larger than a predetermined threshold. Location and scale estimates could be the sample mean and the standard deviation of the z_k values. Such estimates are nonrobust because they include some of the potential outlier values. Including all units in the computations reduces their probability of being declared as outliers. This "masking" effect is avoided by computing robust estimates of m_z and σ_z. Robust outlier-resistant estimates of m_z and σ_z are the median $Q_{2,z}$ and interquartile distance $(Q_{3,z} - Q_{1,z})$ of the z_k values respectively, where the $Q_{j,z}$ values are the j-th ($j = 1, 2, 3$) quartiles of the population (or the sample). Note that we could have used the nonweighted variable $\delta_{y,k} = (y_k - m_y)/\sigma_y$ as well. Units are declared as outliers if their z_k values fall outside the interval $(m_z - \delta_{\text{Low}}\sigma_z, \ m_z + \delta_{\text{High}}\sigma_z)$, where δ_{Low} and δ_{High} are predetermined values. These bounds can be chosen by examining past data or using past experience.

6.2.3. Hidiroglou–Berthelot method

The standardized distance can be used to detect whether the ratio of two variables y and x for a given sampled unit differs markedly from the ratios of the remaining units. Such comparisons do not account for size differences between units. Incorporating a measure of size (importance) with each unit places more emphasis on small ratios associated with those larger values. Hidiroglou and Berthelot (1986) extended the standardized distance procedure by incorporating a size component, and transforming the ratios to ensure symmetry. The extended method has been adapted by several national agencies to detect suspicious units. The procedure consists of six steps: (i) Ratios $r_k = y_k/x_k$ are computed for each unit k within the sample s. (ii) Data are transformed to ensure outliers can be detected at both tails of the distribution. The transformed data are given by $s_k = 1 - (\text{med } r_k)/r_k$ if $0 < r_k < \text{med } r_k$, and $r_k/(\text{med } r_k) - 1$ otherwise. (iii) The data's magnitude is incorporated by defining $E_k = s_k \max(x_k, y_k)^{\phi}$ where $0 < \phi < 1$. These E_k values are called effects. The parameter ϕ provides a control of the importance associated with the magnitude of the data. It controls the shape of the curve defining upper and lower boundaries. (iv) The first (E_{Q1}), second (E_{Q2}), and third (E_{Q3}) quartiles of the effects E_k are computed. (v) The interquartile ranges $d_{Q1} = \max(E_{Q2} - E_{Q1}, |a \, E_{A2}|)$ and $d_{Q3} = \max(E_{Q3} - E_{Q2}, |a \, E_{Q2}|)$ are computed. The quantity $|a \, E_{Q2}|$ reduces the tendency of declaring false outliers, and "a" is usually set to 0.5. This problem may arise when the E values are clustered around a single value and are one or two deviations from it. (vi) Units are declared to be outliers if their associated E_k value is outside $(E_{Q2} - cd_{Q1}, E_{Q2} + cd_{Q3})$. The parameter c controls the width of the acceptance region. Belcher (2003) suggested a procedure to determine the values of the different parameters entering the Hidiroglou–Berthelot method. Figure 4 illustrates how these steps lead to identifying outliers.

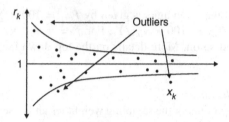

Fig. 4. Hidiroglou–Berthelot method.

6.3. Imputation

Edited records have either passed or failed the edits. A subset of these records may have been declared as outliers as well, regardless of their edit status. Records also considered as having failed edits are those that have not responded to the survey (or unit nonresponse), or that have provided incomplete data (partial response). Furthermore, some data items may have been manually deleted if they have been considered in error as a result of the editing process. The overall impact of edit failure is that it results in *missing data*.

There are several options available for dealing with missing data. The simplest one is to do nothing. That is, missing values are flagged on the output data file, leaving it up to the data user or analyst to deal with them. This "solution" to missing data is usually adopted when its reveal to be too difficult to impute values with sufficient accuracy. For example, this occurs for variables that have no direct relationship with any other collected variable. In farm surveys, for example, livestock and crops cannot be used to impute each other, and it is then preferable to leave the missing values not imputed.

Another option is to adjust the survey weights for nonresponse. Although this procedure is mainly meant for unit nonresponse, it can be used for partial nonresponse. However, the drawback is that there will be as many weight adjustments as there are missing fields across the records. Methods such as calibration, or mass imputation, are used to insure consistency between the resulting tabulations. These approaches have been considered by Statistics Netherlands for the construction of a consistent set of estimates based on data from different sources (see Kroese and Renssen, 2001).

The preferred option for survey users is to impute missing data within individual records. The imputation procedures should be based on the Fellegi–Holt principles (Fellegi and Holt, 1976) which are as follows: (i) data within individual records must satisfy all specified edits. (ii) The data in each record should be made to satisfy all edits by changing the fewest possible variables (fields). (iii) Imputation rules should be derived automatically from edit rules. (iv) Imputation should maintain the joint distribution of variables.

In business surveys, because there are usually strong accounting relationships between collected variables, manual imputation is often considered, especially for small surveys where the resources to be devoted to imputation systems are minimal. This approach is not reasonable as it can lead to different tabulations, thereby yielding inconsistent results. The resulting manual imputation may not be the best method to use if the imputation is based on incomplete knowledge.

Imputation methods can be classified as *deterministic* or *stochastic*. A deterministic imputation results in unique imputed data. Examples of deterministic imputation methods often used for business surveys include: logical, mean, historical, sequential (ordered) hot-deck, ratio and regression, and nearest neighbor imputation. A stochastic imputation results in data that are not unique, as some random noise has been added to each imputed value. Stochastic imputation can be viewed as being composed of a deterministic component with random error added to it. Stochastic imputation, unlike deterministic imputation, attempts to preserve the distribution of the data. Examples of stochastic imputation include random hot deck, regression with random residuals, and any deterministic method with random residuals added.

Imputation of plausible values in place of missing values results in internally consistent records. Imputation is also the most feasible option for partial nonresponse. Good imputation techniques can preserve known relationships between variables, which is an important issue in business surveys. Imputation also addresses systematic biases, and reduces nonresponse bias. However, imputation may introduce false relationships between the reported data by creating "consistent" records that fit preconceived models. For example, suppose it is assumed that $x < y$ for two response variables x and y. If this constraint is false, imputing the missing variable x (or y) will result in incorrectly imputed data.

Imputation will normally use reported data grouped within subsets of the sample or population. Such subsets are known as *imputation groups* or *imputation classes*.

6.3.1. Deterministic imputation methods

6.3.1.1. Logical (or deductive) imputation. Missing values are obtained by deduction, using logical constraints and reported values within a record. Typical examples include deriving a missing subcomponent of a total. This type of imputation is often used in business surveys if there is a strong relationship between variables (especially financial ones).

6.3.1.2. Mean value imputation. Missing data are assigned the mean of the reported values for that imputation class. Mean value imputation should only be used for quantitative variables. Respondent means are preserved, but distributions and multivariate relationships are distorted by creating an artificial spike at the class mean value. The method also performs poorly when nonresponse is not random, even within imputation classes. It is often used only as a last resort.

Note that the effect on the estimates of mean value imputation corresponds exactly to a weight adjustment for nonresponse within imputation groups. Weighting is useful for unit nonresponse as relationships between variables are preserved.

6.3.1.3. Historical imputation. This is the most useful method in repeated economic surveys. It is effective when variables are stable over time, and when there is a good correlation between occasions for given variables within a record. The procedure imputes current missing values on the basis of the reported values for the same unit on a previous occasion. Historical trend imputation is a variant of the procedure: previous values are adjusted by a measure of trend, based on other variables on the record. Historical imputation is heavily used for imputing tax data for incorporated businesses at Statistics Canada (Hamel and Martineau, 2007). Historical imputation can be seen as a special case of regression imputation (see later).

6.3.1.4. Sequential hot-deck method. This method assumes that the order of the data items is fixed, even though they might have been sorted according to some criterion (measure of size or geography). Missing data are replaced by the corresponding value from the preceding responding unit in the data file. The data file is processed sequentially, storing the values of clean records for later use, or using previously stored values to impute missing variables. Care is needed to ensure that no systematic bias is introduced by forcing *donors* (reported data items) to always be smaller or larger than *recipients* (missing data items). Sequential hot-deck is used for the United States Current Population Survey.

Sequential hot-deck uses actual observed data for imputation. The distribution between variables tends to be preserved, and no invalid values are imputed, unlike with mean, ratio, or regression imputation. This is ensured if the imputed variables are not correlated with other variables within the record. If the variables are correlated (as it is the case with financial variables, for example), then imputing the missing variables by those from the preceding unit will not preserve the distribution. This problem is resolved, in practice, by imputing complete blocks of variables at a time: a block being a set of variables with no relationship with variables outside the block (see Hamel and Martineau, 2007).

6.3.1.5. Nearest neighbor imputation. Nearest neighbor imputation uses data from clean records to impute missing values of recipients. It uses actual observed data from recipients. Donors are chosen such that some measure of distance between the donor and recipient is minimized. This distance is calculated as a multivariate measure based on reported data. Nearest neighbor imputation may use donors repeatedly when the nonresponse rate is high within the class. This method is the second most heavily used one, after historical imputation, for imputing tax data for incorporated businesses at Statistics Canada (Hamel and Martineau, 2007).

6.3.1.6. Ratio and regression imputation methods. These imputation methods use auxiliary variables to replace missing values with a predicted value that is based on a ratio or regression. This method is good for business surveys when auxiliary information is well correlated with the imputed variable. However, accounting relationships between variables need to be respected, and this leads to the need for some adjustment of the predicted values. For example, suppose that the relationship $x + y = z$ holds for three variables x, y, and z . If x and y are imputed, we might need to adjust the imputed values (e.g., by prorating) to insure that this relationship still holds.

The response variable needs to be continuous for these methods to be effective. For regression, the independent regression variables may be continuous or dummy variables if they are discrete. A disadvantage of this method is that the distributions of the overall data set may have spikes.

6.3.2. Stochastic methods
Deterministic methods tend to reduce the variability of the data, and to distort the data distributions. Stochastic imputation techniques counter these negative effects by adding a residual error to deterministic imputations that include regression imputation. Another approach for stochastic methods is to use some form of sampling in the imputation process, as it is the case of hot-deck imputation. Whether or not stochastic

methods are used, it is possible to compute variances that take into account the effect of imputation.

7. Estimation

Business survey data are collected for reference periods that are monthly, quarterly, or annual. The resulting data are usually summarized as *level* and *change*. Level is measured as a total for a given variable of interest y. Change is defined as the difference between, or the ratio of, two estimated totals at two different time periods.

Factors that need to be taken into account for estimation include the sample design, the parameters to be estimated, domains of interest, and auxiliary data. The sample design is usually straightforward for business surveys. These surveys mostly use one-phase or two-phase stratified simple random sampling or Bernoulli sampling without replacement at each phase. Domain estimation is used in three ways for business surveys. First, the classification (i.e., geography, industry, or size) of the sampled units may differ from the original one. Second, the classification associated with domains for tabulating the collected data may differ from one used for the stratification purposes. Third, a unit originally sampled in-scope for a given survey may become out-of-scope either by ceasing its business activities or changing its classification to one that is not within the target population (e.g., a unit sampled within the retail sector becomes a wholesaler, which is out-of-scope to the survey).

The increasing use of auxiliary data for business surveys is associated with the wider availability of sources outside the survey, such as regularly updated administrative sources or annual totals from a larger independent survey. Auxiliary data yield several benefits. They improve the efficiency of the estimates when the auxiliary data (say x) are correlated with the variable(s) of interest y. Given that there is some nonresponse, the potential nonresponse bias is reduced if the variables of interest are well correlated with the auxiliary data. A by-product of using auxiliary data is that their weighted totals add up to known population totals. We limit estimation to one-phase stratified Bernoulli sampling in what follows.

7.1. Estimation for level

Let $U = \{1, \ldots, k, \ldots, N\}$ denote the in-scope population of businesses. The population of businesses is stratified by geography, industry, and size as U_h, $h = 1, \ldots, L$. The population size of U_h is N_h, and a probability sample s is selected from U with inclusion probability π_k for $k \in s$. If Bernoulli sampling has been used, the inclusion probabilities associated with stratum U_h are given by $\pi_k = f_h = n_h/N_h$. The expected sample size within the sample stratum s_h is n_h, and the realized sample size will be n_h^*. Each sampled unit $k \in s_h$ will have sample design weights given by $w_k = 1/\pi_k = N_h/n_h$.

Suppose that we wish to estimate the total of y for a given domain of U, say $U_{(d)}$, $d = 1, \ldots, D$. This population total is given by $Y_{(d)} = \sum_{h=1}^{L} \sum_{k \in U_h} y_{(d)k}$ where $y_{(d)k} = y_k \, \delta_{(d)k}$ with $\delta_{(d)k}$ is one if $k \in U_{(d)}$ and zero otherwise.

If no auxiliary data are used, the population total $Y_{(d)}$ is estimated by the expansion estimator $\hat{Y}_{(d)}^{(\text{EXP})} = \sum_{h=1}^{L} \hat{Y}_{(d),h}^{(\text{EXP})}$ where $\hat{Y}_{(d),h}^{(\text{EXP})} = \sum_{k \in s_h} w_k y_{(d)k}$. Although this estimator is unconditionally unbiased, it is conditionally biased given the realized samples size n_h^* (see Rao, 1985). Given that Bernoulli sampling has been used, the estimated variance of $\hat{Y}_{(d)}^{(\text{EXP})}$ will be $v(\hat{Y}_{(d)}^{(\text{EXP})}) = \sum_{h=1}^{L} \sum_{k \in s_h} (1 - \pi_k) y_{(d)k}^2 / \pi_k^2$, which does not compare favorably with the corresponding expression for stratified *srswor* given the same (expected) sample sizes at the stratum level.

Consider the estimator

$$\hat{Y}_{(d)}^{(\text{HAJ})} = \sum_{h=1}^{L} \frac{N_h}{\hat{N}_h} \sum_{k \in s_h} w_k y_{(d)k} \tag{4}$$

where N_h are known population strata counts, and $\hat{N}_h = \sum_{k \in s_h} w_k$ is the estimated population strata counts (Brewer et al., 1972). This estimator, also known as the Hájek estimator, "adjusts" for the discrepancy between expected and realized sample sizes, assuming that $P(n_h^* = 0)$ is negligible for $h = 1, \ldots, L$. This estimator has the following two desirable properties. First, it is conditionally nearly unbiased given n_h^*. Second, its variance estimator is approximately equal to the one that we would get with *srswor* with realized sample sizes n_h^* selected from populations of size N_h, $h = 1, \ldots, L$. That is,

$$V\left(\hat{Y}_{(d)}^{(\text{HAJ})}\right) \doteq \sum_{h=1}^{L} \frac{N_h^2}{n_h} \left(1 - \frac{n_h}{N_h}\right) S_{h(d)}^2 \tag{5}$$

where $S_{h(d)}^2 = (N_h - 1)^{-1} \sum_{k \in U_h} \left(y_{(d)k} - \overline{Y}_{h(d)}\right)^2$ and $\overline{Y}_{h(d)} = \sum_{k \in U_h} y_{(d)k} / N_h$. This variance can be estimated using

$$v\left(\hat{Y}_{(d)}^{(\text{HAJ})}\right) \doteq \sum_{h=1}^{L} \frac{N_h^2}{n_h^*} \left(1 - \frac{n_h}{N_h}\right) \hat{S}_{h(d)}^2 \tag{6}$$

where $\hat{S}_{h(d)}^2 = (n_h^* - 1)^{-1} \sum_{k \in s_h^*} \left(y_{(d)k} - \hat{\overline{Y}}_{h(d)}\right)^2$ and $\hat{\overline{Y}}_{h(d)} = \sum_{k \in s_h^*} y_{(d)k} / n_h^*$.

Estimator (7.1) is reasonable if the realized sample size is sufficiently large within each stratum U_h; if it is not, strata with insufficient realized sample sizes need to be combined with others to reduce the relative bias. The estimator $\hat{Y}_{(d)}^{(\text{HAJ})}$ can alternatively be written as $\hat{Y}_{(d)}^{(\text{HAJ})} = \sum_{h=1}^{L} \sum_{k \in s_h} w_k g_k y_{(d)k}$, where $g_k = N_h / \hat{N}_h$ for $k \in U_h$ is known as the *g-weight* (see Särndal et al., 1992).

The separate count ratio estimator is the simplest example of an estimator that uses auxiliary data. Multivariate auxiliary data can be incorporated into the estimation process via the well known regression estimator, $\hat{Y}_{(d)}^{(\text{REG})} = \hat{Y}_{(d)}^{(\text{EXP})} + \left(\mathbf{X}_d - \hat{\mathbf{X}}_d\right)' \hat{\mathbf{B}}_{(d)}$, where $\hat{\mathbf{B}}_{(d)}$ is obtained by minimizing the variance of $\hat{Y}_{(d)}^{(\text{REG})}$. The regression estimator can also be written as $\hat{Y}_{(d)}^{(\text{REG})} = \sum_{k \in s} \widetilde{w}_k y_{(d)k}$ where $\widetilde{w}_k = w_k g_k$ is known as regression weights. The regression weights are the products of the original design weights w_k with the g_k-weights given by $g_k = 1 + \left(\sum_{k \in U} \mathbf{x}_k - \sum_{k \in s} w_k \mathbf{x}_k\right)' \left(\sum_{l \in S} \frac{x_l x_l'}{\lambda_l \pi_l}\right)^{-1} \frac{\mathbf{x}_k'}{\lambda_k}$: the λ_k term incorporates the optimality of the estimator. In the case of the ratio estimator, we have that $\lambda_k = c_k x_k$. The Huang and Fuller's (1978) iterative procedure is implemented in Bascula (Nieuwenbroek and Boonstra, 2001).

The calibration procedure of Deville and Särndal (1992) minimizes distance measures between the original weights and final weights $\widetilde{w}_k$ subject to $\mathbf{X} =$

$\sum_{k \in s} \tilde{w}_k \mathbf{x}_k$. They propose several such distance measures, and the one defined by $\sum_{k \in s} (\tilde{w}_k - w_k)^2 / w_k \lambda_k$ corresponds exactly to the regression weighs given by $g_k = 1 + \left(\sum_{k \in U} \mathbf{x}_k - \sum_{k \in s} w_k \mathbf{x}_k \right)' \left(\sum_{l \in s} \frac{x_l x_l^t}{\lambda_l \pi_l} \right)^{-1} \frac{\mathbf{x}_k'}{\lambda_k}$. It should also be noted that Deville and Särndal's procedure allows for bounding the g-weights. A good comparison of these two approaches in given is Singh and Mohl (1996).

The $\sum_{k \in s} \tilde{w}_k \mathbf{x}_k = \sum_{k \in U} \mathbf{x}_k$ constraint can be applied to subpopulations $U_p \subseteq U(p = 1, \ldots, P)$ of the population U, where $U_p \cap U_{p'} = \emptyset$ for $p \neq p'$ and $U = \bigcup_{p=1}^{P} U_p$. These subpopulations are also referred to as poststrata. The previous constraint translates into $\sum_{k \in s_p} \tilde{w}_k \mathbf{x}_k = \sum_{k \in U_p} \mathbf{x}_k$ where $s_p = s \cap U_p$. For this case, the g-weights are of the form:

$$g_k = 1 + \left(\sum_{k \in U_p} \mathbf{x}_k - \sum_{k \in s_p} w_k \mathbf{x}_k \right)' \left(\sum_{l \in s_p} \frac{w_l \mathbf{x}_l \mathbf{x}_l'}{c_l} \right)^{-1} \frac{\mathbf{x}_k}{c_k} \text{ for } k \in U_p \quad (7)$$

Hidiroglou and Patak (2004) show that domain estimation efficiency (in terms of variance) is improved when auxiliary data are available for poststrata that are close to the domains of interest. This holds when there is a constant term in the auxiliary data vector, and when the variance structure associated with the model linking the variable of interest y to the auxiliary data $\mathbf{x}$ is constant. It does not necessarily hold, however, that the incorporation of unidimensional auxiliary data at the poststratum level into $\hat{Y}_{(d)}^{(HAJ)}$ will increase the efficiency of the resulting estimator.

If the poststrata sizes are too small, raking on margin variables (counts or continuous quantities) will also improve the efficiency of the expansion estimator $\hat{Y}_{(d)}^{(EXP)}$. Hidiroglou and Patak (2006) displayed how gross business income incorporated into raking ratio estimation could improve the efficiency of the estimates of total sales for the Monthly Canadian Retail Trade Survey by raking on margins based on industry and geography.

Two-phase sampling is also used in business surveys to obtain (or inherit) relatively inexpensive first-phase information that is related to the characteristic of interest. This information may be in the form of discrete variables that yield estimated counts used in poststratified estimation, or continuous x variables for improving the efficiency of the estimators of interest. A number of surveys use a two-phase sample design at Statistics Canada. An example is the Quarterly Canadian Retail Commodity Survey. This sample design was chosen to reduce collection costs by using as the first-phase sample the Canadian Monthly Retail Trade Survey. Auxiliary information (annualized sales) from the first-phase sample is used in all of the Canadian Retail Commodity Survey design steps to maximize the efficiency. More details of sample design are given in Binder et al. (2000).

7.2. Estimation for change

Let $U(t) = \{1, \ldots, k, \ldots, N(t)\}$ denote the in-scope population of businesses at the t-th survey occasion. Note that because of births and deaths of units, we expect $U(t)$ to be different from $U(t')$ for $t \neq t'$. As in the previous section, we consider the case where Bernoulli sampling has been used, with inclusion probabilities associated with stratum $U_h(t)$ given by $\pi_k(t) = f_h(t) = n_h(t)/N_h(t)$.

At survey occasion 1, a sample $s(1)$ of size $n^*(1)$ has been selected, using the selection intervals defined by the starting point $\alpha(1) \in [0, 1]$. Recall that a population unit k falling in stratum h is included in the sample if $\alpha(1) < u_k \le \alpha(1) + f_h(1)$, provided $\alpha(1) + f_h(1) \le 1$ (see Section 5.1). For survey occasion 2, a rotation of $(100 \times r)\%$ of the sample has been performed by shifting the parameter $\alpha(1)$ by $r \times f_h(2)$ in all strata of survey occasion 2. The starting point $\alpha(2)$ is then given by $\alpha(1) + r \times f_h(2)$ (assuming, for simplicity, that $\alpha(1) + rf_h(2) + f_h(2) \le 1$). This rotation results in a selected sample $s(2)$ of size $n^*(2)$, with $n^*(c)$ units in common with $s(1)$. Note that the inclusion probability $\pi_k(1, 2)$ of unit k being selected in both survey occasions is given by $\pi_k(1, 2) = \min [\max [0, f_h(1) - rf_h(2)], f_h(2)]$.

Change between the survey occasions can be defined in a number of ways. A simple definition is to consider the difference $\Delta_{(d)} = Y_{(d)2} - Y_{(d)1}$, where $Y_{(d)t}$ is the population total of domain d at survey occasion t. This difference can be estimated using $\hat{\Delta}_{(d)}^{(HAJ)} = \hat{Y}_{(d)2}^{(HAJ)} - \hat{Y}_{(d)1}^{(HAJ)}$, where $\hat{Y}_{(d)t}^{(HAJ)}$ is defined by (4) for $t = 1, 2$. The variance of $\hat{\Delta}_{(d)}^{(HAJ)}$ is given by

$$V\left(\hat{\Delta}_{(d)}^{(HAJ)}\right) = V\left(\hat{Y}_{(d)2}^{(HAJ)}\right) + V\left(\hat{Y}_{(d)1}^{(HAJ)}\right) - 2\mathrm{Cov}\left(\hat{Y}_{(d)2}^{(HAJ)}, \hat{Y}_{(d)1}^{(HAJ)}\right) \tag{8}$$

The variance $V\left(\hat{Y}_{(d)t}^{(HAJ)}\right)$ is given by (5), and it can be estimated using (6). The covariance $\mathrm{Cov}\left(\hat{Y}_{(d)2}^{(HAJ)}, \hat{Y}_{(d)1}^{(HAJ)}\right)$ is approximately given by

$$\mathrm{Cov}\left(\hat{Y}_{(d)2}^{(HAJ)}, \hat{Y}_{(d)1}^{(HAJ)}\right) \doteq \sum_{h=1}^{H}\sum_{g=1}^{G} N_{gh}(c) \frac{N_h(1)N_g(2)}{n_h(1)n_g(2)}\left(f_{gh} - \frac{n_h(1)n_g(2)}{N_h(1)N_g(2)}\right)S_{gh(d)} \tag{9}$$

where, to simplify the notation, we denoted the strata of survey occasion 1 by $h(h = 1, \ldots, H)$, and those of survey occasion 2 by $g(g = 1, \ldots, G)$. The quantity $N_{gh}(c)$ is the size of the common population $U_{gh}(c)$ crossing stratum g of survey occasion 2 and stratum h of survey occasion 1. The quantity f_{gh} is given by $f_{gh} = \min [\max [0, f_h(1) - rf_h(2)], f_h(2)]$, and $S_{gh(d)} = \left(N_{gh}(c) - 1\right)^{-1}\sum_{k \in U_{gh}(c)} \left(y_{(d)k}(2) - \overline{Y}_{gh(d)2}\right)\left(y_{(d)k}(1) - \overline{Y}_{gh(d)1}\right)$ with $\overline{Y}_{gh(d)t} = \sum_{k \in U_{gh}(c)} y_{(d)k}(t)/N_{gh}(c)$ for $t = 1, 2$. The covariance (10) can be estimated using

$$\mathrm{Cov}\left(\hat{Y}_{(d)2}^{(HAJ)}, \hat{Y}_{(d)1}^{(HAJ)}\right) = \sum_{h=1}^{H}\sum_{g=1}^{G} \frac{n_{cgh}^*}{f_{gh}}\frac{N_h(1)N_g(2)}{n_h^*(1)n_g^*(2)}\left(f_{gh} - \frac{n_h(1)n_g(2)}{N_h(1)N_g(2)}\right)\hat{S}_{gh(d)} \tag{10}$$

where $\hat{S}_{gh(d)} = \left(n_{gh}^*(c) - 1\right)^{-1}\sum_{k \in s_{gh}^*(c)}\left(y_{(d)k}(2) - \hat{\overline{Y}}_{h(d)2}\right)\left(y_{(d)k}(1) - \hat{\overline{Y}}_{h(d)1}\right)$ and $\hat{\overline{Y}}_{h(d)t} = \sum_{k \in s_h^*(t)} y_{(d)k}(t)/n_h^*(t)$ for $t = 1, 2$. If the realized sample size n_{cgh}^* is not sufficiently large in some cross-strata gh; then some collapsing must be done within the strata g of survey occasion 2, or the strata h of survey occasion 1, to reduce the relative bias.

Essential Methods for Design Based Sample Surveys
ISSN: 0169-7161
© 2009 Elsevier B.V. All rights reserved
DOI: 10.1016/B978-0-444-53734-8.00007-8

7

Sampling, Data Collection, and Estimation in Agricultural Surveys

Sarah M. Nusser and Carol C. House

1. Introduction

Surveys that provide information about land use, land stewardship, agricultural production, and the economics of managing both the agricultural industry and our natural resources are of necessity complex, but vitally important. Agriculture is a $240 billion industry in the United States alone. It forms the foundation of an even larger food and fiber industry that contributes 12.3% of the U.S. gross national product and over 17% of the total employment. This industry provides $71 billion in U.S. exports, whereas the U.S. imports more than $65 billion in food and fiber products from other countries. Many developed countries in the world have similar economic dependence on their agricultural industry, and the economies of developing countries are tied even more tightly to their agrarian infrastructure.

Information obtained from surveys of agricultural production and the economics of that production allows the market economy to function efficiently. "With accurate information, individuals can make sound decisions that allow them to adjust their actions to the situation at hand. The value of publicly provided information [to these markets] is often underestimated." (Roberts and Schimmelphennig, 2006). Apart from producers and buyers, this information is used by such entities as businesses supplying inputs to production, to those deciding where to build processing facilities, and those directing train car distributions. These surveys further provide the information needed by policy makers, local governments, academic researchers, and other stakeholders to extract knowledge needed for making informed decisions.

Farming and ranching are closely tied to the use and management of land. There are 1.9 billion acres of land in the United States, with nearly 1.4 billion acres in rural (non-Federal) areas. Non-Federal rural areas include about 400 million acres in active cropland and cropland set aside via conservation programs, more than 500 million acres of pastureland and rangeland used to raise livestock, and 400 million acres of forestland (Natural Resources Conservation Service, 2007a). The quality of this land is important to both agriculture and natural resources. For example, programs to reduce soil loss on cropland and increase wetlands on agricultural land have helped to generate reductions of 43% in soil erosion since 1982 (Natural Resources Conservation Service, 2007b) and

161

net increases in agricultural wetlands of 98,000 acres during the last decade (Natural Resources Conservation Service, 2007c). Thus, surveys to obtain information about land utilization and stewardship are linked closely to those of agricultural production, and this chapter will include these in the discussions. These surveys provide information on such topics as the changing use of land, the spread of urbanization, the conservation practices employed on cropland, and the effectiveness of environmental policies.

From the perspective of applying statistical methods, conducting agricultural surveys requires the use of a distinct mix of methodology. Farms are businesses and much of the data that are collected on agricultural surveys are facts related to the operation of those businesses. During the past three decades, much of production agriculture in developed countries has become consolidated into larger operations. Now in the United States, less than 10% of the farms produce almost 75% of the total production. Many farming enterprises are part of vertically integrated, multinational corporations. Thus, survey methodology developed for business surveys is relevant to agricultural surveys in developed countries (see Chapter 6).

Although many farms have become corporate in nature, many more remain small and family operated. Within the United States, 90% of all farms are still small, family owned operations. In developing countries, small family run farms are the predominant type of production agriculture. Often, there is an interest in the farm household as well as the farm enterprise. Surveys of small farms, and particularly those collecting information about the farm household, may be similar to household surveys of other population groups. Thus, the survey methodology created for household surveys also applies to agricultural surveys (see Chapter 5).

Finally, some surveys of agriculture involve assessments of conditions on the land. These surveys consist of sampling land units and then collecting information and making direct measurements on those land units. This may be done on-site or remotely via imagery and without any input from a human respondent. Thus, methodological strategies that are designed to minimize error in measurements made directly from the field or other resources are also important. In addition, statistical approaches used in environmental surveys are relevant (see Chapter 8).

Thus, agricultural surveys present many challenges, most of which are related to the complex and diverse nature of the target population and the types of information acquired through these surveys. These challenges include:

- Using the appropriate mix of methodology for businesses, households, and direct measurements;
- Choosing a sampling frame that has both reasonable completeness and reasonable efficiency;
- Handling changes in the population, such as the consolidation of production agriculture into fewer population units from which more and more information is demanded, or changes in boundaries of land types;
- Coping with decreasing response rates;
- Creating estimates that are consistent with multiple independent administrative data sources and previously released estimates;
- Addressing the emerging need for all information to be geo-referenced;
- Serving the increasing need for small population estimates related to small geographic areas and specialized agricultural commodities that have been developed and marketed;

- Providing for the growing demand of analysts and researchers for more data, more access, greater accuracy, and to be able to import data into sophisticated tools; and
- Ensuring confidentiality of respondent data.

This chapter provides an overview of methods used for surveys of agricultural producers and land areas that focus on agriculture and natural resource concerns. In doing so, we emphasize emerging methods and future challenges in sampling, data collection, and estimation.

2. Sampling

In surveys of agriculture and the land, a careful review and understanding of the population inferences that are needed from the survey are critically important. This is certainly the case for all surveys, but these issues seem to be easily obscured in agricultural and land surveys. This is illustrated by the following example. There is a current need in the United States and Europe to understand how quickly farmland may be disappearing to development, and to make policy decisions to slow down that development. The first question to ask is "what is farmland." Is it land that is currently used for growing a crop or raising livestock? Is it more broadly any land owned or managed by a farming operation, whether or not it is currently used for producing crops or livestock? Is it any undeveloped land? Are woods and forests included? What if cattle are kept in those woods and forests? Do you include subdivisions of 10 acre homesites (perhaps with horses), or is this land already considered "developed"? The answers to these types of questions are essential to defining the target population, sampling units, reporting units, and observations for a survey.

Sampling methods for agricultural surveys are tied to land, either directly or indirectly. Not surprisingly, area frames are utilized for many agricultural surveys. An area sample frame allows for the direct sampling of units of land, usually through a multistage sampling process to improve efficiency in sampling and/or data collection. A major advantage of the use of an area frame is that it provides complete coverage of the targeted land area. It can be an efficient frame for drawing samples to collect data and make measurements that are highly correlated to that land area, such as land use, production of major crops, general conservation practices, etc. An area frame, although a complete frame, will generally provide very large sampling errors when used for collecting data on uncommon items. For example, the acres of corn in Iowa can be estimated very effectively with an area frame. The estimation of acres of apples in that same state would be problematic.

Area frames may be constructed in different ways. Several important area frames are in use for agricultural statistics, and we will describe in some detail three that have distinctly different designs. One has been built and maintained by the U.S. Department of Agriculture for the estimation of agricultural production. Land area in the United States is divided into primary sampling units (PSUs), which are then stratified by general land use classifications, and sampled at different rates. Strata definitions differ by state, but usually include the following breakouts: >75% cultivated, 51–75% cultivated, 15–50% cultivated, <15% cultivated, agri-urban, urban, nonagriculture, water. Sampled PSUs are further broken down into secondary sampling units of approximately one square mile in size, and sampled for data collection. Usually one secondary unit is selected within each PSU. Breakdown of all units are made along visible and natural boundaries such as roads, streams, ditches, etc. This approach to the development and use of an area sampling frame for agricultural statistics has been used in a number of developing countries.

Another important area frame is used by the U.S. Department of Agriculture (USDA, 2005) to monitor conditions on the land via the National Resources Inventory (NRI). The NRI is a large natural resource monitoring survey conducted by the USDA Natural Resources Conservation Service in cooperation with the Iowa State University Center for Survey Statistics and Methodology. This inventory captures data on land cover and use, soil erosion, prime farmland, soils, wetlands, habitat diversity, selected conservation practices, and related resource attributes. It does so primarily via remote sensing but is periodically augmented by a subsample of field visits (Nusser and Goebel, 1997). The NRI area frame encompasses all land in the United States and its territories. The sample for the first 1982 inventory formed the basis of subsequent inventories. The first-stage area sample was stratified within counties (or equivalents) using political divisions or geographic coordinate systems without regard to visible features or natural resource boundaries (Figure 1a). The use of political or geographic stratum boundaries avoids problems associated with changing natural resource boundaries. Small strata were used in each county to ensure geographic spread and control over sampling rates for diverse geographic domains. Area segments within strata are generally a square of land one-half mile on each side, although segments may be smaller or larger as heterogeneity of land conditions warrant (Figure 1a). Sampling rates within strata generally range from two to four percent. For nearly all sampled area segments, three points are selected within the segment using restricted randomization to encourage geographic spread. Sample points are used to obtain detailed trending information for most survey variables.

A third area frame is one developed for the USDA Forest Service's Forest Inventory and Analysis (FIA) program, designed to estimate conditions over time for private forests in the United States (Bechtold and Patterson, 2005). The frame design consists of a system of grid points that are the center coordinates of a hexagonal tessellation of the land surface of the United States. The FIA uses a three-phase sample design. Remote sensing methods are used in the first phase to classify these grid points into forest and non-forest strata. FIA's second and third phases are used to obtain field measurements. The second phase sample is used to obtain traditional inventory field measurements at a systematic subsample of approximately 125,000 first phase forest points. The third phase is a subsample of approximately 8,000 second phase points (about 1 in 16 phase two points) to collect more intensive forest health field measurements (USDA, 2007b). One-fifth of the Phase 2 and Phase 3 samples are observed each year. A complex plot design has been developed to facilitate collection of a wide variety of biological and environmental properties at different scales (Figure 2; USDA, 2007a). The layout at a Phase 2 point consists of four circular subplots with a 24 foot radius, one centered at the sample point and the other three surrounding the center point at regular intervals 120 feet from the center plot. A circular microplot with a 6.8 foot radius is also established within each circular subplot.

The integration of field and remote sensing surveys via two-phase sampling to extend the depth of information available for individual sampling units is also used with the NRI. Subsamples of NRI photo-interpretation survey samples are selected

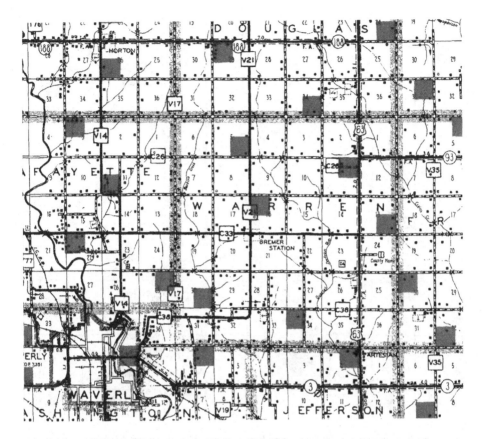

Fig. 1a. Area sample examples for the National Resources Inventory: (a) standard Public Land Survey sample in the central U.S. with a 4% area sample using third townships as strata and quarter sections as area sampling units (each square on the map is a 1 mi × 1 mi section, each solid sample unit is a 0.5 × 0.5 mi quarter section, each stratum is 2 mi × 6 mi).

to investigate field conditions for specific kinds of land, such as the health of rangeland or the quality of cropland soils. Over the last few years, the Conservation Effects Assessment Program has involved a collaboration among USDA agencies in which interviews were conducted with producers whose land was associated with a subsample of NRI points. This approach has led to more detailed practice information at a point than would be available via remote sensing and administrative databases, and has been used to develop simulation models that predict outcomes under alternative practices for policy evaluation.

In Europe, designs that involve remote sensing and field components are also being used to generate timely information on agricultural production for implementation of its Common Agricultural Policy. The MARS (Monitoring Agriculture by Remote Sensing) project is sponsored by the Directorate General Joint Research Center of the European Commission. This project provides crop and yield monitoring by utilizing agrometeorological models, low resolution remote sensing methods, and area estimates using high resolution data combined with ground surveys. Echoing features of the FIA design, Italy's agricultural survey (AGRIT) also relies on satellite imagery to stratify

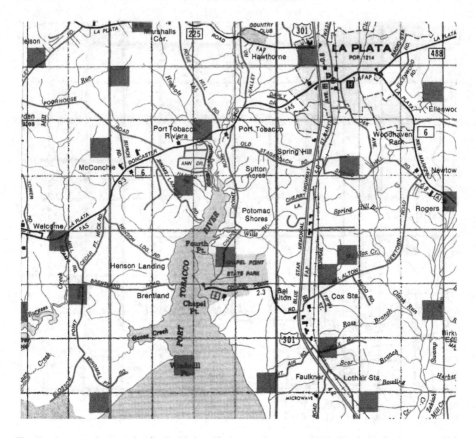

Fig. 1b. Area sample examples for the National Resources Inventory: (b) similar design in the eastern U.S.
with latitude and longitude boundaries for strata and sampling units.

points for agricultural field surveys (Carfagna and Gallego, 2005). This is especially
helpful in fragmented landscapes such as those in Europe.

Technologies used in area frame sampling have evolved considerably over the last
50 years. Initial frames relied on paper products—county road maps and aerial photography to define strata and area segments. Today, geographic information systems
(GIS) are used to assist in stratification and in selecting and managing samples. Utilizing satellite imagery and digital aerial photography, the steps in the process involve
computer-assisted delineation of sampling structures against a digital image. Delineation is improved by the capacity to view different scales. Further, digital storage of
strata and segments enables more efficient production of sampling lists and field materials for the survey. For land-based surveys, the ability to quickly generate area sampling geometries on the earth's surface via GIS greatly increases flexibility in choice
of design features. For example, the FIA program redesigned their survey based on a
tessellation of hexagons for their first phase of sampling, an approach that would be
difficult without GIS. Regardless of the type of sample units, the use of GPS (global
positioning system) receivers enables field staff to find sample points created in GIS
frames when they are not associated with visible features.

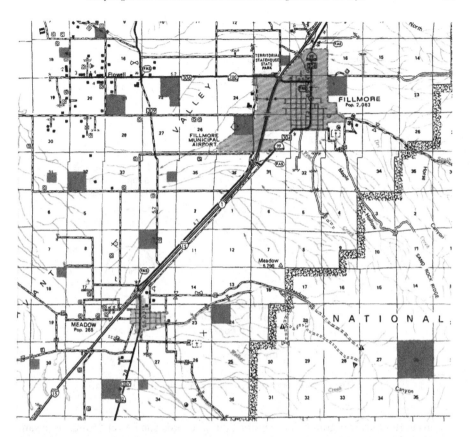

Fig. 1c. Area sample examples for the National Resources Inventory: (c) area sample in variable land use area of the western U.S. with smaller area segments for heterogeneous areas and larger segments in homogeneous areas.

List sampling frames are also very important for agricultural surveys. Generally, these lists are of land owners or of the operators/managers of the agricultural enterprises. These lists can be efficient sampling frames for agricultural statistics particularly if they contain useful information about the agricultural enterprises. Thus, a list of agricultural enterprises that have produced or sold hogs in the past could be an effective sampling frame for a survey to measure hog inventories. If that list also had information about how many hogs each enterprise produced in the past, it would allow stratification of the list for increased precision in the estimate.

Lists of agricultural enterprises may be available through producers associations, producer oriented magazine publishers, or certain publicly available administrative record systems. Most such list sources are likely to have considerable "out-of-scope" records, and will also suffer from serious under-coverage of the target population. Local land and tax offices may have information on land owners that could be used to build a list frame of such owners.

Many governments provide financial and technical support for agricultural production and land stewardship through various programs, and create administrative record

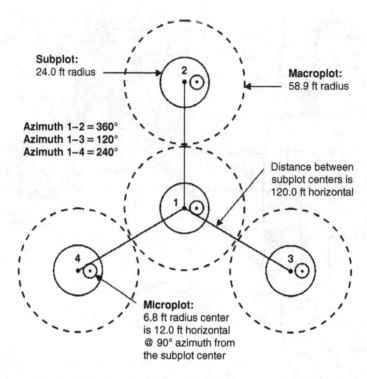

Subplot:
24.0 ft radius

Macroplot:
58.9 ft radius

Azimuth 1–2 = 360°
Azimuth 1–3 = 120°
Azimuth 1–4 = 240°

Distance between
subplot centers is
120.0 ft horizontal

Microplot:
6.8 ft radius center
is 12.0 ft horizontal
@ 90° azimuth from
the subplot center

Fig. 2. The Forest Inventory and Analysis plot design (USDA, 2007a).

systems to run these programs. If accessible, these records can be very valuable in sampling for agricultural surveys. (They can also be extremely valuable in direct estimation, as will be discussed later in this chapter.) Administrative records are created when farmers sign up for production support payments, when they receive government assistance for land or waste management enhancements, or when they apply for a license to build new facilities or to spray certain pesticides.

List sampling from administrative records is perhaps one of the most widely used methodologies for agricultural surveys worldwide. Canada conducts its Census of Agriculture every five years in conjunction with its Census of Population. The database generated from this activity creates a list sampling frame for agricultural surveys. This sampling frame may be updated for births and deaths, and augmented by specialty lists. Countries with centrally controlled economies (including those which have had such economies in the past) usually have extensive administrative systems of records. The Russian Federation, for example, bases its frame for surveys of rural agricultural households on its "Land Taxpayer Register." The only ancillary variable for stratification is total land area. South Africa historically used administrative data from its marketing board to generate information on agricultural production. As that country has moved to a free market economy, that administrative source has become much less complete as a sampling frame. Many of these countries are experimenting with area frames, remote sensing, and other methodological innovations. For example, South Africa is experimenting with a point sample area frame survey in which land use and crops are identified from an ultra-light aircraft.

Sampling from a list frame for general purpose agricultural production surveys can be very complex, even when a good sampling frame with ancillary data with size of operation and commodities produced is available. Often the intent is to measure agricultural production for numerous crop and livestock commodities within a single survey. For efficiency, larger producers (who can report for a larger percentage of the total commodity) should be sampled more heavily than smaller producers. Because farmers do not produce every commodity, care must be taken to allocate the sample so that sufficient responses are received to estimate all targeted commodities. Stratified sampling can be used for this purpose. Strata would be defined first by commodity, and then by the size of operation for those producing the commodity. Because many farming operations will generally produce more than one commodity, each could be eligible for inclusion in more than one stratum. A priority system may be used to assign each population unit to one and only one stratum. Such a system should give higher priority to the less common commodities.

The stratified sampling design described earlier can be effective when only a few commodities are targeted. When there are many commodities, a large number of strata are needed and the prioritization becomes more complex. An alternative sampling procedure for general purpose agricultural production surveys is discussed by Kott and Bailey (Bailey and Kott, 1997; Kott and Bailey, 2000). Their approach selects a Poisson sample (Ghosh et al., 2002) from multiple list frames and then uses calibration estimation. Kott and Bailey propose independently assigning a Poisson Permanent Random Number (PRN) to each population unit. They then create a separate list frame of farming operations for each commodity, creating multiple partially overlapping frames. A specific population unit would be in a commodity frame if ancillary information indicated that farming operation produced the commodity. Many population units would be in more than one commodity list frame. A targeted minimum sample size is determined for each commodity, and a probability proportional to size sample simultaneously drawn from each frame. Use of the permanent random numbers forces overlaps between the commodity samples, thus minimizing the total sample size. Kott and Bailey describe the process of how this overlap is forced, as well as how to calculate the probability of inclusion in the overall sample. In practical application, instead of having a separate sampling frame for each commodity, it may be more appropriate to group similar commodities (such as major row crops) within the same commodity frame.

Using a list frame and an area frame in a multiple frame design can be very effective for agricultural surveys. A well-developed list of enterprises with appropriate ancillary variables can provide a very efficient sample, but may also have considerable incompleteness. An area frame sample may be less efficient for collecting certain types of data, but offers complete coverage. Using standard multiple frame methodology, estimates from the overlap domain are weighted by the inverse standard errors for each frame. In practice, the standard errors using the area frame are often considerably larger than those of list frame and thus the overlap domain is estimated almost exclusively from the list frame.

Repeated observations over time in surveys that target units of land are possible, and often desirable, but will increase the complexity of sample designs (Fuller, 1999). Survey objectives that involve estimating both status at a particular point of time and trends over time are in conflict and need to be balanced. For surveys that target agricultural enterprises directly, longitudinal designs are problematic because individual farming enterprises change their structure over time. They go in and out of business, take on or dissolve partnerships, change their land holdings, and change names. Any longitudinal survey that requires human response needs to take into account respondent burden and the effect of frequent contact on both response rates and the quality of response.

Land surveys with area sampling frames and longitudinal data sets generate a different set of concerns, which are well expressed through the current design for the NRI. Pressures for more frequent NRI data led to a redesign in 2000 involving a shift to an annual survey. This replaced a design that involved observations on 300,000 area segments once every five years from 1982 to 1997. Research evaluating the tradeoffs between status and trend estimates resulted in a supplemented panel design in which 40,000 segments are observed every year along with a rotating panel of approximately 30,000 segments that augments the continuous panel (Breidt and Fuller, 1999). The rotation period varies across types of segments. Segments with points that have land/cover uses that are more likely to change or are of primary interest are given shorter rotation periods. Longer rotation periods are defined for other segments. The 1997 NRI is viewed as a first phase sample for subsequent annual samples for the purposes of estimation (Legg et al., 2006).

Land surveys may seek to produce a time series of information about land-based sample units for constructing tables of gross change (Fuller, 1999). Gross change tables provide estimates of change into and out of various categories of land, and offer a detailed understanding of the dynamics of change. Area data can be used for this purpose, but it is quite difficult to track polygons that represent changes in the extent of area features over time. A more practical method is to use repeated observations at a specific point on the land, which generates direct time series information to support gross change estimates.

3. Data collection

Data collection methods vary with the type of observation unit, the type and complexity of information to be collected, the time frame for data collection, and budget considerations. Many surveys require producer inputs and thus involve collecting data from farm operators. Responses may be collected either in person, by telephone, through the mail or via the web, to collect data on crop and livestock production, conservation practices, and/or pesticide use. As with the other household or business surveys, mail collection can be very economical, but may not yield the required response rates. Multimode surveys are common, using less-expensive mail or web surveys coupled with nonresponse follow-up on the telephone or with in-person visits. Web surveys have an important future for surveying agricultural producers and agri-businesses, especially as internet access becomes more consistent in rural areas. In 2005, 51% of all farmers in the United States, and 72% of commercial farmers, reported having internet access. However, most reported still using dial-up rather than high-speed service, and as yet most appear not to be inclined to respond to surveys over the web.

Most in-person agricultural surveys in the United States are completed using paper survey instruments, in part because of the difficulty of conducting interviews in outdoor farm environments. However, recent research indicates that as technologies for tablet computers improve, it will be feasible to conduct computer-assisted interviews for production surveys (Nusser, 2004). In-person interviews may focus on one or more parcels of land operated by the producer, and thus often rely on an aerial photograph to identify and document parcel boundaries. With the use of computer-assisted interviews, it will be possible to use GIS-based capture of parcel boundaries on digital photographic backdrops (Nusser, 2004). Such an approach also provides the opportunity for computer-based measurement of areas associated with fields and buffers, which may be more accurate than operator-reported areas and less burdensome to obtain.

Land-use surveys often involve field observations of attributes at points, along transects or within areas, including possibly the location and extent of geographic features. However, remote sensing and image analysis have become very effective and efficient means of collecting information on land usage and management. Data collected via remote sensing generally involve variables that can be directly observed using geographic information sources (e.g., aerial photographs, satellite images, topographic maps, soils maps). These may be the extent or surface area of geographic features on the land, or conditions at a specific line segment or point on the land. To supplement directly observable data, some elements, such as conservation program participation, may be obtained from administrative data, again not requiring contact with a human respondent. These approaches save respondent burden and data collection resources. However, even though the costs of remote sensing materials are far less than in the past, these costs still may be substantial, particularly if high-resolution materials are desired for detailed observations.

Computer-assisted data collection methods for land features are considerably more complex than standard questionnaires, but surveys are beginning to take advantage of advances in geographic information technologies. In recent years, the NRI has moved to directly capturing boundary delineations of features such as water bodies and building structures using custom-developed geographic data tools (Nusser, 2005). The geographic interface is integrated with more traditional computer-assisted forms to record land classifications. Prior to the availability of geographic tools in the survey instrument, data collectors manually measured abstracted summaries of features, such as areas and lengths. The geographic survey instrument now enables collection of raw information on the feature, and algorithms process the geographic data.

Although land use surveys rely more heavily on remote sensing imagery and analysis, they may still require on-site field measurements that are tied to remotely sensed delineation and classification of land areas. In 1996, the NRI began using handheld computers with formal computer-assisted survey instruments in the field with coordinates from the photo-interpretation survey displayed on a GPS receiver (Nusser and Thompson, 1997). Today, GPS can be provided as a resource, integrating easily with a mobile computer and digital imagery.

Repeated observations in longitudinal surveys introduce special problems for remotely sensed or field observations at a sample point. The key concept of repeated measurement is to return to the same physical location as data were collected in the past. It is tempting to consider the stored GPS coordinate as the gold standard for a sample or reporting unit's location, but positional error exists in GPS receivers and the coordinate systems of imagery and other geospatial data layers. These errors may exist even after the coordinates have been orthorectified (a process that aligns a geospatial data source to a standard coordinate system that adjusts for the presence of three dimensional terrain features). Material from the previous observation is required to ensure that the new observation is taken in the same location as the prior observation.

4. Statistical estimation

Statistical estimation encompasses data processing, imputation, weighting, and estimation of parameters and variances of parameter estimates. As with any survey, estimation approaches depend greatly upon the objectives of the survey. The choice of methods for surveys that release one-time results are generally less constrained than those that can be used in longitudinal panel surveys in which a series of related estimates are released over time. If inference for small areas is an objective, more complex estimation techniques are needed. Similarly, methods for creating public release data sets are more involved than those used for surveys where only estimates are released.

Even somewhat static estimation of agricultural production and related activities is complex because it generally involves the careful integration of information from various surveys (often collected at different times over the marketing year) and from administrative sources of varying quality. The goal of these endeavors is to estimate current year supply of an agricultural commodity, which can be used with information on demand for that commodity to provide appropriate transparency for proper functioning of the marketplace.

Vogel and Bange (1999) discuss the complexities of crop production estimates and forecasts produced by the U.S. Department of Agriculture and the need for a careful integration of survey and administrative data, both from domestic and foreign sources. The components of production are measured separately, and span the production year. First, planted acreage is estimated based on a multiple frame survey conducted in June. This initial estimate of planted area is subject to revision later in the season when administrative data is available from an USDA production support program. This administrative data provides the area planted to major crops by farming operations signed up for the support program. Because most field crop producers enroll in these programs, these estimates from administrative data must be integrated with survey estimates of planted area. As the season progresses, additional surveys are conducted to estimate area that will be harvested. This amount may vary greatly from year to year, and change throughout the production year, based on both weather and market conditions. Yield is forecasted monthly during the growing season using a combination of estimates from surveys of producers and from direct measurements of the crop in the field. Data from both the producer surveys and the direct measurement surveys are used in models that forecast yield per acre. At harvest, additional surveys (both surveys of producers and those involving direct measurements only) are conducted to estimate harvested area, yield and production.

Following harvest, additional administrative or survey data become available showing utilization of the crop production. For example, all cotton will be ginned. There is no other utilization for this crop. A monthly census of cotton gins around the country provide a critically important estimate of total cotton production that must be integrated with estimates from earlier production surveys to improve the overall estimate of cotton production. Corn is even more complex because some corn enters the market (for food, feed, or ethanol production) and some is fed on-farm to livestock. Corn may be stored for some time without further processing. Administrative and survey data providing estimates of corn exports, corn imports, corn storage, and corn fed to animals are all utilized in an estimation "balance sheet" (see Figure 3) with the initial corn production

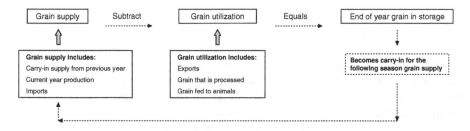

Fig. 3. Balance sheet for grain production and use.

estimates to ensure that the total estimates of corn production, marketing and utilization are consistent for the marketing year.

An even more complex assemblage of methods is required when surveys release related estimates over time and/or of microdata that can be further explored by policy analysts and researchers. The NRI survey program periodically releases estimates of conditions and trends for natural resources as well as public use data sets that have a complete time series for all points in the data set. Numerous issues are addressed in the estimation process. These include ensuring sensible temporal patterns for observed and imputed point data, integrating known trends from external data sources such as program participation data and land areas, appropriately reflecting these trends in small area estimates, and promoting consistency with estimates generated from previous releases. These problems require sophisticated estimation methods and if data are to be released, estimation methods must also create a data set that is simple for analysts to generate routine statistics and their estimated variances.

To address these issues, the NRI uses several strategies. In data review, historical time trends are given special attention to evaluate whether rare or impossible time series have been created in the classification sequences for a point. However, it can be difficult to identify all unusual or unexpected time series, or even inconsistent values for suites of related variables. A recent advance was made by Wang and Opsomer (2006), who describe the use of cluster algorithms to identify unusual combinations of values or unusual trends over time. This approach may reduce the need to create specific computer checks or manually review data for unusual trends. Small changes can also occur in individual time series for continuous measures that are entirely due to measurement error and do not represent meaningful changes. Smoothing procedures have been developed to minimize the impact of such fluctuations.

To improve small area estimates, an imputation approach is used to ensure area data known to correspond to segments or watersheds within counties are geographically allocated to these same locations for subpopulation estimates (Breidt et al., 1996; Nusser and Goebel, 1997). NRI sample area segments can be viewed as representing the first phase in a two-phase sample with points as the second phase. However, instead of using segment data in a standard two-phase weighting approach, segment area data are used to impute points that reflect observed segment changes when such patterns are not observed in the segment's points. Weights that reflect the size of segment areas corresponding to specific time trends are assigned to imputed and observed points that match conditions observed in the segment data. Local attribution of these areas improves the small area properties of estimates relative to using a standard two-phase estimator (Breidt et al., 1996).

As mentioned earlier, it is important that NRI–generated estimates are consistent over time. It is equally important that they are consistent with information from certain administrative record systems (such as that tabulating area enrolled in the Conservation Reserve Program). In NRI, consistency with prior estimates or administrative totals is handled via raking and ratio estimation methods.

Recent research in the NRI has considered an estimated generalized least squares estimator (EGLSE) to incorporate correlations across sample panels and the full 1997 NRI sample (Legg et al., 2005, 2006). These estimators are consistent and asymptotically as efficient as generalized least squares estimators, which have minimum variance among the class of linear unbiased estimators. EGLSEs are used in a ratio step near the end of the weighting process to improve the properties of estimates for key variables.

Because of the complexity of the NRI estimation process, replication methods are used for variance estimation. Replicates are created for delete-a-group variance estimation (Kott, 2001; Lu et al., 2006) that attempt to reflect both sampling variance and variance due to estimation methods such as point imputation. Another approach being investigated, but not yet implemented, is fractional imputation (Kim and Fuller, 2004). In fractional imputation, multiple imputation outcomes are generated for a point and the weight associated with the point prior to imputation is divided equally among the imputation realizations for the point. This approach is expected to provide a better estimate of variance due to imputation than the current method.

The National Agricultural Statistics Service also uses satellite data to improve the precision of estimates of crop acreage, especially at the county level (Allen et al., 2002). Studies have shown that relative efficiencies of 3.0 or more have been achieved over the area frame estimate of planted acreage. However, Allen et al. (2002) point to complexities in this process: "There is a common misunderstanding that crop type signatures are so unique that they could be determined once and for all. Then later classifications would be a matter of running a new satellite data file against known parameters. This is called signature extension. Satellite-based crop classification is based on the measurement of energy emitted or reflected by plants. Those readings do differ somewhat from one crop type to others in different wavelengths. However, that pattern differs throughout the growing season of a particular crop. There can also be considerable differences between healthy plants and plants of the same crop but under serious stress. The density of crop planting and the presence or absence of weeds and recent precipitation also can affect crop response. On top of all other factors, the atmosphere through which the crop response is being measured is not the same from one day to the next."

Because most agricultural surveys are inherently linked to the land, there is great pressure to use the geospatial locations of sample points, even if they are not publicly available. Research agreements can be used as one vehicle to enable point-specific modeling. Agencies are also looking towards the possibility of using image classification in creating map-based products. Supervised image classification involves using ground truth (or reference) data in combination with satellite imagery to produce a data layer depicting a specific theme, such as land cover or crop cover. For example, in selected states, NASS has created a popular data product called the Cropland Data Layer, which is a public use GIS data file with crop specific categorization (at 30 m resolution) for each crop season. This data layer is used in conjunction with other GIS layers to aid in watershed monitoring, soils utilization analysis, agribusiness planning, crop rotation practices analysis, animal habitat monitoring, and prairie water pothole monitoring (Allen et al., 2002).

5. Confidentiality

Maintaining the confidentiality of individual survey responses is a critical part of the survey estimation process for agricultural and land based surveys, as it is for surveys of other populations. Disclosure avoidance issues for farms mirrors many of the issues with surveys of businesses (see Chapter 4). Namely, there are large or specialized producers that are easily identifiable even when their responses are combined with other responses. Thus disclosure avoidance programs and processes must ensure an appropriate number of responses in an estimation cell, but also must ensure that one or two responses do not individually account for a predominant portion of the cell total.

Confidentiality methods for surveys that involve observations of the land are simpler than for surveys that involve respondents. Coordinates of sample points are not publicly released in order to protect land owners and to preserve the integrity of plots that are revisited over time. Thus, the primary concern in these surveys is to evaluate disclosure risks within geographic polygons created with classification variables such as county, watershed or eco-region. The NRI combines polygons with small sample sizes with adjacent polygons to reduce disclosure risk. As the time series at a point becomes more extensive, additional disclosure limitation methods may be needed.

6. Concluding remarks

Agriculture and land utilization issues sit at the vertex of many global concerns and conflicts. The disagreement over governmental subsidy programs for agricultural production is a major barrier to negotiating international free trade agreements. There are ongoing concerns about shortages of food in drought stricken parts of the world. Global concerns about human health extend to food-borne diseases such as bovine spongiform encephalopathy (BSE) in cattle, avian influenza, and *E. coli*. There is global concern about the clearing of forestland for agriculture and about the conversion of agricultural land to development. One common thread through all of these concerns is the need for high quality information with which to make policy and trade decisions. Agricultural and land-based surveys will continue to be an important tool for information gathering in the foreseeable future.

There are a number of methodological opportunities and challenges ahead for these surveys. From the data user viewpoint, there are several critical needs. Because a single survey will never encompass the breadth of interacting components of the agricultural sector, there is an increasing need for linking data from surveys and other administrative sources together in appropriate ways to support more global analyses. There is an increasing need to link data of all types back to a land base, and then to be able to layer that data within a GIS to draw inferences based on increasingly smaller geographic areas. There is an increasing need for improved methods of small area estimation in general. This extends not only to facilitate analysis of small geographic areas, but also because agricultural commodities are becoming so specialized that finer and finer breakouts of production are required. There is an increasing need to look at change over time, and to create data and techniques that are appropriate for longitudinal analysis. Finally, data users want more access to disaggregate data, they want online access to that data, and they want online tools available with the data sets that will allow them to perform a wide range of analysis.

From the data providers' viewpoint, there are equally critical needs ahead. In developed countries there continues to be fewer enterprises engaged in commercial agriculture. Thus, the population base from which to sample for agricultural surveys is shrinking while the need for data from that population is growing. Methodological challenges before us are to minimize the reporting burden by improving electronic reporting capabilities, increasing the utility of other sources of information, and developing and improving various modeling techniques that require less input data. Another critical need for data providers is to maintain the confidentiality of their individual responses and of remotely sensed data associated with a producer's land. There are significant methodological changes ahead as we try to meet data user needs of more online access to disaggregate data from a shrinking population that is becoming more and more concerned about privacy. From a statistical perspective, these requirements ultimately lead to more complex sampling and estimation methods, particularly for surveys conducted over time, which must be balanced with the need to provide easily accessible and usable data to researchers and the public.

Acknowledgments

Some of the research discussed in this chapter was supported in part by Cooperative Agreement No. 68-3A75-4-122 with the USDA Natural Resources Conservation Service, U.S. Department of Agriculture. Any opinions, findings, and conclusions or recommendations expressed in this material are those of the authors and do neither necessarily reflect the views of the Natural Resources Conservation Service, the National Agricultural Statistics Service, nor the U.S. Department of Agriculture.

Essential Methods for Design Based Sample Surveys
ISSN: 0169-7161
© 2009 Elsevier B.V. All rights reserved
DOI: 10.1016/B978-0-444-53734-8.00008-X

8

Sampling and Inference in Environmental Surveys

David A. Marker and Don L. Stevens Jr.

1. Introduction

In this chapter, we focus on surveys of environmental resources, which we loosely define as the air, water, soil, and associated biota that sustain our environment. The objective of the surveys we consider will generally be an assessment of status, condition, or extent of a resource. The target population of the survey may be discrete and finite, for example, small lakes or wetlands, with well-defined population units; or may be a one-, two-, or three-dimensional continuum, for example, a stream network, a forest, or the volume of water in a large lake. Each of these calls for different types of frames and sampling techniques. The survey may be a one-time assessment or may include a long-term monitoring objective to assess change or trend. Addressing both objectives requires a balance of revisiting sites to assess trend and adding new sites to assess status.

Traditionally, the focus of sampling in the environmental sciences has been on relatively small and well-delimited systems, e.g., at the scale of a lake or watershed or forest stand. However, some current environmental issues, such as global warming, contamination of surface and ground water by pesticides and other pollutants, and extensive landscape alteration are not localized. Quantifying the extent of symptoms of widespread concerns requires large-scale study efforts, which in turn needs environmental sampling techniques and methodology that are formulated to address regional, continental, and global environmental issues.

Survey design is a well-developed and established area in the statistical literature. There are many textbooks that provide excellent accounts of the essential attributes of good survey design, such as the necessity of clear definitions of the population of interest, the sample units, the sample frame, and how the sample is to be drawn (Cassel et al., 1993b; Cochran, 1977; Kish, 1967; Lohr, 1999; Särndal et al., 1992; Thompson, 2002; Yates, 1981). However, designs for environmental sampling often present additional challenges which we identify below. These include the need for broad population description; spatial context of the population; availability of ancillary information; inadequate frames; difficult access; multiple objectives, including status and trend; evolving objectives; and the need to satisfy multiple stakeholders.

The focus of most survey methodology is estimating the mean value or total of a population. In contrast, an environmental survey often has a more general object, such as estimating the distribution function or the proportion of the population in various classes, for example, the proportion of lakes that meet designated use criteria. There may be many environmental responses of interest that are interdependent. A common objective of environmental surveys is to characterize the status of some resource as well as the change or trend in that status. These two objectives have somewhat conflicting design criteria: status is generally best assessed by sampling as much of the resource as possible, whereas trends are generally best detected by observing the same resource locations over time. Frequently, a secondary objective is the evaluation of relationships between attributes, both measured at the site and available on the frame.

A characteristic of overwhelming importance for environmental populations is that they exist in a spatial context. The response will have spatial pattern and structure. Sites near to one another will tend to have similar physical substrate and be subjected to similar stressors, both natural and anthropogenic. Response can be influenced by topography, hydrology, and metrology. All these influences will tend to induce spatial patterns in the response.

Another important characteristic of environmental populations is that some ancillary information (in addition to location) is almost always available. Currently, there is a wealth of remotely sensed information available from satellites or aerial photography that may be used to structure the sampling design or used in analysis.

Environmental resources are often expensive and time-consuming to sample. Logistics can be difficult; often, the population of interest includes sites in remote locations, for example, lakes in wilderness areas. There can be considerable time and money expended in traveling between sites. Laboratory costs for analyzing individual samples may be nontrivial. Some environmental metrics can be time-consuming to evaluate, for example, quantifying the species richness, and abundance of a macroinvertebrate sample requires the services of a skilled benthic taxonomist. For a large program, it may be a year or more after data are collected in the field before laboratory analyses are available. These may also be subject to substantially more measurement error than routinely found in other types of surveys. Nonresponse in environmental sampling can be substantial for reasons such as ease of physical access, safety, or permission.

A practical complication frequently encountered in environmental sampling is the difficulty in obtaining an accurate sampling frame. In many instances, available sampling frames include a substantial portion of nontarget elements or fail to cover the entire population. The frame problem is aggravated by the sheer difficulty of collecting and analyzing samples.

Environmental sampling almost always occurs with a backdrop of political, economical, and societal considerations so that statistical considerations represent only one aspect of a sampling design. Furthermore, because environmental issues can impact human populations, there are often multiple groups, agencies, and organizations that have an interest in the products of the survey. The interests of the multiple stakeholders are not always perfectly aligned. Meeting the interests of multiple stakeholders, while maintaining a scientifically and statistically rigorous design, can be a challenge.

In many instances, the need for an environmental sample will be driven by the need to assess the condition of an environmental resource because of concern over potential degradation. The design needs to address current environmental issues but that is not sufficient. The current issues will eventually be resolved, but new, presently unrecognized issues will emerge. These issues will manifest themselves in unforeseeable ways, and they will affect resources that cannot now be identified. An environmental assessment program with the dual objectives of status and trend must be able to accommodate regrouping, recombining, expansion and contraction of the sample to permit such emerging issues, and evolving objectives to be addressed. The issues of inadequate frames, nonresponse and missing data, and evolving objectives drive a need for sampling designs with the flexibility to add, remove, or reallocate samples.

Below, we review some of the sampling methodology that has been developed to meet the challenges that sampling environmental populations present: focus on broad population description; spatial context; ancillary information; inadequate frames; difficult access; evolving objectives; and the need to satisfy multiple objectives and stakeholders.

2. Sampling populations in space

Historically, many environmental samples have been chosen for convenience or subjectively to be representative. Both of these selection methods have severe shortcomings (Paulsen et al., 1998; Peterson et al., 1999). There are two widely accepted, statistically and scientifically rigorous approaches to selecting an environmental sample: probability-based and model-based. These two approaches begin from different theoretical bases but can both address the common working objective. In probability-based sampling, the response is viewed as fixed but unknown. A model-based approach views the response as one realization of a random process. In environmental sampling, we are usually interested in an attribute of an environmental resource, and a probability-based design objective is to estimate that attribute. A parallel model-based design objective is to predict the outcome of the process and to calculate the attribute from the predicted outcome. The focus in this chapter is on probability-based design and inference. However, there are some insights from model-based optimal design that are relevant to probability-based sampling.

Because space has a central role in environmental sampling, much of the relevant theory and practice has dealt with spatial sampling. A number of authors have investigated designs for sampling in space. The papers that have a sampling theoretic orientation tend to consider only finite populations; some of those with a spatial statistics model-based orientation consider continuous populations. Overton and Stehman (1993) compared three designs (systematic (SYS), simple random sampling (SRS), and random tessellation stratified (RTS)), then contrasted them in terms of their precision and variance-estimation properties. They concluded that the designs ranked RTS<SYS<SRS in order of increasing variance. Many investigations of two-dimensional sampling have taken a superpopulation approach (Cochran, 1946; Das, 1950; Quenouille, 1949). Matérn (1986) investigated sampling in continuous two-dimensional space and derived some comparisons of stratified and systematic sampling, using several systematic arrangements and spatial covariance functions. He concluded that a systematic sample on a triangular grid was optimum for a wide class of nonincreasing, isotropic covariance functions. Olea (1984) compared several variations on systematic designs that give good sample dispersion yet avoid the potential problems with periodicity that strict systematic designs have. Iachan (1985) derived some asymptotic comparisons of two-dimensional sampling designs and extended Cochran's (1946)

result that under some restrictions on the covariance function, systematic sampling is more efficient than stratified sampling, which in turn is more efficient than simple random sampling. Dalenius et al. (1961) showed that with some restrictions on the spatial covariance, sample designs using a triangular grid are optimal, a result supported by later work by McBratney et al. (1981) and Yfantis et al. (1987).

3. Defining sample frames for environmental populations

Probability-based sample designs were defined in Chapter 1. The first characteristic of a probability sample is that each unit in the population must have an explicit definition. That definition is used to develop a frame for the population, that is, a construct from which population elements can be selected via a random process. The construction of a frame is a first step for any probability sample, but environmental frames can take some different forms than that usually found in survey methodology.

In the case of discrete environmental resources with distinct sampling units (such as lakes), it would be possible in concept to develop an exhaustive list of all elements of the resource. The Eastern Lake Survey (ELS) (Linthurst et al., 1986) and the Western Lake Survey (WLS) (Landers et al., 1987), which were conducted by the EPA as a part of the National Acid Precipitation Assessment Program (NAPAP), took this approach. The frame for these surveys was developed by listing each lake in the target region on U.S. Geological Survey (USGS) 1:250,000-scale topographic maps.

A serious drawback to a list frame is the amount of time required to construct the list. Even in those fortunate circumstances when a nearly ready-made frame exists, considerable effort must be expended to verify that the frame completely covers the population of interest without excessive inclusion of nontarget populations. In the ELS, for example, investigators discovered that many bodies of water represented as lakes on 1:250,000-scale maps were in fact bogs, intermittently flooded areas, or wide spots in a stream. These nontarget units can be eliminated in the sampling, but they do complicate population estimation procedures. There is no easy way to compensate for units in the target population that were omitted from the list.

Another strategy for sampling environmental resources is to develop an area sampling design based upon a single-area sampling frame. In such a frame, the entire region to be monitored (e.g., the conterminous United States) is partitioned into a set of mutually exclusive and exhaustive areas. These areas are frequently designated primary sampling units (PSUs). The partition can be based on arbitrary geometric figures or on some characteristic of the landscape, such as the USGS hydrologic cataloging units. Commonly, PSUs are chosen with boundaries that are easily discernible in the field, such as permanent roads, railroads, or rivers. A sample is selected from these PSUs according to a probability-based protocol, such as selecting a PSU with a probability proportional to its size. Usually, some restriction is imposed on the sample selection to ensure spatial dispersion of the sample. The resources occurring in each sample PSU are identified, characterized, and measured.

Land use/land cover databases are available that cover the entire United States. These databases can provide ancillary information, such as land use, land form, soils classification, vegetation cover, hydrology, and human-induced modifications that can be used to structure a sampling design. Additional information is available from existing meteorological databases.

The National Hydrography Database (NHD; USGS, 2000) is arguably the best information available for water bodies in the United States, but it is not an ideal frame. Although attributes within NHD can be used to identify a subset of NHD that more closely matches the target, the subset may still include many nontarget entries. For example, suppose the original goal was to describe the population of lakes in California that support fish populations; the NHD has over 9000 entries identified as lakes in California. Many of these are very small (over 1000 are less than 1 ha, and over 5000 are less than 5 ha). Many of those smaller entries identified as lakes are in fact not lakes; they are farm ponds, sewage-treatment holding ponds, or intermittent seepage basins that are dry most of the year.

Conversely, the best available frame may miss some of the target resource. A geographic information systems (GIS) coverage was most likely based on aerial photos, and the utility of the coverage as a population frame is dependent on the skill of the photo interpreter and cartographer. The best source of frame information for wetlands in the United States is the National Wetland Inventory (NWI; USFWS, 2002; http://www.fws.gov/nwi/). However, it can be very difficult to identify wetlands in forested areas because of canopy cover. As many as of 50% of wetlands can be missed in the NWI (Brooks et al., 1999).

A grid can be used to frame a population that is distributed over some spatial extent by superimposing a grid over a representation of the spatial extent and then sampling at or around each grid point. Randomized systematic grids have a long history of application in environmental sampling. At the national level, they have been used extensively by the National Forest System (NFS) and Forest Inventory and Analysis (FIA) (Bickford et al., 1963; Gillespie, 1999; Hazard and Law, 1989) to sample forest growth and production. The National Stream Survey (Kaufmann et al., 1988; Messer et al., 1986) also used a grid frame to locate stream segments for sampling.

With the availability of GIS, electronic representations of maps are becoming more common for environmental populations. The NHD is the most complete information to be had on the extent and location of aquatic resources in the United States. The U.S. EPA's Environmental Monitoring and Assessment Program (EMAP; Messer et al., 1991; http://www.epa.gov/emap/) uses the NHD as a preferred frame (Angradi, 2006). With a GIS representation of a frame, the population can be viewed as finite or as a continuum of points. For example, a list of coordinates of lakes can be obtained from the NHD. Conversely, a forest, a large estuary, or a stream network may be treated as continua. Sample sites can be identified by choosing points from the GIS representation.

The use of a GIS can facilitate the preservation of the spatial context of the sample points. At a minimum, spatial context is the information required to locate a sample point on the landscape, for example, latitude and longitude. However, there is a richer connotation in all the available landscape information that is also attached to geographic coordinates: ecoregion, land use, soil topography, and so on. Knowing the spatial context of a sample from a resource, that is, knowing where the samples are located, and knowing their spatial relationship to one another provides the link of proximity to admit the joint evaluation of multiple resources and to evaluate the effects of stresses with known spatial properties.

A cautionary note on the combining of GIS data from multiple sources is that different levels of accuracy can result in inconsistencies in the derived sampling frame. It is important when combining multiple data sources to check for unacceptable combinations and clean these before proceeding.

4. Designs for probability-based environmental samples

The simplest and easiest to implement sampling method is simple random sampling (SRS) (Cochran, 1977). Simple formulae apply to population attribute and variance estimation. However, SRS is rarely appropriate because auxiliary information and prior knowledge about the population are not used in the selection. SRS samples will be inefficient compared to methods that do utilize knowledge about population characteristics or structure.

4.1. Multistage designs

In a multistage cluster design, an area selected at a given stage is further split into subareas, and a sample is selected from the subareas. Complete characterization and measurement take place only at the lowest order set of areas. This is essentially the design used by the National Agricultural Statistics Service (NASS http://www.nass.usda.gov/research/AFS.htm; Cotter and Nealon, 1987; Mazur and Cotter, 1991) in their June Enumerative Survey of national agricultural production, where each sampled PSU is split into secondary sampling units called segments. Field visits are made to a sample of segments.

The National Resource Conservation Service (NRCS) (formerly the Soil Conservation Service (SCS)) also uses a two-stage design in several national resource surveys (Goebel, 1998; Goebel and Schmude, 1982; Nusser and Goebel, 1997). The 1958 Conservation Needs Inventory (CNI) used a frame based on 100-acre squares of land in the northeastern states and partitions of public land survey sections (approximately 640 acres) in the rest of the country. The 1967 CNI treated the 1958 sample areas as PSUs and subsampled within them at specific points. The 1977 National Resource Inventory (NRI) also used the 1958 CNI area frame and a two-stage sample. The 1982 NRI also used a two-stage area frame based on public land survey sections in most cases. A similar design was used in 1987 and 1992 (Goebel, 1998).

4.1.1. Spatially constrained designs
Some prior knowledge about environmental populations is always available. We may have reason to believe that the response is influenced by or is related to a variable for which we have complete information, for example, from remote sensing techniques. One important item of information that is always available for environmental populations is location. As noted in the introduction, environmental populations invariably exhibit spatial structure and pattern.

The advantage of spatial control accrues from the tendency of elements of an environmental population that are near one another to be more similar than elements that are far apart. Observations of elements that are near one another contain redundant information. Thus, samples that are well dispersed over the population domain tend to lead to more precise estimates of population attributes than samples without spatial control. The advantage of a spatially dispersed sample has long been recognized; accordingly, there are many techniques for achieving that dispersion, including area sampling, spatial stratification, systematic and grid-based sampling, spatially structured list frames, and spatially balanced designs.

4.1.2. Spatially balanced designs

The notion of a balanced sample was introduced by Yates (1946). A sample of Z is *balanced* over an auxiliary variable X if the x-values (which are known beforehand) are chosen so that the sample mean of the x-values is exactly equal to the true population mean of X. A stricter version of balance was suggested by Royall and Herson (1973), who required that the first several sample moments of the x-sample match the population moments. The intuition behind balancing is that the auxiliary variable is correlated with the unknown response to be assessed. By balancing over the auxiliary variable, we hope to get approximate balance over the unknown response and hence to get a more precise estimate than SRS would give. Kott (1986) noted that an option intermediate between random sampling and strict balancing can be obtained by splitting the range of X into n quantiles and picking one sample element in each quantile. Although this option does not achieve balance in the strict sense of having sample moments match population moments, it does guarantee that the sample distribution function of X will be close to the true distribution function for every sample draw. This is the idea behind all stratified sample designs. Because of the correlation between X and Z, the hope is that the sample of Z will be more precise.

If the ancillary variable is location, then we define a sample to be *spatially balanced* if the spatial moments of the sample locations match the spatial moments of the population. The first two spatial moments are the center of gravity and the inertia. The center of gravity for a region R is given by the ordered pair (μ_x, μ_y), where μ_x, the moment about the y-axis, is given by $\mu_x = \int_{-\infty}^{\infty} x \nu_y(x) dx$. The function $\nu_y(x)$ is the extent of the cross section of R at the point x and is given by $\nu_y(x) = \int_{-\infty}^{\infty} I_{\{w|(x,w)\in R\}}(y) dy$. Similar definitions hold for μ_y and ν_x. The second spatial moment is analogous to the covariance matrix and measures the regularity of the shape of R or of the point pattern formed by the sample points.

In general, a probability sample will not achieve exact spatial balance, but approximate spatial balance is a worthwhile goal. Also, the discrepancy between the sample moments and the population moments can be used as a measure of spatial balance of a sample. The techniques described below for achieving spatial control all do better at achieving spatial balance than does SRS.

4.2. Stratified designs

Sample designs can almost always be improved by introducing stratification. As discussed in Chapter 1, frames can be stratified to assure representation in the sample for units with particular characteristics or to improve precision of estimates. Stratification by analytic domains of interest can assure representation of each domain, thereby improving small area estimation (see Marker, 2001). If units within strata are more homogeneous than the population as a whole, then stratified designs (with corresponding estimators) can improve survey precision. While many environmental surveys use stratified list sample designs similar to those discussed in earlier chapters, the remainder of this section describes other types of stratified designs more common in environmental settings.

4.2.1. Systematic sampling

For some kinds of environmental resources, systematic sampling is an attractive means of achieving spatial dispersion. For a two-dimensional, extensive resource, for example, a forest, a systematic sample can be obtained by placing a grid over a map of the resource and selecting the center points of grid cells or intersections of grid lines as sample points. Olea (1984) discusses several alternate ways of picking points in grid cells so that strict alignment is avoided. Randomness can be achieved by random placement of the grid. For a stream network, sample points could be picked at regular intervals along the network, starting at the outflow and working upstream. A rule for how to proceed at confluences would be needed.

A potential drawback of systematic samples is that they can align with natural or anthropogenic features of the landscape. If those features also influence the response, then high variability of estimators can result. This phenomenon is usually cited in relation to periodic or near-periodic responses but can also occur in responses with a mosaic structure. Another shortcoming with a systematic sample is its inflexibility. Frame errors or inaccessible sites are not easily accommodated, nor is variable probability. A sample can be locally intensified, say by halving the grid spacing, but there are a limited number of intensification factors available (Dacey, 1964; Hudson, 1967). Finally, systematic samples do not yield unbiased variance estimation formulas.

4.2.2. Spatially stratified designs

One of the most popular means of achieving a spatially dispersed sample is through the use of spatial stratification. Strata are defined to be disjoint polygons that tile or tessellate the target domain. Strata can be regular geometric figures such as grid cells; arbitrary polygons such as ecoregions; political boundaries such as state or county borders; or natural boundaries such as drainage basins. Maximal spatial balance will generally be achieved by maximal dispersion over the domain, which in turn will be obtained by choosing strata with few samples per stratum. The aim of defining the strata should be to have an equal amount of the resource (number, length, area) and equal number of samples in each stratum, resulting in an equiprobable design. Commonly, samples are selected within strata using SRS, but other techniques could be used. If the design is not equiprobable, then the aim should still be to have a constant number of samples per stratum, but the amount of the resource per stratum will vary.

Maximal stratification achieves good spatial control, but having only a few samples per stratum limits flexibility. Given the difficulties of environmental sampling, it would be quite possible to lose all the samples from a stratum because of inaccessibility. If lost samples were replaced, as could be done if SRS is used for within stratum selection, then the inclusion probability could be substantially different for that stratum.

Forming strata with equal size and equal number of samples is usually straightforward for equiprobable designs and two-dimensional target resources but can be problematic for finite resources of unequal size. For example, lakes are often treated as finite populations for sampling purposes. The size distribution of lakes is heavily skewed toward small lakes (Larsen et al., 1994; Stevens, 1994). An equiprobable sample would result in mostly very small lakes, which are not likely to be the lakes that are of most interest (e.g., the ones that are accessible, support recreational or commercial fisheries, support other recreational use, have developed shorelines, or are subject to development impacts). Also, the spatial pattern of lakes tends to be clumped rather than uniform. To reap a benefit from spatial stratification, the strata should encompass a more or less homogeneous area of spatial influence, for example, land use/land cover, terrain, ecoregions, and anthropogenic impacts. This suggests strata that are spatially compact, with small perimeter to area ratio. Forming such strata (spatially compact with equal number of samples) can be difficult. Stevens (1994) describes an algorithm used by EMAP to form spatial strata of lakes.

4.2.3. Random tessellation stratified designs

A compromise between SRS and SYS designs is a RTS design. An RTS design is implemented by randomly placing a grid over the population domain and selecting one point at random in each grid cell. See Olea (1984) or Overton and Stehman (1996) for a discussion of RTS designs. For a linear resource, the design is implemented by systematically dividing the resource into units with equal length and then picking a point at random in each unit. It can also be applied to a finite resource by picking one unit at random from the units covered by a grid cell. Although the RTS design does give good spatial dispersion, it also suffers from the same lack of flexibility and unbiased variance estimator that a systematic design does.

4.2.4. Spatial address techniques

One method that has been used to disperse points in space is to induce a linear order on points in two-dimensional space, apply that order to the population elements, and then use systematic sampling along the ordered population. For example, NASS (Cotter and Nealon, 1987; Mazur and Cotter, 1991) has used a serpentine order to arrange the PSUs. Saalfeld (1991) discussed a method for sampling a connected tree structure, such as a stream network, by starting at the base of the network, tracing up one side (following all tributaries to their source) and then down the other side to the point of beginning. The resulting path traces each stream segment on both sides and is thus twice as long as the total length of the network so that every point on the network is mapped to two points on the path. A systematic sample along the path will have good spatial dispersion.

Some methods for creating spatial addresses are related to the concept of space-filling curves, such as first constructed by Peano (1890). Wolter and Harter (1990) have used a construction similar to Peano's to construct a "Peano key" to maintain the spatial dispersion of a sample as the underlying population experiences births or deaths.

The Peano key is an example of a spatial address created via a quadrant recursive function (Mark, 1990). Without loss of generality, we can assume that the two-dimensional population domain has been scaled and translated into the unit square. A quadrant recursive (q-r) function maps the unit square onto the unit interval and has the property that subquadrants of any order are mapped onto subintervals. This property preserves some two-dimensional proximity relationships in the one-dimensional image (Mark, 1990). As the name implies, a q-r function is defined recursively. We illustrate the construction of a q-r function with the Peano key. First, divide the unit square into four quadrants, which are labeled 0 through 3, beginning in the lower left, proceeding up, diagonally down, and then up to end in the upper right (see Fig. 1). The second step then divides each quadrant into four subquadrants labeled in the same order. Successive steps continue the subdivision process to smaller and smaller subquadrants. Figure 1 illustrates the process, with the subdivision carried out only in the first subquadrants. A spatial address is constructed by joining the labels attached to the subquadrants, beginning with the first division and proceeding down the chain, and treating the resulting number as a base 4 fraction. Thus, every point in the crosshatched subquadrant in Fig. 1 will get an address beginning with 0.001_4. If this process was carried out indefinitely, then the limit is a measurable, 1-1, onto function from the unit square to the unit interval (Stevens and Olsen, 2004).

The basic quadrant recursive function is made into a random map by randomly and independently permuting the order in which labels are attached to the quadrants, at every possible opportunity. This randomization, termed hierarchical randomization (Stevens and Olsen, 1999), preserves the quadrant recursive nature of the map.

Stevens and Olsen (2004) use recursive partitioning to develop a very general technique, the Generalized Random Tessellation Stratified (GRTS) method, for selecting approximately spatially balanced designs. The concept underlying GRTS is to apply recursive partitioning to create a spatial address. At each step in the recursion, the total inclusion probability for each cell is computed as the sum or integral of the inclusion probability of all population elements within the cell. The inclusion probability need not be constant and very general variable probability designs can be accommodated.

The recursion is continued until every cell has total inclusion probability less than one and then hierarchical randomization is applied. The process is illustrated in Fig. 2. Part (A) of the figure shows the q-r address of the first 16 cells, with a line connecting the cells following the q-r order. In Fig. 2(B), the path connecting the cells follows a

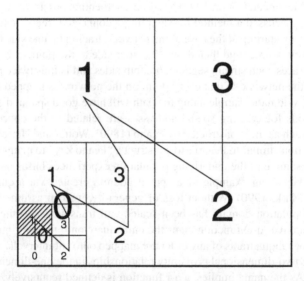

Fig. 1. Illustration of the recursive partitioning steps in construction of the Peano key.

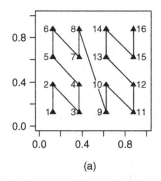
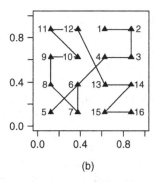

(a)   (b)

Fig. 2. Example of GRTS q-r addressing and sample location. (a) Nonrandomized q-r address for the first two levels. (b) Hierarchically randomized q-r address.

hierarchically randomized order. Each cell is assigned a length equal to its inclusion probability, and then the lengths are strung together, forming a line with length equal to the total sample size. A systematic sample is selected along the line. Because of the 1-1 nature of a quadrant-recursive map, every point on the line corresponds to some population element, so the selected points on the line can be mapped back to specific population elements. Details are given in Stevens and Olsen (2004).

One additional step gives tremendous flexibility to the GRTS technique. The samples, as selected, will appear in the proximity-preserving order inherited from the randomized q-r address. Stevens and Olsen (2004) show how to order the sample points so that any consecutive subsequence of the sequence has good spatial balance. This property allows adjustment of sample size based on field experience or changing priorities, adjustment for nontarget sites, and formation of interpenetrating temporal panels.

4.3. Other sample designs

4.3.1. Adaptive sampling

Some environmental populations have spatial structure that makes them difficult to sample efficiently, even when using stratagems for spatial balance. For example, natural populations frequently exhibit clustering: individuals of the same type or species tend to group together. One potential technique for improving sample efficiency for clustered populations is adaptive sampling (Thompson, 1990, 1991b). Adaptive sampling allows one to modify the sample based on information as it is collected. The basic idea is best illustrated with an example. Suppose that a regular square grid has been placed over the domain of some clustered population. Further, suppose that the clusters tend to be of a size that covers several grid cells so that the grid cell area is substantially smaller than the average cluster size. An initial sample of grid cells is selected, the cells in the initial sample are visited, and the response (e.g., the number of individual members of the target species in the grid cell) is recorded for each cell. If the response meets some criteria (e.g., number of observed individuals is positive, or greater than some number), then adjacent cells are added to the sample. This sequence of observation/ augmentation is continued until no newly observed cell meets the criteria of triggering augmentation.

The resulting sample presents some analysis difficulties because the inclusion probability of a cell is impossible to calculate without complete knowledge of the population structure, which is not available. Thompson (1990) shows how to obtain some modified weights that permits unbiased estimates of the total using an estimator similar to the Horvitz–Thompson estimator. Christman (1997) compares the efficiency of several designs for sampling clustered populations and concludes that adaptive sampling is an efficient sampling scheme for rare, tightly clustered populations (Chao and Thompson, 2001).

There are also some difficulties in applying adaptive sampling in the field. The rule for adding to the sample must be formulated prior to beginning sampling and must be followed in the field. In particular, new neighboring sites must be added so long as the site just observed meets the criteria. Some investigators have reported that has lead to unmanageable sample sizes (Hanselman et al., 2003; Kimura and Somerton, 2006). Thompson (2006) has recently extended the allowable stopping criteria to permit more control over the evolution of the sample. In particular, the new methodology allows the investigator to ensure a fixed sample size. However, the procedure can be computationally intensive.

4.3.2. Mark/recapture studies

Estimation of the size of a wildlife population is a frequent need in environmental studies. Many fish and wildlife populations in the United States are listed under the Endangered Species Act as being either threatened or endangered, and there is a consequent legal requirement to track the abundance of those species. Additionally, most state fish and wildlife agencies use population size information to manage harvest levels and set fishing and hunting seasons.

One of the most popular methods of estimating the size of a wildlife or fish population is known as mark/recapture. In a basic mark/recapture study, an initial sample of individuals is collected, tagged with some permanent mark, and then released. A subsequent sample records the number of marked individuals recaptured from the first sample as well as the total number of individuals. The simplest mark/recapture models for estimating population size assume a closed population (there are no additions to or removals from the population during the observation period), and that each individual has a constant and equal probability of capture at each trapping occasion (Otis et al., 1978; White et al., 1982). For a single recapture event, with data consisting of the number marked (M), the total number in the second sample (C) (including recaptured), and the number recaptured in the second sample (R), Chapman's (1951) estimator of the population size is

$$\hat{N}_C = \frac{(M+1)(C+1)}{R+1} - 1$$

However, in practice, basic mark/recapture studies are rarely used because the required assumptions are not likely to hold. Open population models for mark/recapture studies, known as Jolly-Seber models, were introduced by Jolly (1965) and Seber (1965). The closed-population assumptions have been relaxed (Link and Barker, 2005; Pradel, 1996; Schwarz, 2001; Schwarz and Arnason, 1996, 2000) in a variety of ways to permit estimation of apparent survival and recapture probabilities, the population size at each trapping occasion, and the number of individuals entering the population at each occasion. Seber and Schwarz (2002) provide a recent review of the state of capture–recapture studies.

4.3.3. Designs for assessing trend or status and trend

Powerful and sophisticated statistical techniques are available to identify and test changes in population parameters, for example, tests for change in the mean. We also have techniques for identifying and quantifying trend at a single site or in a single parameter, for example, we can quantify trend in some chemical concentration at a particular sampling station or trend in the average value of several sampling stations, by fitting a regression model that includes a time-dependent term. However, the notions of regional change and especially regional trend are much less well understood. Duncan and Kalton (1987) identified several types of change that might be addressed by sampling a population over time. Although their discussion was oriented toward human populations, the following kinds of change that they identified are relevant for environmental populations:

Gross change is the change at the *site* between two time periods.

Average change over several time periods can refer to the rate of change or trend at a *site* (as opposed to regional change or trend).

Individual instability is a measure of the variance at a *site*, possibly corrected for trend.

The traditional statistical concept is of *change in a population parameter*, where the usual population parameter of interest is the mean. Change is described by sampling the population and estimating parameters at distinct points in time. The resulting estimates are then analyzed for change/trend, for example, with time series or regression methods, or tested for significant difference.

Duncan and Kalton also described *net change* at the aggregate level. They use the example of change in unemployment rate between two months; however, a more general concept is implicit in net change. One can view the unemployment rate as the mean value of a dichotomous population, coded 1 for unemployed and 0 for employed. From this viewpoint, net change is merely a change in a population parameter. However, defining net change as a change in the population distribution, for example, population cumulative distribution function, captures a more general concept of allowing elements of individual change to counterbalance one another. Thus, it is quite possible for individual elements of a population to change, yet for there to be no net change in the population. A related concept occurs in forestry, where change is sometimes broken down into components consisting of growth of existing trees, mortality, and in-growth of new trees. Each of these components of change could be positive, yet the age and size population distribution could be invariant.

Generally, the most precise information of change (trend) comes from sites that are revisited, whereas the most precise information of status comes from visiting more sites. A critical point in designing a survey with the dual objective of status and trend is the allocation of visits to new sites versus revisits, attempting to describe current status and to detect trends in a set of ecological indicators. Observing the same sites over time eliminates the between-site component of variation. If the sites maintain their identity through time, this can greatly increase the power of trend detection methods. For some environmental resources, this is clearly not an issue, for example, for forested sites. For others, for example, lakes or estuaries, there may be little advantage in returning to the same set of site coordinates. Moreover, even if a site retains its identity, there is potential impact of previous visits on the site stemming from both perturbation due to sampling activity at the site and differential management of the site. That impact, sometimes referred to as "time-in-sample bias" (Bailar, 1989), can be substantial. The gain in precision may be more than offset by loss of representativeness.

Skalski (1990) recommended the use of rotating panel designs for the dual objective of status and trends. These designs partition the total sample into several subsets or panels. Each panel is then revisited on a different schedule. Fuller (1999), McDonald (2003), and Chapter 2 of this volume provide details and nomenclature for a wide variety of panel designs. Stevens and Olsen (1999) show how to use GRTS to form interpenetrating temporal panels so that each panel is spatially balanced as well as the composite. The Oregon Department of Fish and Wildlife (ODFW) uses a panel design to monitor the size of Coho salmon populations on the Oregon Coast (Stevens, 2002). Coho spawn in fresh water, migrate to salt water to spend their adult lives, and then return to spawn in about a three-year cycle. The design ODFW uses is tied to the Coho life cycle. One panel is visited every year. There are three panels visited on a three-year cycle and nine panels visited on a nine-year cycle. Four panels are visited every year: the annual panel, one of the three-year panels, one of the nine-year panels, and one panel of new sites.

The power of panel designs to detect change or trend is addressed in Fuller (1999), Urquhart et al. (1993, 1998), and Urquhart and Kincaid (1999). Their insight is that some frequently visited sites are important (e.g., a small annual panel), and the revisit schedule should be tied to the level of change relative to background noise. Thus, to detect a small but persistent trend, a design with a long revisit cycle will be more powerful for the same level of effort than a design with frequent revisits.

5. Using ancillary information in design

Ancillary data can be used in both sampling and inference to improve the accuracy of estimates. In sampling, it can be used to stratify the sampling frame to assure representation of all types of units. It can also be used as a basis for oversampling certain types of units to improve estimates for subdomains of interest. This is true for environmental and other sampling applications. We focus on two applications of ancillary information, which are frequently discussed in environmental applications: additional dimensions to the sampling frame and the appropriate use of ranked set sampling (RSS).

Spatial strata in environmental sampling typically are defined in four dimensions: three geographical and one time dimensions. Sampling bays and streams requires defining the geographical sampling frame in three dimensions. This is also true of sampling land, whether hazardous waste sites or downstream from potential pollution sources. Air pollution monitoring also has three geographical dimensions. Sampling wildlife, on the other hand, is more likely to only have two geographic dimensions. But in all these examples (with the possible exception of nonmoving pollution in the ground), the population of interest, whether wildlife or pollution, is moving, introducing a temporal dimension to the stratification.

The sampling units comprising the sampling frame are usually not difficult to delineate, but ancillary data on these units are often sparse. This is particularly true across time and height/depth. Ancillary data are generally known for a few time periods, so its consistency across time is subject to doubt. Similarly, many historical measurements represent a vertical slice (whether water, land, or air) that has been composited before analyzing, providing limited information on variation across this dimension.

In some environmental applications, it can also be very difficult (or costly) to move data collection locations. For example, only a few monitoring wells are likely to be drilled downstream of a potential pollution source. Once these sampled locations are selected, they can monitor across time, but all the measurements will be at the same location with respect to the other three dimensions of the sampling frame. This physical clustering of the measurements can dramatically reduce the effective sample size and resulting precision of estimates.

As with other traditional sample allocations (see Chapter 1), one frequently over-samples rare domains of interest and parts of the sampling frame, where the outcome measure is thought to be more variable. The four-dimensional nature of environmental sampling can make this more difficult. Variability has to be considered across space and time. The basic principle that one should not oversample too heavily if it is not clear which units are to be oversampled, applies to environmental samples. The resulting design can be very inefficient if the units that are oversampled do not increase the frequency of the rare subdomains or have consistent measurements of the domain of interest.

Ranked set sampling is a particular type of two-phase (double) sampling. In RSS, a small number of units, m, are not measured but are simply ranked, and then the measurement is taken on one unit based on its rank. This is repeated for m sets, each time selecting a different order statistic to be measured. To select a sample of nm measurements, it is necessary to rank nm^2 units by taking n cycles of m sets. This method was introduced by McIntyre (1952) to estimate pasture yields but has received renewed interest in recent years (Patil et al., 1994; Takahasi and Wakimoto, 1968). The advantage of RSS is that it does not require you to know how to stratify the sampling frame in advance, nor do you have to take the initial less-costly ranking information on all first-phase units before beginning second-phase measurement. By being able to rank multiple units and measure one immediately, RSS is attractive to the field operations of environmental measurements.

As an example of the possible use of RSS, consider wanting to select a sample of stream riffles (where the water moves roughly across a series of rocks) to measure fish stocks. Stocks are quite possibly correlated with riffle size. Rather than just to take a random sample of one-third of riffles, it is preferable to walk a stream and rank each set of $m = 3$ riffles, selecting the largest of the first set to measure, the middle of the second set, and the smallest of the third set, then repeating this process. (It is possible to modify this balanced approach to oversample units of particular interest. (Patil, 2002b)) With remote sampling locations and a lack of stratifying ancillary information in advance, this process can provide increased precision.

Unfortunately, much of the research on RSS has compared it with SRS. As demonstrated by the earlier chapters of this volume, SRS is rarely appropriate and the correct comparison is against other complex sample designs that might be used. Mode et al. (2002) compared RSS with three other sampling designs: (1) SRS; (2) weighted double sampling with cut points; and (3) double sampling using ratio estimation. They showed that RSS is appropriate when inexpensive (and possibly qualitative) auxiliary data are available for ranking, for which little distributional knowledge exists. If the general distribution of the auxiliary data is known in advance, then determining which to sample by comparing the auxiliary information to the cut points can achieve improved precision. If the auxiliary data are known to be highly correlated and linearly related to the variable of interest, then ratio estimation is preferable.

In situations where the available covariates make RSS a reasonable data collection method, it is important to consider the cost implications (Mode et al., 1999). Ranked set sampling requires ranking nm^2 units in addition to sampling nm of them. If ranking has minimal costs relative to measurement, RSS can be used. The relative cost of measuring a single unit compared to ranking can vary depending upon the application. Mode et al. provide examples of cost ratios of 5.3 for crude oil in contaminated sediment, 20 for estimating fish abundance, and 50 for detecting radiation. They found that depending on the shape of the distribution and the accuracy of the ranking, cost ratios exceeding 6–11 were sufficient for RSS to yield improvements for a fixed total cost.

6. Inference for probability-based design

The analysis of a probability survey is often called *design-based* because the validity of the population inference rests on the design rather than on an assumed statistical model. The randomness is explicitly included in the sample-selection process and forms the basis for estimating population characteristics. The key quantity in the estimation is the inclusion probability for a population unit, which is the probability that that unit is included in the sample. It must be positive for every unit. In the case of a continuum, the inclusion probability is defined by an inclusion density, usually denoted by $\pi(s)$. In contrast to a probability density, the inclusion density has units. For example, an inclusion density for a point sample from a map might have units of $(number\ of\ sample\ points)/km^2$. In the case of a finite population, the inclusion probability sums to the sample size; in the continuous case, the integral of the inclusion density over the target domain gives the sample size. The importance of the inclusion probability for a sample element is that its reciprocal is a measure of the portion of the population represented by that element. Thus, for example, in a SRS of size n from a finite population with N total elements, the inclusion probability for each sample element is n/N, and each sample element represents N/n population elements. If a SRS of n sites were selected in a wetland with area A km^2, then the inclusion density would be $\pi(s) = n/A$ and each site would represent A/n km^2 of wetland.

The basic analysis tool is the Horvitz–Thompson or π-weighted estimator (Horvitz and Thompson, 1952; Thompson, 2002). The continuous version of this estimator is given in Cordy (1993) or Stevens (1997). The concept of the π-weighted estimator is that estimates of totals are obtained by weighting individual observations with a weight inversely proportional to their inclusion probability.

Let n be the number of sample plots, z_i the response for the ith sample plot, and π_i be the inclusion probability (or density) evaluated at ith sample point. Note that z_i could be a numeric score (e.g., per cent forested land cover) or a binary classification, for example, $z_i = \begin{cases} 1, & \text{if } i\text{th plot in degraded condition} \\ 0, & \text{otherwise} \end{cases}$. The Horvitz–Thompson estimate of the total of z is given by $\hat{z}_T = \sum_i^n \frac{z_i}{\pi_i}$ and the estimate of the mean value by $\bar{z} = \frac{\hat{z}_T}{A}$, where A, the population size, is the total area of the target population. These formulas are the same for both finite and infinite populations. Note that in the case of z_i being a binary classification, $\bar{z}$ estimates the proportion of the resource in the condition class, for example, the proportion of the watershed in degraded condition.

An alternative estimator of the mean value uses the estimated population size $\hat{A} = \sum_1^n \frac{1}{\pi_i}$ as a divisor in place of A. In some circumstances, use of the estimated population size in place of a known population size can lead to a more precise estimate of the mean because of positive covariance between $\bar{z}$ and $\hat{A}$. If the size of the target population is not known, for example, the imperfect frame case described below, then the alternative estimator must be used. Also, if some plots were not accessible, say because access permission was not obtained, then an estimate of the average condition of the *accessible* wetlands is $\bar{z} = \frac{\hat{z}_T}{\hat{A}}$, where both $\bar{z}$ and $\hat{A}$ are computed using only those sites for which a response was obtained. An alternative is to use a nonresponse adjustment to compensate for the nonaccessible locations.

A spatially balanced sample will normally be more precise than a SRS of the same size because its spatial balance capitalizes on the spatial structure of the response. However, because of the restricted randomization inherent in the spatial balance, variance estimation can be an issue. Technically, the variance depends on pairwise or joint inclusion probabilities (the probability that a pair of points are both included in the sample). The restricted randomization implicit in spatial balance makes some of those joint probabilities very small or zero. The joint probabilities appear in the denominator of the usual variance estimators, so the estimators are undefined if joint probabilities are zero and unstable if small. A commonly used approach is to ignore the spatial constraint in the design and apply the SRS variance estimator. The resulting estimator will almost always be biased high. Horvitz and Thompson (1952) derived an unbiased variance estimator to accompany their estimator of the total, but the joint inclusion probability appears as a divisor in the estimator, so it is unsuitable for spatially balanced designs.

Wolter (1985) identified eight one-dimensional variance estimators for one-dimensional systematic sampling. D'Orazio (2003) extended three of these to two-dimensional systematic sampling. A general purpose technique that provides reasonably good results is to apply a postselection spatial stratification with at least two points per stratum. The strata can be selected arbitrarily but the points in a stratum should be close together. The usual stratified sample variance estimator is then applied. Stevens and Olsen (2003) developed a variance estimator specifically for spatially constrained designs that is based on a similar concept. Instead of explicitly forming strata, a local variance is computed at each sample point. The local neighborhood of a point is defined as a region containing the point's four nearest neighbors and then expanded to satisfy a symmetry constraint (if a is in the neighborhood of b, then b must be in the neighborhood of a). The overall variance estimate is a weighted average of the local estimates.

7. Model-based optimal spatial designs

The development of statistical theory or methodology is often driven by the search for optimality, that is, to find a new procedure that is "best." Design optimality involves two choices: which estimator or predictor to use, and which population elements to select, or, in a spatial context, where to place design points. In a statistical context, the standard is usually some measure of closeness of the estimator or predictor to the population attribute. Thus, our working objective needs to combine an estimator and a criterion that can be optimized. This requires the specification of an optimality criterion, which in statistics is usually minimum variance. This is not the only possibility; minimax criteria are also used, where one tries to minimize the maximum unfavorable outcome, for example, minimize the maximum loss. In situations where bias is a major concern, minimum mean square error is often the criterion. Unless otherwise stated, optimal should be interpreted to mean minimum variance.

Statistical models may be used to describe the underlying environmental process that generates the response. The statistical models usually applied in this setting are models of a mean process, possibly depending on ancillary variables, plus models of a spatial random process. The mean process $\mu(s|X, \beta)$ may be a constant, a function of location s, or a function of location and ancillary variables X with parameters β. The spatial random process $Z(s|\theta)$ with parameters θ is frequently taken to be *intrinsically stationary* so that $E[Z(s + h|\theta) - Z(s|\theta)] = 0$. The spatial covariance of $Z(s|\theta)$ is usually described by the *variogram* $2\gamma(h)$, where $2\gamma(h|\theta) = \text{Var}[Z(s + h) - Z(s)|\theta]$. The quantity $\gamma(h)$ is then called the semivariogram. In this discussion, we will consider a spatial random field given by $Y(s) = \mu(s|X, \beta) + Z(s|\theta), s \in R$ for location s and domain R. Frequently, the semivariogram is also assumed to be isotropic so that it depends only on distance and not direction, so that $\gamma(h|\theta) = \gamma(|h||\theta)$.

Some of the early insights on optimal design (Dalenius, 1961; Iachan, 1985; Matérn, 1986) were derived by assuming a known covariance, using the sample mean as an estimator, and by optimizing a variance rate, that is, a variance per unit area. This approach sidesteps the influence of a domain boundary. In practice, the presence of a boundary, especially an irregular boundary, influences the optimal site locations. The results were consistent in suggesting that a systematic sample was better that a stratified sample, which was in turn better than a SRS. Moreover, the compactness property of a triangular grid was also shown to lead to favorable designs.

For the random field model, the sample mean is not the optimal estimator of our working objective. For the case when $\mu(s)$ is an unknown constant or a linear combination of explanatory variables, the optimal (in the sense of minimum squared error loss) predicted value for a new location s_0 is given by the kriging or best linear unbiased prediction estimator $\hat{Y}(x_0) = \sum \lambda_i Y(x_i)$, where λ_i are the kriging weights and are described in many textbooks on geostatistics such as Cressie (1993) or Schabenberger and Gotway (2005). The variance of the prediction at location s_0 is given by $\sigma^2(s_0|S, \gamma) = 2\sum_{i=1}^{n} \lambda_i \gamma(s_i - s_0) - \sum_i^n \sum_j^n \lambda_i \lambda_j \gamma(s_i - s_j)$. Note that the prediction variance depends on the location of the sample points and the semivariogram. There is no dependence on the actual values at those points.

To get an optimal design for our working objective, it makes sense to use the optimal estimator and to choose the sample S to minimize the total prediction variance $V_T(S, \gamma) = \int_D \sigma^2(s|S, \gamma)ds$. In most cases, this integral is very difficult to work with. It is intractable analytically and must be dealt with numerically.

As an alternative, Yfantis et al. (1987) evaluated square, triangular, and hexagonal grids, assuming a known covariance. Their optimality criterion was to minimize the maximum mean square prediction error. Their conclusion was that a triangular grid was optimal. McBratney et al. (1981) reached a similar conclusion using the average prediction variance.

The concept that the optimum location of sampling points for prediction will be some sort of regular arrangement is well established. One approach to optimizing design is to maximize some measure of regularity of the point pattern of the sample locations. The underlying assumption is that a highly regular design will also be a low variance design. An algorithm for locating sample sites that has been used with known domain boundaries or the presence of existing points is spatial simulated annealing (SSA) (Di Zio et al., 2004; Lark, 2002; Stevens, 2006; Van Groenigen, 2000; Van Groenigen and Stein, 1998). Sample points are selected to optimize some criterion that reflects the study objective, for example kriging variance, or a measure of regularity of the resulting spatial point process. The SSA begins with a set of arbitrary locations, and cycles through the points, perturbing each one in turn. At each step, the optimality criterion is calculated. If the new configuration resulting from the perturbation is better than the prior optimum, it is retained as the new optimum configuration. If it is worse, it is retained with a probability that decreases with the number of cycles. The concept behind retaining the suboptimal configuration is to bump the iteration away from a local optimum. Letting the probability of (temporarily) accepting a suboptimal configuration decrease helps to ensure eventual convergence to the global optimum.

Another approach is to modify the criterion somewhat. For example, instead of attempting to optimize over all possible designs, limit the space of potential designs. One way of limiting the design space restricts attention to sequentially optimal designs. In this method, an initial design with m points is chosen, arbitrarily or at random. Then s_{m+1} is chosen at an optimal location conditional on the locations of the previous points (Cressie et al., 1990). The process is then repeated until all n points have been chosen. Another way to do this is to discretize by replacing the two-dimensional continuous domain with a finite point set, say with a regular grid that covers the domain. In principle, then, one can evaluate all possible designs and pick the optimal one. This has been tried by Di Zio et al. (2004) and Wiens (2005). Even then, the computational burden can be overwhelming unless the design space is severely limited. Other authors have used SSA in conjunction with discretization (Wiens, 2005; Zhu and Stein, 2006).

In most applications, the covariance structure will not be known and must be estimated. Some papers have considered optimal designs solely for estimating the covariance function without regard to prediction. Warwick and Myers (1987) develop a search algorithm for achieving particular distributions of point pair distances, by which they take sums of squares of discrepancies in the realized and desired distributions and select a point pattern with a minimum sum of squares. Müller and Zimmerman (1999) consider generalized least squares fit to the empirical variogram to estimate variogram parameters. They use the determinant of the information matrix as design criteria. They compare several techniques, including the Warwick and Meyers method (1987). Their results show that a more irregular design with some points placed close to each other is better for variogram estimation. Zhu and Stein (2005) use maximum likelihood to estimate covariance parameters. They use minimax and Bayesian criteria to select an optimal designs. The design space is restricted to a fine grid, and SSA is used to locate optimal designs.

The more realistic case where the objective is prediction and the covariance structure is unknown and must be estimated has been considered by several authors, who attempt to consider the impact of covariance parameter estimation on the prediction variance. Zimmerman (2005) notes that the design objectives for efficient prediction assuming known dependence and efficient estimation of spatial dependence parameters are largely antithetical and often lead to very different optimal designs. Zimmerman introduces a hybrid design that emphasizes prediction but accounts for the uncertainty in the covariance parameters. His approach is to choose the design to minimize an approximation to the variance of the empirical kriging (empirical-BLUP) prediction error. Note that the empirical kriging/BLUP predictor involves evaluating the covariance matrix at the estimated $\hat{\theta}$ rather than the assumed known θ. He makes some empirical comparisons between designs to optimize parameter estimation, prediction variance, and a hybrid design.

Zhu and Stein (2006) compare the designs for (1) prediction using covariance parameters estimated from an existing data set, (2) estimating covariance parameters, and (3) prediction with estimated parameters. They use SSA to locate optimal design configurations. Consistent with previous work, the optimal designs in case (1) are highly regular and approximately triangular grid structure, subject to perturbation because of irregular boundaries. For case (2), the optimal designs consisted of multiple clusters of points. Their case (3) gives a pattern that is mostly regular, with several clusters of closely spaced points.

Diggle and Lophaven (2006) described a Bayesian approach to spatial design that balances the design for parameter estimation with spatial prediction. The designs are efficient for spatial prediction and make an appropriate allowance for parameter uncertainty. They also compare the efficiency of designs based on a regular grid plus extra close pairs to a regular grid with in-filling. Ritter and Leecaster (2007) also evaluate several designs that combine regularly spaced points with clusters of points. They conclude that the clusters are valuable for estimating the semivariogram and offer several recommendations for a design.

8. Plot design issues

Environmental measurements are frequently taken as an average over a three- (or four-) dimensional space. Water, land, or air samples are collected from a small physical area rather than a point. This area is referred to as the physical support. Complications arise in inference from environmental samples when the analytic units do not match with the physical support of the samples or when units of different size (or composition) support are combined. These are referred to as change of support problems (Gotway Crawford and Young, 2006).

Combining units of different size does not effect mean estimation, but it can cause significant problems in estimating precision and correlation. This in turn effects estimation of significant differences and distributional percentiles. Cressie (1996) points out that if the physical units are positively autocorrelated, the collapsing of the units into larger physical support will have less effect on the variability of the mean than when this correlation is absent.

In general, the larger the physical support, the lesser the variability in the measurements. This averaging of smaller units into larger ones shrinks the variation among units. This can be vital in many environmental situations. Polluted areas are often defined as those exceeding a set level. The determination of whether or not a site is polluted can be completely determined by the size of the physical support used for collecting the sample, not the underlying amount of contaminant.

Although not an environmental application, Openshaw and Taylor (1979) provide an excellent example of how the size and shape of the physical support can determine the estimate. They examined the correlation between the percentage of Republican voters and elderly voters in Iowa counties. Depending on which groupings of counties were used as the physical support, the correlation varied from -0.99 to $+0.99$.

Another plot design issue that can have important implications for analysis is when the physical support overlaps analytic domain boundaries. This situation can arise, for example, when plots are based on watersheds or river reaches, but analyses are planned by political boundaries such as states or counties. When the sampled plots cross the analytic boundaries, it makes analyses very difficult, with the accuracy of the estimates a function of the model assumptions that have to be made to allocate the support across domains. To minimize this problem, it is important to try and identify key analytic domains before the sampling frame is determined. It is then possible to define sampling units as the intersection of logical geographic units and these planned domains.

Composite sampling (Patil, 2002a) is a tempting methodology for measurements that are much more expensive (or time consuming) to analyze than they are to collect. Common examples are sampling for pesticides in soil, air monitoring, and contaminants in fish. Composite sampling is a logical method if when the analyte is present, it is likely to be in large quantities. For example, when conducting exploratory measurements around a suspected hazardous waste dump site, it is reasonable to take multiple samples from around the site, composite them, and then do the chemical analyses. If the analyte was really dumped on this location, it is assumed that the diluting resulting from combining the different samples, some of which have high levels of the analyte and others having none, will still result in detectable levels. Once the presence has been identified, more careful, noncomposited sampling can be conducted. (Alternatively for fish, the initial field sample collects 10 fish: 5 are composited and 5 are archived. If the composite analysis raises a flag, then the individual fish is analyzed. This avoids the expense of multiple field visits. Sample storage is cheap compared with travel to a remote site.)

There are two dangers associated with composite sampling. First, if the detection limit for the analyte is high relative to the expected levels in relative hot spots, then a composited sample might be below the detection limit, even if a hot spot has been included. For example, if the detection limit is only one-quarter the concentration found in the hot spot and five or more physical locations are composited, it is possible for the composite analysis to be nondetectable, even when it included the high value.

Second, composite samples are very good at producing estimated mean values for the area being composited. Thus, it can be used to produce a daily average air pollution level or an average exposure from digging up soil on a site. Unfortunately, composite samples underestimate the variability about that average. That is, the variation in samples with physical support of size equal to that of individual samples will be much greater than the variation observed from the composites. The difference is proportional to the square root of the number of composited samples, so if sets of four individual samples are composited, the variability will be 50% that of the individual samples. This is particularly important if one is interested in measuring percentiles of a distribution far away from the mean. For example, if one is interested in estimating the 90th percentile of individual soil samples, this number will be quite a bit higher than the 90th percentile of composited samples. This is similar to the fact that the 90th percentile of hourly airborne (or waste water) emissions is much larger than the 90th percentile of daily emissions.

Also note that composite samples may not estimate the average that one is really interested in. Going back to the fish example, the result from compositing is a weighted mean of the fish that went into the composite, with the weight being the actual weight of the individuals. This is not necessarily the same as the average body burden of the five fishes. Unless the weight distribution in the sample matches the weight distribution of the target population, there could be enough discrepancy to be of concern.

9. Sources of error in environmental studies

There are a number of sources of error that, if not unique to environmental studies, are more commonly observed in environmental situations than with other types of data. One particularly problematic error source is that the only sources of data are frequently not located at the site of interest. Frequently, measurement requires installation of expensive equipment, which already exists in preset locations. Water pollution measurements are frequently taken near outfalls from industrial facilities, but the concern is the effect on drinking water in peoples' homes. Air pollution monitoring stations are often located near the manufacturing plants producing the pollution, but the concern is with pollution levels in the air breathed by people where they live, often far away from the pollution source. Unlike most other data collection situations, the physical sampling locations are presets based on decisions having nothing to do with optimal statistical sampling. It is left to analysts to model how the data observed in one set of locations migrates to the locations of interest. Madsen et al. (2007) develop a regression model utilizing spatial correlation where the predictor variables are observed at different locations than the response. Zhu et al. (2003) apply a Bayesian hierarchical model to relate incidence of asthma to traffic density data, where the response and the stressor are misaligned in both space and time. Mugglin and Carlin (1998) also use a hierarchical model to interpolate disease incidence counts using spatially misaligned covariates. This migration may be subject to air and water currents, seasonality, and a host of other complicating factors.

Even when the environmental study designer gets to identify their sample locations, they may not be able to gain access to the sites. Often sampled locations are identified from large-scale maps based on GIS or other methods. Although in theory it is possible to go to all such locations, it is not necessarily true in practice. The location might be in middle of river rapids, on a steep slope, or on private property where the landowner refuses to provide permission. While the latter is analogous to refusals in household surveys (Groves and Couper, 1998), environmental data collection introduces additional situations in which it will be impossible to collect the data from the sampled location.

Seasonality affects many types of data collection and analysis. But in most environmental surveys, seasons affect the location being sampled; while in household surveys the person being measured is associated with a specific location. (Migrant populations are an exception to this generalization and in this situation are more like environmental samples than other surveys of people (Kalton, 2003).) Environmental surveys of living species have the extra source of variability due to the fact that many move locations across seasons. For example, the location of fish varies greatly by time of year. Surveys of salmon in rivers of the Northwestern United States can only be conducted in those locations during spawning season. At all other times of year, there will be no fish to measure. If a single species is being studied, it can be timed appropriately for the migration patterns of the species; but if a general survey is being conducted, the results for specific species may be largely dependent on the time of year at which data are collected.

The two just-discussed sources of error can interact to create additional difficulties. During certain seasons, it may be impossible to collect data from locations that are available other times of year. For example, in a survey of domestic well drinking water quality, it may be necessary to sample from the well pipes before treatments are added. This requires accessing the pipes outside of the home. It is impossible to collect this data in rural Alaska during the winter. The data must be collected during the rest of the year, even though the water is consumed all year long.

Measurement error can come from a great variety of sources, including data collection instruments, laboratories, staff collecting and/or measuring the data, inconsistent physical materials, and detection limits. In addition, measurement errors that are unbiased can result in bias for statistics of interest.

Data collection instruments may not measure accurately. This can result in both extra variability and bias. Frequent recalibration of instruments can reduce bias but with an increase in data measurement costs. For example, X-ray fluorescence (XRF) machines are used to measure lead content of painted surfaces. The XRF machines might regularly underestimate the amount of lead or it might inconsistently measure depending on the underlying substrate. If multiple machines are used at the same time by different data collectors, inconsistencies across machines can introduce more error. Machines in laboratories can be similar sources of measurement error. In addition, there may be practices at laboratories (e.g., cleanliness or data tracking) that can affect data quality as well. Again, if multiple laboratories are used to measure the same analyte, then measurement errors can be compounded.

As with other types of surveys, data collectors can be sources of both variability and bias. In environmental surveys requiring data collectors to use machines, the varying skills of the people using the machines can cause increased levels of error. This makes it very important to develop protocols for data collection that will minimize error, are easy to follow, and are easy to monitor for quality.

The need to collect physical samples from inconsistent physical material is a source of error unique to environmental surveys. For example, collecting samples from municipal dump sites to measure the presence of toxic materials requires developing procedures to assure a representative sample of materials from a combination of computer parts, lawn mowers, furniture, and miscellaneous waste.

Detection limits are another difficulty unique to environmental surveys. Detection limits are "defined as the lowest level of the measurand where the probability of a positive result is at least 95 percent" (Van der Voets, 2002, p. 504). Frequently, samples with values of the analyte that are not detected are assumed to be 0 or possibly one-half of the detection limit. A more sophisticated approach is to model the measured data and then distribute the nondetected values between 0 and the detection limit according to the model. Lambert et al. (1991) describe how it can be improper to assume that all values that are nondetected are really below the detection limit.

The fact that measurement error can cause bias in estimated regression cocfficients is well known across all analytic situations (see Fuller, 1987; Stefanski and Buzas, 1995) In environmental analysis, the effect of measurement error is increased because the parameters of interest are frequently not means and totals but rather extreme percentiles or the percent of a distribution that falls above a preset limit. The estimated distribution that results from naive analysis of observations with measurement error is an estimate of the convolution of the true distribution of the parameter and the measurement error distribution. Central location may remain unchanged, but the tails of the convoluted distribution will be spread out relative to the true parameter distribution. In these situations, unbiased measurement error can result in biased estimates of these parameters of interest. Cook and Stefanski (1994) introduced the SIMEX estimator (for SIMulation_EXtrapolation) as a way to remove bias from estimates from estimators that may be complicated, nonlinear functions of the data. The simulation step of SIMEX consists of generating multiple sets of pseudodata by adding known levels of error to the observed data and by calculating the estimator for each set of pseudodata. For the extrapolation step, the set of estimates is then regressed against the known level of error contamination. The regression is then extrapolated back to a zero level of error contamination. Stefanski and Buzas (1996) show how to apply the SIMEX estimator to deconvolute the estimated distribution function for a finite population, and Stefanski and Cook (1995) explore the relationship between the SIMEX estimator and the jackknife method (Quenouille, 1956) for reducing bias in nonlinear estimators.

Many environmental measures are right skewed, such as the log-normal distribution shown in Fig. 3. If the goal is to estimate the percent of the distribution above a preset limit (L in Fig. 3), it is clear that a greater percentage of the data are just below L than just above it. Thus, unbiased random measurement error will cause more values that

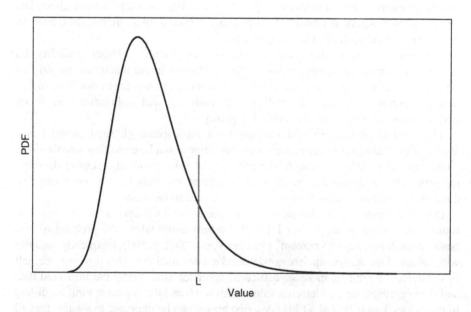

Fig. 3. Log-normal distribution with preset limit L.

are truly below L to be measured above L than the reverse. Measurement error will therefore bias upwards the estimated percent above the limit L.

A second source of bias that interacts with this measurement error is that environmental measurements often are concerned not with the percent of measurements above the cutoff L, but with the percent of physical units that have any values above L. Examples include hazardous waste sites that are declared superfund sites if they have contamination levels above L anywhere on the site or homes that are considered to have a lead hazard if levels above L exist anywhere in the house. Comprehensive measurements of all locations are never taken, rather a sample of locations is measured and then a determination is made as to whether the site or home is contaminated. While the sample of measurements provides an unbiased estimate of the average level of contamination, they provide an underestimate of the highest level of contamination. (Similar sources of bias arise when trying to estimate life cycle exposure to contamination by only collecting physical samples at selected interview times during a longitudinal survey.)

An example of how these two sources of bias interact is provided in Clickner et al. (2002). As part of the National Survey of Lead and Allergens in Homes, a representative sample of homes in the United States was selected and measurements were taken to determine how many had lead hazards. One source of a lead hazard is having any floor dust with lead loadings of greater than $40 \mu g/ft^2$. Samples from four rooms were collected and the maximum was computed. Figure 4 (Figure C.8 from

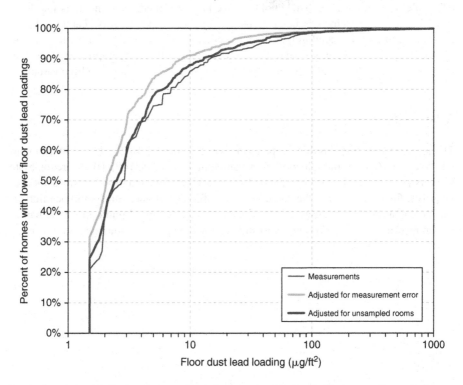

Fig. 4. Cumulative distribution of the maximum floor dust lead loading for Homes sampling and inference in environmental surveys.

Clickner et al., 2002) shows three curves. The thin black line shows the maximum of the measured dust levels. The gray line adjusts the maximum for the measurement error bias described above. This adjustment lowers the percent of homes that exceed the cutoff. The thick black line adjusts these maximums for the fact that only four rooms per house were measured. Since the maximum cannot go down if data from additional rooms were included, this adjustment increases the percent of homes exceeding the cutoff. Using the estimates after both adjustments yields an estimate of 4% of homes having floor dust lead loadings of $40\,\mu g/ft^2$ or more in one or more rooms. This is about 1% (one million) fewer homes than estimated using the actual measured floor dust measurements.

10. Conclusions

Sampling and inference in environmental surveys have much in common with other surveys. Thus much of what is contained in the earlier chapters of this book is relevant to environmental surveys as well. We have discussed some aspects of populations that need particular attention when designing and analyzing an environmental sample: focus on a broad population description, make use of the spatial context, use ancillary information, inadequate frames, difficult access to sampling locations, responses that are difficult and expensive to measure, evolving objectives, and the need to satisfy multiple objectives and stakeholders. We have outlined some methods that have been developed to address issues engendered by these aspects. This chapter is by no means a complete compendium, as there are other issues and methodology that we did not touch upon. The methods we did discuss are constantly being improved as we gain experience using them and software implementations become more readily available.

Acknowledgements

The research described in this chapter was partially funded by the U.S. Environmental Protection Agency though cooperative agreement CR83-831682-01 to Oregon State University. It has not been subjected to the Agency's review and therefore does not necessarily reflect the views of the Agency and no official endorsement should be inferred. The conclusions and opinions are solely those of the authors. Mention of trade names or commercial products does not constitute endorsement or recommendation for use.

Essential Methods for Design Based Sample Surveys
ISSN: 0169-7161
© 2009 Elsevier B.V. All rights reserved
DOI: 10.1016/B978-0-444-53734-8.00009-1

9

Survey Sampling Methods in Marketing Research: A Review of Telephone, Mall Intercept, Panel, and Web Surveys

Raja Velu and Gurramkonda M. Naidu

1. Introduction

Survey sampling methods play an important role in marketing research. The discipline of marketing itself draws its techniques from various social and physical sciences and any advances made in sampling methods in these areas almost always find an application in marketing research. Recognizing the importance of the topic, the first special issue (August 1977) of the *Journal of Marketing Research* was devoted to survey research. The articles in that special issue addressed three aspects of survey research, namely, sampling design, questionnaire preparation, and data collection. In an article that has appeared in an earlier volume, Velu and Naidu (1988) provided a survey of these aspects. Our objective there was to briefly review and update the aspect of sampling design with special focus on telephone, mall intercept, panel, and internet surveys. Because the design issues related to telephone sampling in particular the random digit dialing (RDD) methods are covered in a separate chapter 3 in this volume, we will focus on the other forms of surveys. A bibliography follows and, while not exhaustive, the listing of books and other references should provide a starting point for an iterative search. Although the coverage of topics is more relevant for United States, to the extent possible we provide information about the practices in other countries as well. To begin with, we have compiled a list of select institutions that actively engage in marketing survey research (Table 1).

Marketing researchers have been aware that subjective sampling procedures must be avoided in favor of probability methods of selection to make valid inferences about the target segments. Because of the inherent diversity of the marketing discipline, there has been a growing demand for all types of data necessitating more complex marketing surveys. Also, during the past three decades, the household (the nucleus of most consumer surveys) has undergone dramatic changes in terms of its composition and size. More women have joined the workforce, have become economically independent, and are making buying decisions. As the environment was changing, techniques for sample surveys were also changing. The high cost of personal household interviews has led to the development and use of more efficient sample designs and less expensive

Table 1
A list of marketing research institutions

URL	Institution	Marketing Research Services Offered
www.surveysampling.com	Survey Sampling International (SSI)	Leading provider of superior samples for mail, telephone, Internet, panel, B2B, B2C, surveys, RDD in 20 countries and internet panels in 17 countries. With global partners, and a reach of over 50 countries. Some 1500 organizations and 43 of the top 50 research organizations use their services.
www.createsurvey.com	Create Survey	Create Survey is a Web-based survey software that lets you build and run online surveys in the Internet. You may start using it right now or read more below.
www.surveysystem.com	The Survey System	The Survey System is the most complete software package available for working with telephone, online, and printed questionnaires. It handles all phases of survey projects, from creating questionnaires through data entry. The Survey System was designed specifically for questionnaires; so their software saves you time.
www.greenfield.com	Greenfield Online	Greenfield Online helps marketing research companies and consultancies connect with their consumer insights by programming and executing online surveys using their Internet-based online panel of prerecruited respondents. They couple access to survey respondents with executional excellence and quality.
http://us.lightspeedpanel.com	LightSpeed Consumer Panel	By taking the time to participate in Lightspeed Panel surveys, you have the power to let companies know exactly what you think. This helps you, the consumer, to develop and improve the products and services offered.
www.tnsglobal.com	TNS in North America	TNS is a market information group. It is the world's largest custom research company and a leading provider of social and political polling. It is a major supplier of consumer panel, TV audience measurement, and media intelligence services. It provides market information and measurement, together with insights and analysis, to local and multinational organizations.
www.e-focusgroups.com	e-focus Groups	e-FocusGroups offers solutions for all market research needs. It brings the benefit of more than 20 years of market research experience in a wide variety of industries, including consumer products, advertizing, pharmaceuticals, e-commerce, computer hardware, computer software, telecommunications, and banking, among others.

(Continued)

Table 1
(*continued*)

URL	Institution	Marketing Research Services Offered
www.forrester.com	Forrester	Forrester Research, Inc. is an independent technology and market research company that provides pragmatic and forward-thinking advice to global leaders in business and technology. For more than 23 years, Forrester has been making leaders successful every day through its proprietary research, consulting, events, and peer-to-peer executive programs
www.zoomerang.com	Zoomerang	Zoomerang pioneered online survey software in 1999 to give organizations like yours a powerful self-service alternative to conduct accurate comprehensive surveys with a minimum of cost and effort. Today, Zoomerang is the world's No. 1 source of online surveys, helping thousands of organizations in more than 100 countries.
www.web-surveyor.com	Web Surveyor	They are empowering people to make informed business decisions using their online data collection solutions. It provides online survey services that enable their customers to easily collect real-time feedback to drive their businesses. They ensure data security and confidentiality, a reliable survey hosting service, dependable survey software, and a responsive team of survey experts.
www.web-online-surveys.com	Web Online Surveys	This is an all in one service designed for people who are not computer experts and have the need to conduct surveys by themselves.
www.synovate.com	Synovate–Research Reinvented	Synovate is the world's most curious company. Their job is to learn what people like, and why they like the things they like. That knowledge helps product designers and manufacturers give people what they want. The work they do at Synovate is continuously stretching the definitions of conventional research. They operate across six continents, in 50 countries.
www.gartner.com	Gartner	They deliver the technology-related insight necessary to make the right decisions, every day. Gartner serves 10,000 organizations, including chief information officers and other senior IT executives in corporations and government agencies, as well as technology companies and the investment community.
www.vnu.com	Nielsen	The Nielsen Company is a global information and media company with leading market positions and recognized brands in marketing information, media information, business publications, and trade shows. The privately held company is active in more than 100 countries, with headquarters in Haarlem, the Netherlands, and New York, United States.

(*Continued*)

206 R. Velu and G. M. Naidu

Table 1
(*continued*)

URL	Institution	Marketing Research Services Offered
www.imshealth.com	IMS Intelligence Applied	IMS is the one global source for pharmaceutical market intelligence, providing critical information, analysis, and services that drive decisions and shape strategies.
www.kantargroup.com	Kantar	Kantar is one of the world's largest research, insight and consultancy networks. They help clients to make better business decisions through a deeper understanding of their markets, their brands, and their customers. They help clients find *better ways to answer business questions.*
www.harrisinteractive.com	Harris Interactive	In an increasingly chaotic and competitive world, Harris Interactive can provide clarity and confidence. They believe that market research helps our clients understand the drivers of decision making and can strengthen enterprise equity. Providing clients with this accurate knowledge will help them achieve measurable and enduring performance improvements.
www.jdpower.com	J.D. Power Consumer Center	Since 1968, J.D. Power and Associates has been conducting quality and customer satisfaction research based on survey responses from millions of consumers worldwide. It has developed and maintains one of the largest, most comprehensive historical customer satisfaction databases for various products and services.
www.opinionresearch.com	Opininon Research Corporation	At Opinion Research Corporation, they provide objective, fact-based decision support, they earn their confidence with our fresh ideas and perspectives, grounded in rigorous research methods and business savvy.
www.dentsuresearch.co.jp	Dentsu Research On-line	Dentsu Research, a specialist in market research, has served as the eyes and the ears of team Dentsu, collecting and analyzing the latest in consumer information. Now, over 30 years later, marketing research remains the core of their work, providing any and all services clients require.
www.infores.com	IRI—Information Resources Inc.	Driving the transformation of the consumer packaged goods (CPG), retail, and healthcare industries, only IRI provides a unique combination of real-time market content, advanced analytics, enterprise performance management software, and professional services.
www.npd.com	NPD Group	The NPD Group, founded in 1967, is the leading global provider of consumer and retail market research information for a wide range of industries. They provide critical consumer behavior and point-of-sale information and industry expertise across more industries than any other market research company.

Note: There are several other vendors and due to space limitations, they are not listed here.

data collections methods such as the use of telephone and mall intercept interviews and the use of Web surveys. At the same time, the public has become increasingly concerned about invasion of privacy and the maintenance of confidentiality of the information obtained.

Some developments (see Frankel and Frankel, 1977) of interest to marketing researchers include: (i) techniques related to the manipulation of sampling frames, (ii) techniques related to respondent selection, (iii) methods for minimizing the total survey error, and (iv) improving the quality of nonprobability sampling. Broadly we organize the discussion of various forms of surveys around these areas. We shall focus briefly in Section 2 on sample frames and procedures for telephone household surveys, a topic that received a great deal of attention over the last several decades. Judgment or nonprobability sampling procedures, still viable in marketing research, are convenient to carry out and are less expensive compared to other methods. In Section 3 we shall comment on mall intercept surveys, which are used increasingly by several marketing research firms. The consumer panel studies are reviewed in Section 4. With the advent of the Internet in the mid 1990s, the Web survey has become quite popular because of its ease of implementation as well as its cheaper cost. We briefly review this area in Section 5.

2. Telephone surveys

The telephone is an important tool for the collection of marketing survey data in the United States. Although it has been used in the past mainly for short follow-up interviews, usually for clarifying the information provided in personal or mail interviews, marketing researchers had resorted to using the telephone due to the increasing cost of other forms of surveys. A distinct advantage of the method is accessibility to the respondent. Some major disadvantages are the limited time a respondent may want to spend with a physically absent interviewer and the inability of the respondent to actually "see" the product in question as in surveys where the interviewer can display the product and obtain observational data.

In the United States telephone numbers have three parts: a three-digit area code, a three-digit central office code or prefix, followed by a four-digit suffix. The list of all area code–central office code combinations currently in service can be obtained from the telephone companies. With the introduction of mobile phones, these combinations have exponentially increased in recent times. Numbers to exclude from such a list are those of (i) the telephone company central offices (such as 555 used for directory assistance) and (ii) other central offices used solely by government or businesses (such as 866). We shall refer to groups of consecutive numbers starting with 0, 00, or 000 within the suffix as "banks of numbers." For the operational convenience of the exchange, only certain banks of numbers are assigned to users.

2.1. Sampling frames for telephone households

The sampling unit for most marketing investigations has been primarily a household and it is implicitly assumed when telephone sampling is used that a single telephone serves a single household. This is not necessarily the case in practice. Some households have more than one telephone and more than one number. With the call forward option, business calls are sometimes automatically transferred to home phones. It is estimated that more than 94% of the households in the United States can be contacted either via land line or via mobile phone (see Tucker et al., 2007).

Households with mobile telephones are different from households with land telephones, as shown by several demographic and economic variables. These demographic and economic differences are expected to manifest in attitudinal differences as well. We will comment on sampling issues related to mobile phones later in the chapter. What we describe below mainly applies to land lines only.

There are basically two kinds of sampling frames used for telephone surveys. The rest are minor variants of these two frames. One is the list-assisted frame. The list can come from the telephone directories or from previous surveys. The other is the set of all possible four-digit suffixes within the existing central office codes. The latter is used in RDD methods. There are some advantages and disadvantages in both frames. The most important drawback of this frame, however, is that it excludes working telephone numbers that are not listed in the directory. Also, telephone directories are outdated, on the average, by at least 7–8 months. The percentage of unlisted numbers varies by regions with roughly 30% of numbers in large metropolitan areas of the United States unlisted. Households with unlisted telephone numbers tend to differ from households with listed telephone numbers on key demographic characteristics (see Moberg, 1982). Brunner and Brunner (1971) found significant differences between the two groups on certain product ownership, usage, and purchase patterns.

The disadvantage of the second frame is the large number of nonworking telephone numbers that may be sampled with unrestricted random sampling. In the United States only a fraction of dialings will connect with a usable residential household. The effort to identify these numbers adds considerably to the cost of a survey. Waksberg (1978) reports that this spade work is done by marketing research firms, and the more "useful" sampling frames are developed by these firms at considerable expense and are not available to the general public. Most researchers cannot afford to duplicate such a costly task. It is important to narrow the frame used for RDD. The designs to be discussed in what follows are expected to reduce the proportion of unused numbers sharply.

To emphasize the inherent differences between the two frames and their variants, it is useful to mention the problems in determining the status of a given number. Dialing a working number can result in (i) a completed call, (ii) unanswered rings, (iii) a busy signal, or (iv) wrong or no connection because of misdialing or technical problems. Unless the call results in a contact, it is impossible to determine whether the number belongs to a household or a nonhousehold. In RDD sampling, a nonworking number is not always easily determined. Dialing such a number can result in (i) a recorded message stating that the call cannot be completed as dialed, (ii) no connection, (iii) unanswered rings, or (iv) connection with a number other than that was dialed. The last possibility introduces biases in RDD sampling, because the telephone system equipment is not normally designed to receive a nonworking number. Note that the households reached in this manner have a greater probability of inclusion in the sample.

2.2. *Telephone sample designs*

Telephone sample designs can be broadly divided into list-assisted and RDD methods. We shall briefly discuss these designs and finally discuss the concept of dual frame designs.

2.2.1. List-assisted methods

2.2.1.1. Direct selection from directory.

This is the most basic of all the directory assisted methods. A sample of directory lines is selected using either systematic or simple random sampling. One could also use cluster sampling for easy execution. The cluster consists of a randomly selected line and the next k lines. To avoid actually counting lines, directory column inches can be used. This method yields an equal probability sample of all listed numbers with a minimal percentage of wasted dialing due to nonworking numbers. A disadvantage of the cluster sampling method is that names listed together in a directory might belong to the same community, religion, etc., and if they are homogeneous with respect to the variables being estimated, the design is inefficient as compared to simple random sampling of lines. The major disadvantage of the directory method is that it does not give any chance for unlisted working telephone numbers to appear in the sample. The bias may be significant in certain surveys and the following procedures are proposed to correct partially for the bias.

2.2.1.2. Addition of a constant to a listed number.

A number is randomly selected from the directory and an integer, either fixed or randomized (between 0 and 9), is added to the directory number. This gives a chance for inclusion of possibly unlisted numbers in the sample. Some variants of the above mentioned procedure involve randomization of the last r (2, 3, or 4) digits or a directory number. Two drawbacks of these procedures are as follows: (i) when r increases, the number of wasted dial rings will increase, and (ii) all telephone numbers do not have an equal chance of inclusion, because the probability of selection of a number would be proportional to the number of directory listed numbers in the same rth bank. If the numbers are not in the directory, they automatically eliminate the possibility that numbers which follow them will be in the sample. A method suggested by Sudman (1973) to correct for (ii) is described in the following section.

2.2.1.3. Sudman's method.

A random sample of listed numbers is selected and the last (usually $r = 3$) digits are ignored. This results in banks of numbers selected with probability proportionate to the number of listed numbers in the bank. Calls are made using RDD within the bank until a predetermined number of households with *listed* numbers have been reached. The predetermined number is fixed so that the resulting sample is self-weighting. If we let N = total number of household telephones, N_L = number of telephones among N that are listed, n = sample size, m = number of selected banks of working numbers, and N_{L_i} = number of listed telephones in the ith bank, then

$$\text{probability of inclusion of a number in the sample} = \left(N_{L_i} \frac{m}{N_L} \right) \left(\frac{N_{Ln}}{NmN_{L_i}} \right) = \frac{n}{N}. \tag{1}$$

REMARK. This probability is exact (and the sample is self-weighting) only if (a) the proportion of listed households numbers in the ith bank is equal to the overall proportion (N_L/N) of listed household numbers and the predetermined number of sampled listed households in a bank is n/m, or if (b) the predetermined number of sampled listed households in a bank is fixed as nN_iN_L/NmN_{L_i} where N_i is the number of household telephones in the ith bank.

The first bracketed term indicates the probability of inclusion of bank i in the sample and the second term, that of selecting a number within the bank. The procedure is unbiased and self-weighting. As Waksberg (1978) points out, this method also has several problems. Ascertaining whether a number dialed is listed or not can be difficult. For example, in a national survey, the procedure requires the use of a large number of telephone directories. Finally, because the numbers are clustered, a large proportion of them may occur in relatively empty banks, resulting in unequal numbers of households per cluster.

2.2.2. Random digit dialing methods

These methods are used to obtain equal probability samples of all telephone numbers both listed and unlisted. As mentioned earlier, an unrestricted application of the procedure will lead to the inefficient use of survey resources. Therefore, it is important to narrow the sampling frame by eliminating nonworking numbers. If information on nonworking numbers is available (e.g., which banks are not assigned), random digits within these banks could be excluded from the sample. Some telephone companies will provide information about working banks. However, this information is usually not available, forcing researchers to use directories to determine working banks. Typically those banks with less than three listed phone numbers are eliminated. The incidence of telephone households in the sample can be increased by eliminating the business telephones listed in the yellow pages of the telephone directory. It is evident that all these efforts require a considerable investment of time and, unless the frame is used repeatedly, the cost may be prohibitive for a small survey.

2.2.2.1. Waksberg–Mitofsky design.

The (RDD) selection procedure proposed by Waksberg (1978) is as follows. Obtain from the telephone companies all area code–central office code combinations currently in service. Append all possible two digits and treat the resulting eight-digit numbers as primary sampling units (PSU). Randomly select a PSU and the next two digits. If the 10-digit number is for a residential address, the PSU is retained in the sample and if not, it is rejected. If retained, additional pairs of random numbers to identify the two last digits are selected within the same PSU and dialed until a set number of residential telephones are reached. This process is repeated until a predetermined number of PSUs are chosen. This design produces an equal probability sample of working telephone numbers. The procedure of selecting PSUs is similar to Lahiri's (1951) selection procedure for probability proportionate to size (pps), although the latter requires a prior estimate of cluster size. This procedure which selects PSUs with probability proportional to working numbers differs from Sudman's method which selects PSUs proportional to listed working numbers. The stopping rule for the Waksberg–Mitofsky design also refers to working numbers and is not restricted to listed numbers. It is important to note that this procedure uses a cluster size of 100, a practical advantage over a cluster of 1000.

A crucial problem in this procedure is the large value of the proportion of PSUs with no residential numbers. Because all possible choices of two-digit numbers are appended to area code–central office code combinations to arrive at the PSU, it is possible that a large number of PSUs may not contain any residential numbers. It is important to obtain an estimate of the proportion of PSUs with no residential numbers. This can be expected to be smaller for urban than rural areas. An estimate based on a national U.S. study is given by Groves (1978) as 0.65.

2.2.2.2. Stratified element sample. An alternative design is discussed in Groves (1978). The procedure initially groups together all central office codes in the same exchange and then groups together exchanges in the same area code. Size categories of the exchanges are then formed based on the number of central office codes in an exchange with the number of central office codes acting as a proxy to population density. Within each size category, exchanges are ordered geographically within an area code and similarly area codes are then ordered geographically. Given this ordering of the frame, a systematic sample of central office codes is drawn. A four-digit random number is generated and appended to a selected central office code, yielding a 10-digit sample telephone number. Groves (1978) observes that only about one-fifth of the numbers were confirmed a working household numbers, whereas in Waksberg's design a roughly threefold increase in identifying working household number is possible. The main attraction for using this design would be when there is a greater homogeneity among the prefixes. This design can be treated as a simple random sample when the stratification introduced based on the exchange size is rather weak.

2.2.2.3. Dual frame sample design. The two-stage cluster design, proposed by Waksberg (1978), is better than directory-based designs in terms of coverage rates and over stratified element sampling in terms of cost. However, the design requires a new selection from the same PSU for each nonworking number encountered, and thus adds to the cost of screening numbers to identify residences. It is difficult to distinguish nonworking numbers from unanswered residential numbers. Another problem is the low-response rates for telephone surveys attempted without prior contact. It is found that persons with listed numbers are more likely to cooperate than those with unlisted numbers. Groves and Lepkowski (1986) consider dual frame designs as proposed by Hartley (1962) to be useful when the target segment forms a majority of elements in one incomplete list frame (directory listings) but a minority in another complete frame (RDD generated numbers). The poststratified estimator suggested by Casady et al. (1981), which mixes the estimates from each of the two frames, is investigated by Groves and Lepkowski (1986) and Lepkowski and Groves (1986). If we let p denote the proportion of the unlisted telephone population and θ denote a mixing parameter, the estimator of the mean is

$$\bar{y} = p\bar{y}_{\text{UL, RDD}} + (1 - p)\left[\theta\bar{y}_{\text{L, RDD}} + (1 - \theta)\bar{y}_{\text{L, DL}}\right] \tag{2}$$

where $\bar{y}_{\text{UL, RDD}}$ is the estimate for the unlisted population chosen by RDD, $\bar{y}_{\text{L, RDD}}$ is the estimate of the listed population chosen by RDD, and $\bar{y}_{\text{L, DL}}$ is the estimate of the listed population chosen from the directory frames cases. The cost advantage of the dual frame derives from the list frame in identifying the working numbers. Several survey research firms (see Table 1) maintain a computerized data bank of all published directories and in one test for the state of Michigan, Groves and Lepkowski (1986) report 88% of numbers on the list were found to be working numbers as compared to 59% for the selection of samples within the PSU in RDD design. From the form of the poststratified estimator, it can be seen that the crucial parameters are p and θ which depend on the geographical region and the type of marketing research investigation. It is estimated that roughly 64% of contacted RDD sample households arc in directory listings, but the proportion of RDD numbers not contacted but found in the listing is around 66%. At the national level it is not known what proportion of these noncontacted numbers are working residential numbers. This may be influenced by large metropolitan areas where a low rate of list frame coverage is known to exist. Thus, the dual frame design can result in increased coverage (than list frame) and also increased precision (than the cluster RDD) by following simple random/stratified element designs on the list frame, thereby avoiding homogeneity due to clustering. To evaluate the dual design more thoroughly, the marketing investigator must know several cost elements and the relative nonresponse bias. The nonresponse bias is typically measure by the difference between two group means, where only one group receives an advance letter. Based on a simulation study for the U.S. National Crime Survey, Groves and Lepkowski (1986) suggest optimal allocations between 35% and 80% to the list frame.

2.2.2.4. List-assisted RDD methods.

The operational difficulties involved in implementing the Mitofsky–Waksberg method has led to increased use of list-assisted sample designs. Two main issues with the Mitofsky–Waksberg method were in replacing the nonresidential numbers and in variances being larger than a simple random sample or stratified random sample of the same size. The properties of the list-assisted methods were examined in detail by Casady and Lepkowski (1993). But the underlying structure of the telephone system has changed greatly since then. More area codes are now being assigned and there is a gradual decrease in the proportion of numbers that appear in directories. Thus, it has become increasingly difficult to identify the residential numbers. Tucker et al. (2002) evaluate relative efficiencies of list-assisted and Mitofsky–Waksberg designs and conclude that the relative gain in precision from list-assisted design has increased in the past decade.

2.3. Respondent selection in telephone surveys.

There are a number of other issues to be addressed in telephone surveys. Some households have more than one telephone number, making it necessary to obtain this information during the interview so that appropriate estimation weights could be constructed. In any telephone survey, ambiguities exist about no answers, uncertain rings, busy signals, etc. Any stopping rule for classifying these is bound to introduce some bias in sample selection. A more serious problem from a marketing researcher's point of view is that the person answering the telephone is not necessarily the same person who makes the purchase decisions. As shown in the literature on consumer behavior, buying decisions result from an interaction of all family members. To retain the characteristics of a probability sample, the person to be interviewed should be selected at random. We discuss a few approaches to the problem in the following.

A selection procedure suggested by Kish (1967) in the context of area probability samples requires all eligible respondents within a household to be listed by sex and by age within sex categories. The interviewer then selects one respondent using a random number table (see Kish, 1976, Section 11). This procedure is difficult to use in telephone surveys where most refusals to participate occur at the beginning of the interview. The procedure is time-consuming and could present problems establishing rapport. For example, asking for the number of adult males in residence could be perceived as insensitive to single women living alone. Because rapport with the respondent is so vital to telephone surveys, Troldahl and Cater (1964) adapted the Kish format but based the selection on only two easy-to-answer questions: (i) How many adults live in your household, counting yourself, and (ii) how many of them are men? Using four selection matrices rotated randomly over the sample, a respondent is selected. This procedure does not significantly reduce refusals when compared to the Kish strategy (see Frey, 1983, p. 80). Bryant (1975) suggested dropping one of the four matrices every second time it appears in the rotation. This would result in the selection of more male respondents and the procedure takes into account increases in one-person households and households headed by women. It must be noted that these alternative strategies assign unequal probabilities of selection to some eligible respondents such as middle-aged adults. Another variation used by Groves and Kahn (1979) is to modify (ii) "how many of them are women?" A recent investigation by Czaja et al. (1982) reveals no major differences in cooperation rates and demographic characteristics across the three models.

Two procedures reported recently seem to be effective in terms of operational use and eliciting higher response rates. Basically, these two avoid asking household composition questions before beginning the interview. The first procedure is suggested by Hagan and Collier (1983). The designated respondent is predetermined to be one of four possibilities: oldest man, youngest man, oldest woman, or youngest woman. After the initial introduction, interviewers simply ask for the designated respondent (randomly chosen and printed on the interview form a priori) and when a respondent of that designation does not live in the household, the opposite sex is interviewed. In single-person households, the age designation is irrelevant. Based on a national study, the authors suggest that this procedure is an improvement in terms of lower refusal rate. The second procedure given by O'Roourke and Blair (1983) selects the adult who had the "most recent birthday." This is a probability selection method and ascertaining the birthday is considerably easy. Comparing this with Kish's procedure, based on a survey, the authors found the major difference in refusal rate occurred at the preselection stage. Once the respondents agreed to participate, it did not matter which procedure was used to continue the interview.

Rizzo et al. (2004) provide a less intrusive method for selection of within-household members. It uses the fact that about 85% of households in the United States have less than two adults. Thus, this method randomly selects either the screener respondent or the other adult. Other than gathering information on the number of adults in the family, the procedure does not call for any information. The procedure operates as follows. Let N be the number of adults: if $N = 1$, the respondent is selected; if $N > 1$, randomly sample the respondent with probability equal to $1/N$. If $N > 2$ and if the screener respondent is not selected, then use the Kish method. This is a probability sampling method and does not result in self-selection biases.

2.4. Randomized response techniques in telephone sampling

The randomized response technique originally introduced by Warner (1965) to obtain the estimates of behavior that is usually underreported and is found to be useful for personal interviews. A randomizing device is used to choose a statement and the respondent is asked to provide a response to the one selected. The interviewer is neither shown the outcome of the device nor is informed of which statement is answered. The most difficult aspect of a telephone application of the randomized response technique for sensitive questions is the provision of a randomization device. As Stern and Steinhorst (1984) observe, there are two main problems: (i) the device is not readily available to many respondents, and (ii) the complexity of instructions necessary to provide a satisfactory distribution may inhibit respondent cooperation. Also, suggestions from a "faceless" voice to flip a coin may be regarded as foolish by some respondents. However, an advantage of using a respondent-supplied randomizer is that it eliminates the respondent's suspicions that the interviewer has "fixed" the randomizer. A potential disadvantage is that it does not provide a known probability distribution. The technique continues to be used widely in social sciences. See Van der Heijden et al. (2000).

There are several randomizers suggested in the literature including credit card number, street address, occurrence of events, etc. (see Orwin and Boruch, 1982). The one that is tested on a limited basis is the last digit of randomly selected telephone numbers. This provides a known distribution for both the selection of sensitive and nonsensitive questions and the generation of surrogate answers (see Stern and Steinhorst, 1984). Although this method is considered to be successful on the issue of response privacy, the nonresponse is still high. This method also requires both the interviewer and respondent to have access to the same telephone directory. Each geographical area served by a different telephone exchange and telephone directory would be sampled as a separate stratum. At a national level, this may create some operational problems. Other randomizers such as the last digit of street address are supposed to overcome this problem, but in the absence of a known distribution of the last digits, they are not statistically attractive to use.

2.5. Locating a special population using RDD

In many instances, the researcher may be interested in locating a subclass of the total population. Blair and Czaja (1982) show how Waksberg's two-stage cluster design can be modified, if it is known that this special population clusters geographically. This modification takes advantage of the fact that the telephone central office codes are assigned to well-defined geographic locations. It works as follows: select a simple random sample from all possible telephone numbers. These numbers are then called and only those working residential numbers of a household with the appropriate special characteristics are retained. The first eight digits of each retained number are then defined as a PSU. Using each retained telephone number as a random start in the PSU it created, numbers are then sequentially generated and screened. This procedure is continued until a certain cluster size is identified.

As Waksberg (1983) notes, this procedure has some serious statistical implications in which many situations may reduce the efficiency. But in the case of special populations, PSUs could exist in which it is not possible to reach the predetermined cluster size even if the 100 numbers are used. The special population households associated with clusters that are smaller than the specified cluster size have a lower probability of selection than the rest of the special population. Hence, to produce an unbiased estimate for the total population, we must adjust for unequal probabilities which increase the sample variance (see Kish, 1967, p. 430).

2.6. Ring policy in telephone surveys

Each telephone call is composed of 2-second rings followed by 5 seconds of silence. Survey research firms on the average allow six rings per call, thus the amount of time taken to reach a potential respondent is on the average 37 seconds. Smead and Wilcox (1980) questioned how long the phone should be allowed to ring based on a telephone survey using the members of a major university consumer panel. Ten rings and three callbacks were used. The average answer time for the 219 respondents was 8.7 seconds with a standard deviation of 6.3 seconds. The answer times followed a gamma distribution and suggested that only four rings (or 23 seconds) were necessary to reach 97%.

2.7. Telephone sampling: other uses

The use of telephone interviewing is widespread because of its major cost advantages. However, there are still many situations that require face-to-face interviewing, particularly those that deal with special subgroups of the general population. This involves screening, and the rarer the group the more costly the screening. However, in general, telephone screening costs are lower than face-to-face screening. Sudman (1978), based on a realistic cost model, has shown that telephone screening will be an optimum procedure unless (i) the degree of homogeneity is small, (ii) the density of interviews is low, and (iii) locating and screening costs are small relative to interviewing costs. From the discussion in Section 1, it follows that (iii) could be an important consideration in using RDD. However, directory-based telephone screening might be cost effective.

Many survey research firms have databases constructed from the telephone directories supplements with auto registration data. These are useful for mail samples. Information collected from other sources such as census records are sorted by area code and telephone exchange that provides a faster way to reach a target population such as low income families, Hispanic groups, etc. The yellow page listings are used for business samples, because the directory category headings are broad and easy to use by marketing researchers.

Computer-assisted telephone interviewing (CATI) was used first by market research agencies in the private sector. The concept was proposed by the American Telephone and Telegraph Company to measure customer evaluation of telephone services. CATI is now very popular in other types of organizations as well. Interview responses are quickly processed and by accumulating counts of key respondent characteristics while interviewing, quota targets, that is, desired sample sizes in strata in RDD sampling, can be tracked. Adding visual monitoring to telephones from supervisory terminals, CATI provides efficient control in the interview process (see Nichols and Groves, 1986).

2.8. Recent developments in telephones surveys

Cell phones, pagers, faxes, modems, Internet, call forwarding, voice mail, and other convenient services offered to phone subscribers are creating increasing challenges to researchers to contact the public for telephone interviews. The explosion of telephone area codes as a result of these new products creates a much bigger challenge to researchers to draw representative samples from their target population. According to a Lockheed Martin Study, United States will run out of new area codes by 2010. This implies evolving challenges for telephone survey sampling methodology. It is estimated that more than 25% of U.S. households have more than one land line. Households with children, Internet access, home-based businesses, and the difficulty to identify multiple phone line households create new challenges to draw a random sample of households.

Telephone Consumer Protection Act (TCPA) prohibits calls to wireless and assess a penalty for each such call. Land lines are household-based whereas cell (wireless) is population-based. Of the 9% of U.S. households that have no land lines, around 2.5% do not subscribe to a phone, and 6.5% have only wireless service. Cell phone numbers proliferate into RDD samples due to call-forward their land lines to wireless services. Wireless service-only households tend to be young males (less than 35 years), educated, employed, renter, earning less than $40,000 annually, and have no children. Wireless also does not have 911 services and when they get married and have family, they may opt for a land line. Some agencies such as survey sampling international (SSI) use software, wireless ID that reduces sampling risk by identifying potential wireless phones. Merging of two overlapping and incompatible sampling frames and households with multiple phone lines create potential cover bias.

The recent Cell Phone Sampling Summit II sponsored by Nielsen Media Research was convened to discuss how the cell phones are treated in RDD surveys. It is estimated that approximately 70% of the U.S. households have cell phones and it is growing. The telephone frame can be partitioned into three components: (a) land-line telephone exchanges, (b) cellular telephone exchanges, and (c) mixed-use exchanges. It must be noted that cellular telephone numbers are located in all those components of the frame. In addition to the issues discussed in telephone sampling earlier, the design should explicitly consider,

> "Weighing for unequal probability of selection, including whether a cell phone is a personal device reaching only one potential respondent or a household device reaching more than one potential respondent."

Because cell phone usage is on the rise among the teenagers, it is possible to reach ineligible persons when surveying adults and thus RDD cell phone calls may result in a wastage. These and other recent developments are to be carefully studied. We summarize a few studies that have addressed these issues later.

Tucker et al. (2007) report the telephone service and usage patterns in 2004 based on the information obtained from Current Population Survey (CPS). As observed earlier, standard RDD techniques usually exclude the cell phones, thus resulting in undercoverage. It is estimated that 6% of the households have only cell phone service. The percentage of one-person households that are cell-only (8.1%) is somewhat higher than that of large households (5.5%). Cell-only households are more likely to be renters than owners of homes. If the distribution is sliced by age approximately 20% young adults (18–24) are cell-only users. The data indicate that among those households that have both cell and land line, very few receive any calls in cell phones. Tucker et al. (2007) suggest using individuals as sampling units rather than households. But this can cause problems for households with multiple members who may share a single cell phone.

Brick et al. (2007) discuss the feasibility of cell phone surveys in United States. The contact rates across various time periods were the same for cell samples, whereas the rates for land samples were lower during weekdays. The refusal rate for cell sample is generally much higher and efforts to follow up also do not result in success. The text messaging was not effective in raising the cell response rate.

3. Fax surveys

In the mid 90s, there was a growing interest in conducting surveys via fax. Faster delivery was the main reason put forth along with the possibility that it may give the impression to the responder that the matter is important. Dickson and Maclachlan (1996) conduct a study to compare the mail surveys with fax surveys. They estimate that the cost per returned questionnaire in the fax was less than one-fourth of the cost for the mail surveys. The selection bias due to ownership of fax machines was not addressed. It is not known what percentage of households have fax machines and even if they have, what percentage of them keep them on. With the increasing use of scanners and the internet, fax surveys are not likely to take off.

4. Shopping center sampling and interviewing

Interviewing shoppers in shopping malls started in the early 60's when the development of totally enclosed shopping centers provided researchers access to a large number of shoppers from a wide geographic area. Prior to the mall intercept, surveys were mostly conducted in supermarkets, discount stores, train stations, and places where large concentrations of people could be found. More than 170 malls have permanent market research facilities, some of which are equipped with interviewing stations, videotape equipment, and food preparation facilities for conducting taste tests. A large number of malls permit intercepts on a temporary basis but may prohibit interviewing because they see it as an inconvenience to their shoppers.

The two major advantages of a mall intercept interview are cost and control and it has many of the advantages associated with personal interviewing. Also, it is the only way to conduct most taste tests and ad tests requiring movie projectors or videotape equipment. However, there are a number of disadvantages. The important one is that shoppers are frequently in a hurry and may not respond carefully. It may be difficult to maintain a controlled interviewing environment in the presence of the respondent's children, relatives, etc. Despite these problems, mall intercept interviews are increasingly used in market research. It is estimated that, of those who had participated in any form of a survey, 18% were contacted through mall intercept interviews compared to 12% through personal interviews (see Gates and Solomon, 1982). Because of the administrative efficiency, it has some potential for growth.

4.1. Sampling issues

Samples for most shopping center interview are selected haphazardly and do not reflect the general population. The effect and sources of biases are not properly understood and are not taken into account. If the investigation is at the early stages of product development, it may not be necessary to follow rigorous sampling procedures. But if the objective is to generalize to the population, it is important to follow rigorous sampling schemes. Shopping center sampling can be compared to sampling mobile populations. The major interest in studies related to mobile populations has been in estimating the size of the population, but little attention has been paid to sampling time and location. Sudman (1980) provides some procedures that take these aspects into account.

The key assumption in the mall samples is that all households have a nonzero (but not equal) probability of begin found in a shopping center. The assumption may not be realistic and the bias introduced for some special groups such as lower income or older households may be substantial. Second, because the probability of selection is a function of the frequency of visits, that frequency must be estimable. This may strain respondent memory and may introduce some biases.

Sudman's procedure works as follows: First, select the shopping centers using the same basic random sampling procedures used in the selection of locations in a multi-stage area probability sample with probability proportional to a size measure such as total annual dollar volume. The optimum number of shopping centers and the number of respondents can be determined using the formulae for area cluster samples,

$$n_{opt} = \left[\frac{C_1}{C_2} \left(\frac{1-\rho}{\rho} \right) \right]^{1/2} \tag{3}$$

where C_1 is the set up cost at a shopping center, C_2 is the cost per interview and with a total budget $C = C_1 m + C_2 mn$, where m is the sampled number of shopping centers, n is the number of interviews per shopping center, and ρ is the intraclass correlation coefficient between shoppers within shopping centers. Because C_1 is generally much larger than C_2, large samples are selected from each center; but the heavy clustering increases the sampling variance.

The respondents can be selected either when they arrive at the center or as they move around within it. For the latter, we require information on how much time they have spent in the center because persons spending more time shopping have a higher probability of selection. To select an unbiased sample of entrances, it is important to know the fraction of customers the entrances attract from previous counts. This size measure can be used to sample entrances with probability proportional to size and is much more efficient than sampling them with equal probability. Though the less-used entrances will be sampled fewer times than the more heavily used entrances, the sampling rate would be higher at the less-used entrances if a self-weighting sample is desired. Establishing rules for within shopping center sampling is more difficult than entrance sampling. Identical traffic patterns in all parts of the center cannot be assumed because the location of discount stores is more likely to attract customers different from those who shop at fashion centers.

It is important to use careful time sample procedures, to avoid biases against certain types of customers, for example, working women who mostly shop in the evenings and weekends. Selecting an eligible time period with equal probability is not an efficient design. The solution is identical-sampling of time periods with probabilities proportionate to the number of customers expected in the time period. Sudman (1980) suggests forming time–location clusters, based on past data and selecting these clusters with probability proportional to past size.

The above mentioned procedures are far more sophisticated than those procedures used in the past. There are still problems in their implementation and generalizability. We suggest using the dual frame concept. For each shopping center, we may obtain trade area maps showing geographic areas from which stores draw their trade, because shopping centers generally attract those households nearest to it. These maps are sometimes drawn from shopper surveys (see Blair, 1983) intended for a different use by the retail merchants. With such a map, we may have a sampling frame from which we can draw an independent sample by telephone that can be combined with the mall sample. For a related discussion, see Bush and Hair (1985).

It is known that sampling bias may occur when the individuals spend different lengths of time at the survey location. Because most surveys are conducted away from the entrances, individuals who spend more time at the mall are more likely to be sampled. Such samples are known as length-biased (Cox, 1969). Recreational shoppers are likely to be overrepresented in the sample. Nowell and Stanley (1991) report a study on the bias of length of stay and suggest correcting for the bias using the procedures given in Cox (1969). The key factor appears to depend on whether individuals can accurately estimate the time they spend at the mall. Nichols et al. (1995) report that the length of time spent in the mall is different for Hispanics. Generally they spend more time traveling to the mall, but spend less time in it. Thus, both frequency bias and length of stay bias need to be considered for the shopping mall estimates.

5. Consumer panels

The panel has become an important tool for monitoring market factors ever since Jenkins (1938) and Lazarsfeld and Fiske (1938) used them to study brand preferences and reader reactions to a magazine (*Women's Home Companion*). Since then, the use of panels to study the purchase behavior of nondurable consumer goods has gained importance in North America and some Western European countries. See Hardin and Johnson (1971) for various applications of panels in marketing research. Marketing Research Corporation of America (MRCA) followed with a panel of 7500 households in 1941 and focused on the consumer purchase behavior of grocery, health and personal care, and textile products. Today, the use of panels in marketing studies is much more widespread and there are hundreds of consumer and industrial panels mostly located in North America and Western Europe. Nevertheless, some of the initial sampling problems related to panels still remain. This section will briefly review some of these problems from a sample design perspective. The problems related to panel sample design are not usually covered in discussions of sample survey methods. Sudman and Ferber (1979) identified three critical areas likely to induce bias in panel sample design. These are as follows: (i) bias created by initial refusals, (ii) bias created by subsequent mortality, and (iii) bias created through conditioning. A brief discussion of these areas follows. It must be recognized that there are other critical areas, such as aging of the panel and possible changes in the population that are not represented in the sample, which are not discussed here.

A consumer panel measures purchases of a product at any given point over a period of time. This has been used to measure market trends, seasonal effects, and the effects of marketing strategies. Panel data from the *Chicago Tribune*, National Panel Diary (NPD), National Family Opinion (NFO), Marketing Research Corporation of America (MRCA), Intercontinental Marketing Services (IMS), etc. focus on different product lines and industries. The majority specialize on consumer products, mostly nondurables distributed through grocery stores, whereas industrial panels such as those from IMS focus on hospital equipment, supplies, and doctor's prescriptions. Alternatively, store audits are used to estimate market size and trends (A. C. Nielsen) and with the advent of electronic scanners of Universal Product Codes (UPC), purchase data have become much more reliable and offer extensive detail on product/brand purchases as well as profiles of sample buyers. Information Resources, Inc. with headquarters in Chicago provide Infoscan and Behaviorscan services to business clients. Each panel member receives a member identification card that is presented to the store clerk at the time of checkout. All purchases are electronically recorded, eliminating the need for written diaries. This method has distinct advantages, as its popularity is growing both in North America and abroad (Information Resources has operations in Australia, Canada, France, Great Britain, Japan, and West Germany). These sources also study consumer brand preferences and brandswitchings over a period of time. Panel data have been extensively used in the formulation and evaluation of pricing strategies (see Montgomery, 1971). Segmentation by usage, package size, effectiveness of "marketing mix" variables have been studied by, among others, Blattberg and Sen (1976). Models are developed to predict market penetration based on repeated buying rates (see Eskin, 1973). With the information provided by panels on both purchasing and media exposure, efforts were made to estimate the effectiveness of advertizing particularly for new products (see Nakanishi, 1973). Carefoot (1982) and Information Resources, Inc. have used scanners to evaluate the effectiveness of advertizing. MRCA's panel data have been utilized to sense changing food habits leading to the modification of existing products and the development/introduction of new products to better serve the consumer.

5.1. Bias created by initial refusals

Refusals, noncooperation, and nonresponse are to be expected in any survey. The level of cooperation attained is dependent on recruiting methods used and the nature of tasks required by the panel members. Often higher rates of cooperation are achieved if the expected effort from the respondent is lower. Panels recruited by face-to-face contact tend to have higher rates of cooperation than those recruited by telephone or mail. Oversamples are drawn initially to balance demographic variables such as geography, household size, income, education of the head of household, etc. Even if the panel fits all these demographics, there is no assurance that the panel results are bias free if willingness to cooperate on a panel and purchase of a product are related to a variable such as lifestyle. Panel cooperation seems to be closely associated with family size; for example, households with two or more members tend to cooperate more readily than single-person households. From the studies of the U.S. Department of Agriculture (1953) and additional investigations ("Panel bias reviewed," 1976), the following patterns emerge:

- Single-person households have a higher tendency to be noncooperators or "not-at-homes." They have less interest in food purchases and maintain records on an irregular basis.
- The older the housewife (after 55 years), the lower the chances of joining the panel. This may be related to education and the ability to keep records.
- Homeowners are more likely to cooperate than tenants. This again may be related to household size.
- Working wives are less likely to join the consumer panel than nonworking wives.
- Panel cooperators tend to be more "price conscious" than noncooperators.
- The income distribution of panel members and that of the U.S. population tend to be very similar except at the lower end where a smaller percentage of lower income households are represented in the panel.

Except for household size, the differences between cooperators and noncooperators tend to be negligible with respect to demographic profiles. However, the differences could be significant with respect to socio-psychographic characteristics such as organization, record keeping, and price consciousness. With new developments (Infoscan and Behaviorscan) the need to keep records by the panel members is eliminated, reducing potential errors in reporting, recall, and record keeping. Atwood consumer panels in Great Britain and Germany show no significant differences between panel members and the general population with respect to readership of magazines and newspapers and selected psychological and buying variables (Sudman and Ferber, 1979).

In summary, the evidence from the United States and European studies indicates that some biases in consumer panels such as household size, age of the housewife, and level of education of the head of household are possible. In panels requiring less effort, the refusal rate is lower resulting in lower sample bias. Panels that require more effort and those recruited by mail or telephone often tend to have a higher percentage of noncooperators resulting in higher bias.

The ratio method of estimation has often been used to obtain better estimates of the population. Under-representation of smaller households or a specific geographic region is overcome by the application of suitable poststratified weights in deriving the population estimates.

5.2. Bias due to attrition/mortality/formation of new households

A panel should be representative of a target population. Though the population itself may not change drastically from year to year, some changes do occur over time. Dissolution of old households, formation of new households, household moves, etc. are examples of changing population characteristics. Potential problems are as follows: (i) panel member dropouts, (ii) household moves, (iii) household dissolutions, and (iv) new household formations. We will discuss each of them briefly.

(i) Dropouts: Panel dropouts or attrition is often estimated to be 5–10% from one period to the next in the United States. Charlton and Ehrenberg (1976) reported that 88% of their limited sample completed the 25-week panel. Farley et al. (1976) reported a 43% dropout rate from the waves of interviewing spanning 18 months. Personal situations, such as illness in the family, birth of a child, enlistment in the army, etc., are often the reasons for dropout. Two methods have been used to overcome this problem. An oversample could be made in anticipation of an expected dropout rate. However, in practice, it may not be possible to maintain large oversamples (European panel operators tend to follow this procedure). Besides, this would lead to sampling bias. The second method is to replace the dropout household with a new household of similar characteristics by a method of imputation in the field. The problem of noncooperation of a newly selected household is similar to that of initial recruiting. A prepared list of substitute households is searched until a replacement is found. Even if replacements are representative with respect to selected socioeconomic and demographic variables, they could differ on behavioral variables such as purchase quantity, degree of brand loyalty, private brand proneness, etc. Winer (1983) suggested that replacements be made with due consideration to selected behavior variables.

Sobol (1959) and Bucklin and Carman (1976) demonstrated that attrition introduces potential bias in panel-based market research. Hausman and Wise (1979) have designed a model of attrition and proposed a maximum likelihood method of estimation of parameters. They estimated the parameters in the presence of attrition as well as bias due to attrition. Winer (1980, 1983) and Olsen (1980) developed procedures for estimation of attrition bias in the absence of replacement of dropouts.

Maintaining a representative panel is not easy. Most panel operators recognize the importance of suitable compensation and effective communication with panel households as essential factors in keeping morale high and turnover rate at a minimum.

(ii) Household moves: When a household moves, it is a generally accepted principle to follow it. The only exception is if the panel is confined to a specific geographic area and the move takes the household out of that target area. Following the panel wherever they go ensures continuous representativeness of the panel including the patterns of mobility inherent in the population.

(iii) Household dissolutions: In the event that all members of a panel household die, the household is often replaced with a similar household. If one of the spouses dies and the other joins a nursing home, the household is dropped from the panel.

(iv) New household formations: The panels are continuously monitored as to the size of the household. If a new household is formed through marriage, the new household is recruited with probabilities inversely proportional to the number of persons who will constitute the new household. Thus, in the case of new households resulting from a marriage, half the split-offs are recruited. This way the panel recruits younger households to maintain their representatives in the population.

5.3. Bias created through conditioning

The term "conditioning" refers to stimuli in a broad sense and includes all contacts between panel operators and panel households such as initial recruiting calls, instructions/training, diary keeping, compensation, and newsletter or other forms of communication whether personal or mail. Sudman and Ferber (1979) classified the effects of the stimuli into three categories: immediate, short-term, and long-term. These effects could be in terms of purchase behavior affecting brand choice, store choice, quantities purchased, number of shopping trips per unit time, expenditures on a product per unit of time, etc. For example, keeping a "time-use" diary might cause a person to use a different pattern of time utilization than the "usual." Besides changes in behavior, it might also change attitudes and beliefs affecting future behavior.

Studies focusing on the immediate effect of the acceptance of an invitation to join the panel on a household have used "recall" techniques to assess the differences in purchase behavior before and after joining the panel. The results, however, were inconclusive. The effect of short-term conditioning seems to be evident bases on empirical studies. A 1973–1979 study conducted by the Survey Research Laboratory at the University of Illinois on medical diaries found that first month reportings were 14% higher than the subsequent records of the following 2 months. Similarly, Sudman (1962) found that a panel diary method used to collect data on 10 product purchases over an 8-week period reported that first week purchases were 20% higher than the 8-week average and second week expenditures were 8% below the average. The experiences of U.S. Bureau of Census (1972–1973) also support the evidence of the existence of a short-term conditioning effect on the behavior of panel households. As a result, many practitioners ignore the first period as "trial" data or omit it in the trend analysis.

Substantial evidence exists that a special stimulus can result in major changes in reported purchase behavior. A sticker reminder in a diary and a postcard reminder to record all soft drink purchases resulted in an increase or more than 30% in reported purchases. A similar study on reporting purchases of citrus products (special form included for reporting) showed that the experimental group had a significantly higher incidence of purchase records of citrus products during the first month than the control group. However, the initial conditioning effect seemed to have disappeared in later months.

Some researchers have speculated that keeping diary records could sensitize households over time and cause them to be better shoppers. One panel study indicated that an average household made 2.7 trips per week for grocery shopping during the first 3 months of data collection period and 2.6 trips per week in the next 3 months. The differences are not statistically significant, and any conditioning effect is negligible. Ehrenberg (1960) using a British consumer panel and Cordell and Rahmel (1962) using A. C. Nielsen panel for television viewing habits concluded that there may be a slight short-term effect of panel conditioning but it disappears over the long term.

Long-term effects on households serving as panel members is of major concern as they could develop fatigue or become uninterested in keeping diaries. Interestingly enough, there is no evidence to support such a hypothesis. Ehrenberg (1960) described several studies and pointed out that over a 10-year period the Atwood consumer panel compared "old" and "new" panel members and found no significant differences. The general conclusion was that the length of panel membership did not systematically affect the reported results. Any "conditioning" that may exist in the early period of panel membership is likely to wear off or stabilize over a reasonably short time.

Some form of compensation is very common for most continuing panels and is often in the form of money, gifts, or other forms of motivation (participation in lotteries, etc.). The amount or value of compensation seem to vary widely depending on the type of respondent. For most consumer nondurables, the compensation has been in the range of $10–$60 a year. For physician panels, the compensation was several hundred dollars. Both European and Japanese panels seem to receive better compensation than those in North America. Ferber and Sudman (1974) and Sudman and Ferber (1971) reported that the households receiving compensation provided better quality data than those who did not. Their conclusion was that compensation in sufficient amounts is necessary to ensure initial and continuing cooperation as well as quality of reporting. There is no evidence that the form of compensation has any major impact on cooperation (Ferber and Sudman, 1974).

5.4. Consumer panels: other issues

A study by Grootaert (1986) on the estimation of household expenditures in Hong Kong using the panel diary method suggested the use of multiple diaries—each member of the household maintains a separate diary of daily expenditures. This method resulted in more accurate reporting of expenditures particularly on "personal" products such as clothing, shoes, and services. The reporting arrangements depend on family structure, size, and decision making process within a household. As such, the results are not usually generalizable to other countries.

With high-tech electronic methods of data collection using scanners, the need to maintain written diaries is diminishing. As increasingly more retail stores are equipped with UPC scanners, data collection using panel method has become increasingly important for various marketing experiments. This has led to what is called "single source" research where many promotional experiments can be tested out by following the panel members from their TV sets to checkout counters.

It is easy to measure accurately the effect of promotional campaigns via this high-tech research. Information Resources, Inc. (IRI) monitors 3000 households in eight small town markets. The microcomputers record when the television is on and which station it is tuned to. IRI sends out special test commercials via cable channels. The single source research has it drawbacks. The size of the panels is still relatively small because of the high-cost nature of data collection and hence it is doubtful how generalizable the results would be to the entire market. Second, how do we know viewers are actually watching the test commercials. The change in the buying behavior is also questionable when the panel members are probably conscious of being in the panel. Brand loyalties are somewhat difficult to change by a short-term advertizing. But this research may be useful for new products (see Kessler, 1986).

5.5. Recent developments in consumer panels

International household consumer panels are maintained by various commercially oriented survey research companies. SSIs surveyspot (U.S. Panel) covers North America whereas Opinionworld offers collective panel for Europe. SSI offers proprietary panels in more than 40 countries and in early 2007, it added China to the list of countries offering consumer panels. Though the literature on panels initially focused on consumer/household panels, now panels are extended to commercial and professional panels. Commercial panels are often used to track movement of goods and services at different stages of distribution to monitor trends. For example, a panel of pharmacists is used to track or monitor trends in prescription drugs.

6. Web surveys

The online world has become as important to Internet users as the real world (http:// digitalcenter.org). "The internet has been a source of entertainment, information, and communication since the web became available to the American public in 1994." During the past decade Internet has become the primary vehicle for conducting marketing research. Web surveys, Internet panels, E-focus groups, web advertizing research, etc. have replaced traditional methods of conducting marketing research. Internet has also become rich source of secondary data and become universally accessible by anyone from anywhere and brought down the cost of conducting marketing research more effectively, with higher speed, and ever declining costs of unit information. Further developments, as noted in the sections that follow, will have profound effect on tools, and methods employed in marketing research over the next 10–15 years.

6.1. Internet penetration

Since Internet accessibility to the public in 1994, it has made rapid strides (see Table 2) to become a very powerful platform and changed the way we do business, and the way we communicate. It is the universal source of information. In fact, Internet is the most democratic of all mass media. With a very low investment, any business irrespective of its size can have a web page and reach a very large market, directly, fast, and economically. With a small investment almost anybody can have access to the worldwide web. The number of internet users in December 1995 was 16 million representing only 0.4% of the world population. This has grown to 361 million or 5.8% of the world population by December 2000 and then to 1.018 billion or 15.7% of world population by December 2005 and to 1.093 billion or 16.6% of world population in December 2006 (www.internetworldstats.com accessed on Feb. 12, 2007). This represents an annual growth of some 46.8% since 1995 and reaching a moderating annual growth of 25% during the past 5 years.

Although the annual growth of Internet users may moderate from the past rates of growth, it is a fair assumption to forecast 15–20% annual growth between now and 2010. That translates to 1.91–2.27 Bil Internet users by 2010. This tends to imply that Internet research can be representative and effective as other traditional methods with the fast growth of Internet populations. As reported recently by researchers, the problems of conducting Internet research must be effectively addressed and resolved, just as the problems with traditional research. (Ilieva et al., 2002; Kellner, 2004; Mathy et al., 2002; Schillewaert and Meulemeester, 2005)

English is the language of some 30% of Internet users followed by Chinese and Spanish by 14% and 8%, respectively. Japanese and German occupy fourth and fifth ranks with 7.9% and 5.3%, respectively. The top five Internet users languages account for nearly two-thirds of Internet user population. (www.internetworldstats.com) Table 2 presents SSI Internet Samples by country and official language. It can be seen that economically well-developed countries tend to have higher and similar levels of penetration.

Table 2
Internet users by region and penetrations by their respective populations

Region	No. of Internet Users (million)	Internet Penetration (% of Population)
Asia	389	10.5
Europe	313	38.6
North America	232	69.4
Latin America	89	16.0
Africa	33	3.5
Middle East	19	10.0
Australia/Oceania	19	53.5

Source: www.internetworldstats.com accessed on Feb. 12, 2007

6.2. Web surveys: issues

The power of the Web appears to have both positive and negative sides. Because it is relatively inexpensive to conduct surveys on the Internet, any business organization irrespective of its size can avail this opportunity. On the other hand, the proliferation of Web surveys makes it difficult to evaluate the quality of the surveys. Couper (2000) has observed: "It has become much more of a fragmentation than a bifurcation ("quick and dirty" versus "expensive but high quality," as was originally predicted) with vendors trying to find or create a niche for their particular approach or product." The Web surveys must be evaluated like other surveys in terms of their sampling, coverage, nonresponse, and error properties.

6.2.1. Coverage and sampling error

The construction of sample frame for Web surveys that will lead to selecting probability samples is not easy. The sampling frames are often incomplete and the coverage error is probably the most serious as only about 42% of the population have used the Internet even in the United States. Although it is expected to grow, it is not clear how much this percentage will be in the future. It may be constrained by the interest of the population in information sources.

The problem is not only who has access to the Internet but also the demographic and behavioral difference of the population base between those who have access and those who do not. The National Telecommunications and Information Administration (NTIA) report generally identifies that income, race, education, and household composition all play a role in having Internet access. Thus, the challenge for Web survey researchers is to find ways to reach the target population or otherwise, the inferences from survey results could be very restrictive. Because of the coverage issue, the sampling errors are likely to be high and skewed.

6.2.2. Nonresponse error

The nonresponse error depends on the rate of nonresponse and on the difference between respondents and nonrespondents on the variables of interest. When the sampling frame itself cannot be defined, the problem becomes even more acute. Couper et al. (1999) summarize the response rates in e-mail surveys and observe that the response rates in e-mail surveys are lower than the rates for mail surveys. Several reasons are attributed to this gap: lack of personalization as in mail surveys, technical difficulties in using the Internet, and confidentiality concerns.

6.2.3. Measurement error

The Web is more flexible for constructing survey instruments, such as adding visuals, etc., and therefore provides many options in its form and in its content. There is no definite conclusion on the ideal form of the surveys; it is clear that it all depends on the target population. In longitudinal surveys it is possible that the response over time may be biased. Given that the sampling frames are not easily defined for Web surveys, the statistical adjustments that could be made need to be studied more carefully.

6.3. Types of web surveys

Couper (2000) provides a neat summary of various types of Web surveys (see Table 2 of his paper). The surveys could be broadly classified as based on nonprobability and probability methods. Although the nonprobability methods are similar to other media surveys such as telephone and mail surveys, we focus here on probability methods. They take only two forms: restrict the sample to population with web access and thus limiting the generalizability of the survey results; use other methods to reach broader population via RDD-type of tools. These are briefly summarized in the following sections.

6.3.1. Intercept surveys

These are targeted toward visitors to a Web site and are used mainly for eliciting product-related opinions and in general to acquire customer feed back. Typically, systematic sampling is used and cookies are used to track the visits and for the timing of the exposure to the survey.

6.3.2. List-based samples

Here, we begin with a list of households with Web access and invite a select sample to participate in the survey with proper checks for avoiding duplications. Although this may cover only a portion of targeted sampling frames, yet useful estimates of samples error, etc. could be derived.

6.3.3. Prerecruited panels

Panel members are selected using probability sampling methods such as RDD and are recruited for surveys on the Web. Because the selection has a probability basis, the quantification of various types of errors are possible. However, it is recognized that the nonresponse errors, etc. could be still different for Web and further research must be done to better understand the dynamics of nonresponse between the Web and other modes.

6.3.4. Probability samples

The basic approach here is to first define the appropriate sampling frame and then provide Web access to those who are recruited but do not have access. Although this is much more scientific than the other methods, the initial response rate to the recruiter interviews has been somewhat poor. But this approach essentially solves the problem of representative coverage and Web access.

6.3.5. Mixed-mode Design

These designs combine various modes of reaching the targeted group. Mixed-mode surveys provide an opportunity to overcome the weaknesses of each method, but deployment of mixed-modes of data collection raises many challenges including the possibility that some respondents may give different answers to each mode. Dillman (2007) identifies five situations for use of mixed-mode surveys.

- Collection of the same data from different members of a sample.
- Collection of panel data from the same respondents at a later time.
- Collection of different data from the same respondents during a single data collection period.
- Collection of comparison data from different populations.
- Use one mode to prompt completion by another mode.

Use of mixed-mode surveys may enhance the possibility of improving response rates and reduction of nonresponse and coverage errors. Though there is no compelling evidence for choice and sequence of mixed-modes to be employed, it appears prudent to start with a method that is least expensive such as web surveys and then move towards mail, telephone, and if still necessary employ personal interviews. When multiple modes of data collection are employed it is essential to make a deliberate effort to deliver equivalent stimulus regardless of whether it is delivered aurally or visually.

6.4. Online Panels

The conduct of surveys via online panels has gained prominence in recent times because of easy access to the internet. But the recruitment of panel member in some cases is not based on known sampling methods. Harris Interactive for examples recruits panel members on a voluntary basis, but collects extensive demographics for postsurvey adjustments. Knowledge Network uses the telephone RDD methodology to recruit the panel members, but the survey information is collected via the internet. In a study related to health surveys, Baker et al. (2003) found that these online panels are no different from other panels in terms of response rate, attrition, etc.

6.4.1. Some panel sources

Many suppliers of marketing research services offer online panels (www.iri.com; www.acnielsen.com; www.lightspeedresearch.com; www.mra-net.org, and others). Information Resources Incorporated offers a behavior scan system whereas AC Nielsen offer scan track and specialty panels such as the African-American consumer panel (www.targetmarket.com). Foreign vendors such as Marsc panel management (www.marsc.co.uk), and Intage Inc. (www.jmra.net.or.ip) offer consumer panels in the U.K. and Japan, respectively. Several vendors such as Forrester (www.forrester.com), Lightspeed (www.lightspeedresearch.com), Marketmakers group (www.marketmakersgroup.com), Perez (www.perez.com), Robert Thale Associates (www.robertthaleassociates.com), and others offer B2B panels as well as consumer panels. Future Information Research Management (FIRM) (www.confirmit.com) operates a database of global 5000 for use of online B2B surveys and panels.

7. Conclusion

With the advent of Internet, we expect that a great number of market research surveys will be carried out through the Web. Because of the coverage issues we expect increased use of mixed-mode surveys. The surge in the number of surveys is bound to affect both the nonresponse rates as well as the quality of the response. The Web provides a unique opportunity to customize the surveys but this comes with a price of increase complexity to generalize the results of the surveys because, the web universe is still evolving. Several articles that have appeared in the *International Journal of Market Research* emphasize the various issues pointed out in this chapter. It is obvious that until the issue of Web sampling frame is clearly understood, the quantification of sources of survey errors will continue to have bias.

Essential Methods for Design Based Sample Surveys
ISSN: 0169-7161
DOI: 10.1016/B978-0-444-53734-8.00010-8

10

Sample Surveys and Censuses

Ronit Nirel and Hagit Glickman

1. Introduction

For many people, the simultaneous use of the terms census and sample survey seems contradictory. This chapter highlights the past use of sample data in census projects, as well as describes innovative developments in census methodology that accommodates sample data to various degrees. We start with a brief presentation of the main features of a census and the new trends in census methodology.

Censuses provide a core of official statistical data, around which demographic analyses, survey estimates, and administrative data are calibrated. A population census has been defined recently as "the operation that produces at regular intervals the official counting (or benchmark) of the population in the territory of a country and in its smallest geographical sub-territories together with information on a selected number of demographic and social characteristics of the total population" (United Nations Economic Commission for Europe [UNECE], 2006, p. 6, no. 19). The essential features of a census, as specified by the Commission, include universality, simultaneity of information, and individual enumeration. These features imply that there is a (a) well-defined census population, (b) reference date for all census data, which is usually referred to as Census Day, and (c) accurate data pertaining to individuals with regard to place of residence and other sociodemographic characteristics on Census Day are collected. Thus, for a person to be enumerated correctly in a census, nontrivial eligibility criteria must be met.

National eligibility. Each person should have one and only one usual place of residence, which defines the country he or she belongs to. A person living continuously in one country for more than a predefined period of time (e.g., for over one year on Census Day) is considered a "usual resident" of that country and is included in the target census population. A person living outside the country for longer than that period is not eligible for enumeration in the census. An illegal work migrant, for example, may be eligible for enumeration in the census if he or she has resided continuously in the destination country for more than, say, one year. The crucial role of the time reference is also noteworthy. For example, a baby born one day after Census Day is considered ineligible for enumeration, whereas a person in the target population who died one day after Census Day is eligible.

Local eligibility. Within a country, every individual should be counted at his or her usual place of residence on Census Day. The definition of a "usual residence" specifies, for example, criteria for people who divide their time between two places of residence (e.g., people working away from their family's place of residence; children in shared custody). The issue of place of residence also interacts with the time reference. People who moved one day before Census Day are considered ineligible at their former address, even if they lived there for 30 years, and are eligible at their new address.

For detailed recommendations on eligibility issues, see UNECE (2006) and United Nations (2006).

In the past, censuses have involved nationwide area enumeration (door-to-door data collection). In recent decades, however, it has become evident that there is more than one way to conduct a census with the essential features mentioned above. First, various collection methods have been developed within the traditional door-to-door framework, including self-enumeration and mail back and/or mail out options. Another group of new methodologies includes censuses that are based on administrative sources, with or without supplementary fieldwork data. The third group of innovative census-taking methods is based on appropriate accumulation of *sample* survey data that cover the census population over a predefined period of time.

It is not difficult to understand why many countries invest in developing new census methodologies. To begin with, advances in information technology have enhanced the role of administrative data in managing official statistics, including construction of registers such as population and housing registers. Enhancement of record linkage procedures has made it possible to accumulate data from various sources. Concomitantly, the demand for more detailed information has increased, whereas the willingness of individuals to respond to questionnaires has declined. Thus, many new approaches focus on improving quality and timeliness of census outputs while reducing the response burden. Finally, some countries expect the new methodology to reduce costs and enable expenses to be distributed more evenly over time.

Regardless of the methodology used, census counts are subject to different types of errors, which include coverage, content, and operational errors. Of those, coverage errors are the most crucial. There are two types of coverage errors: undercount and overcount. Undercount occurs when an eligible person is omitted from the enumeration, and overcount occurs when an ineligible person is erroneously enumerated. The definition of coverage errors depends on the geographical scale of interest. For example, some countries define a coverage error only at the national level, whereas other countries consider a person who is enumerated incorrectly at the level of a geographical region (e.g., in the wrong province) as contributing to undercount in the correct region and to overcount in the region of enumeration. Errors at various geographic scales within a given country can be of interest as they may relate to different uses of census data.

Because the census is an important and costly operation, evaluation of its coverage errors has become a "state-of-the-art" procedure in census-taking. Thus, many countries conduct a postenumeration survey (PES) immediately after enumeration (whether the enumeration is field-based or administrative), with the objective of estimating census coverage rates. Furthermore, some countries use PES estimates to adjust census counts for coverage errors. Section 2 describes the use of sample surveys to estimate coverage errors. The section begins with a description of model-based undercount estimation (Section 2.1) and extends the approach to estimation of undercount and overcount (Section 2.2). To conclude, we present design-based approaches to coverage evaluation (Section 2.3). In that section, corresponding sample designs are presented together with illustrative examples.

The PESs are generally large surveys possibly comprising hundreds of thousands of households. As such, some countries conduct operations that evaluate the PES. This "second-order" evaluation becomes a "first-order" evaluation of the census output when the PES results are used to statistically adjust the census counts. Although the main source of bias in "raw" census counts are errors pertaining to coverage, the main sources of bias in a PES or in adjusted counts relate to measurement errors, modeling, and processing. Owing to the dearth of comprehensive investigations on this topic, Section 3 attempts to provide a conceptual framework for possible uses of sample data to evaluate statistical adjustment of census counts. In light of the growing diversity of census-taking methods, we focus on broad principles rather than on specific solutions. The first step we propose is to analyze the remaining uncertainty in the adjusted counts (Section 3.1). The second step is to identify potential errors resulting from this uncertainty (Section 3.2), and the last step is to design different evaluating operations, including "Evaluation Follow-Up" (EFU; see Section 3.3).

Section 4 deals with an entirely different types of census that is based on a system of sample surveys. This approach adopts the principles of a rolling sample design, which is briefly described in Section 4.1. The remainder of Section 4 focuses on a description of the rolling census in France, which is the only country that has decided to carry out a sample-based census to date. In the concluding section, we describe sample surveys carried out in conjunction with a census (Section 5). These are the "long-form" surveys, in which comprehensive socioeconomic information is collected in the framework of the census. The U.S. Census is presented as an illustrative example of the main methodological features of such surveys (Section 5.1). A relatively recent development of the long-form concept is referred to by the UN a "traditional census with yearly updates of characteristics" (UNECE, 2006). In this type of census, only short-form data are collected, and the long form is replaced by a set of annual samples from which socioeconomic data are collected during the intercensal years. This idea was implemented in the American Community Survey (ACS), which is described in Section 5.2.

2. The use of sample surveys for estimating coverage errors

A population census is exposed to different types of errors, including coverage, content, and operational errors (UNECE, 2006). Of these, coverage errors are the most serious because the main objective of a census is to provide a full and accurate count of the population. Let C be the census count and N be the true population count. The census net coverage error D is defined by

$$D = N - C.$$

A positive value of D indicates net undercount and a negative value indicates net overcount. Coverage errors arise due to omissions or erroneous enumerations of people in the census. In the past, when enumerators conducted door-to-door enumeration, the most common coverage problem was undercount of dwellings and of individuals within dwellings. In recent years, interest in correcting overcount errors has grown, as several countries base their censuses on administrative registers, and as door-to-door enumeration and form collection has been replaced by various combinations of data collection through the mail and internet. Registers are often subject to inaccuracies in both directions (undercount and overcount) because of delayed updates or lack of reporting. Mailing and internet responses are exposed to duplications and fabrications, as well as to difficulties in understanding census eligibility criteria. Because the census is a large, central, and costly operation carried out once every 5–10 years, evaluation of coverage errors has become a "state-of-the-art" procedure in census-taking. Thus, many countries conduct a PES immediately after the census enumeration activity that estimates the census coverage rate. Furthermore, some countries use PES estimates to adjust census counts for coverage errors. In this section, we describe several coverage models and the corresponding PES sample design, as well as design-based evaluation programs.

2.1. Dual system estimator–based estimation of undercount

To begin with, we consider a census list that is exposed only to undercount. To estimate the extent of undercount in the census list, another source of information is required, namely, another full or sample-based enumeration. The most known model for estimating the size of a closed population using two incomplete enumerations is the capture–recapture model. This model has been used since the 19th century in many disciplines, such as wildlife management, epidemiology, physics, criminology, software testing, and, of course, demography (see, e.g., reviews by Chao, 2001; Schwarz and Seber, 1999). Variants of the problem include the case where two enumerations attempt to count all members of the population (nonsample case) and the case where one of the enumerations is sample-based (census sample). Data may be obtained from field data collection or from a list frame such as a register. In the census-taking literature, these models are referred to as dual system estimators (DSEs). For a comprehensive review of literature on capture–recapture modeling in census methodology, see Fienberg (1992) and Chao and Tsay (1998).

2.1.1. The dual system model

We will begin with a description of the standard DSE (Peterson, 1896; Sekar and Deming, 1949; Wolter, 1986). Consider a closed population Ω, which comprises N individuals residing in a given geographical area at a specific time. Assuming that N is fixed but unknown, the problem is to estimate N. Suppose, for the time being, that we have made two attempts to count the entire population and have obtained two lists of identified individuals. After matching the two lists, a 2×2 table is set up, as in Table 1. Table 1 is commonly referred to as *The Dual System Estimation Table*.

The entries in the table relate to the number of people counted in list A and list B, Y_{11}; the number of people counted in A but not in B and in B but not in A, Y_{10} and

Table 1
The dual system estimation table for the nonsample case

		Census List B		
		Counted	Missed	Total
Census	Counted	Y_{11}	Y_{10}	Y_{1+}
List A	Missed	Y_{01}	Y_{00}	Y_{0+}
	Total	Y_{+1}	Y_{+0}	$Y_{++} = N$

Y_{01}, respectively; and the number of people missed in both lists, Y_{00}. Note that Y_{00}, the marginal totals Y_{+0}, Y_{0+} and the population total $Y_{++} = N$ are unobservable, and therefore need to be estimated. Let p_{ab} be the probability of inclusion in the abth cell, $a, b = 0, 1, +$. The estimation procedure is based on three major assumptions: (A1), lists A and B are created as a result of N mutually independent trials (autonomous independence); (A2), counting probabilities are homogeneous across individuals (heterogeneous independence); and (A3), the event of being counted in list A is independent of the event of being counted in list B (causal independence). With these assumptions, the probability of being counted twice is the product of the marginal counting probabilities, $p_{11} = p_{1+}p_{+1}$, and the maximum likelihood estimators of the probabilities that a person will be counted in list A and in list B, p_{1+} and p_{+1}, respectively, and of the total population N, are

$$\hat{p}_{1+} = \frac{Y_{11}}{Y_{+1}}, \quad \hat{p}_{+1} = \frac{Y_{11}}{Y_{1+}}, \quad \hat{N} = \frac{Y_{1+} \cdot Y_{+1}}{Y_{11}} = \frac{Y_{1+}}{\hat{p}_{1+}}. \tag{1}$$

It can be seen that the total population estimator $\hat{N}$ is expressed as the number of individuals counted in the first list divided by the estimated counting probability of this list $\hat{p}_{1+}$. Given assumptions (A1)–(A3), these estimators are strongly consistent with asymptotic normal distribution (see, e.g., Alho, 1990).

In many applications of the dual system methodology, it is not realistic to assume that both lists are based on a counting procedure that aims to count the entire population. Wolter (1986) provides a detailed description of an alternative model where only one list, for example, the first one, is a nonsample list. The second list is based on a sample of people that is selected from the target population for possible inclusion in the second list. In the context of a census, the first list would be the full enumeration census list and the second list would be provided by a postenumeration undercoverage sample survey.

Suppose that the underlined geographical area is divided into M plots known as enumeration areas (EAs). A simple random sample of m EAs is chosen, and the data collected for list B consist of an enumeration of the population living at the sampled areas only. At this stage, it is assumed that both the census and the survey count only eligible people. For this census-sample case, the maximum likelihood estimators of N and the marginal capture probabilities are

$$\tilde{p}_{1+} = \frac{Y_{11}^{U}}{Y_{+1}^{U}}, \quad \tilde{p}_{+1} = \frac{Y_{11}^{U}}{Y_{1+}^{U}}, \quad \tilde{N} = \frac{Y_{1+} \cdot Y_{+1}^{U}}{Y_{11}^{U}} = \frac{Y_{1+}}{\tilde{p}_{1+}}, \tag{2}$$

where the superscript U indicates sampled EAs. The population size is estimated as the total number of individuals counted in the census list divided by the sample-based estimator of the counting probability $\tilde{p}_{1+}$. The parameter p_{1+} will also be referred to as the undercount parameter.

When a more complex sampling design is used for sampling EAs, predictors of Y_{11}, Y_{1+}, and Y_{+1} are calculated according to the particular sampling scheme. In that case, the population size is estimated by substituting the appropriate design-based predictors in (1), $\tilde{N} = (Y_{1+}\hat{Y}_{+1})/\hat{Y}_{11}$, where Y_{1+} is the census count as before. For example, in the U.S. Accuracy and Coverage Evaluation (ACE) Survey described in Section 2.2.2 (U.S. Census Bureau, 2004), Y_{+1} is essentially predicted by $\hat{Y}_{+1} = \sum_{k \in S} w_k$, where w_k reflects the inverse of the probability of selection of person k in the sampled EAs, as well as adjustments for missing data and other operational problems. A similar predictor is calculated for Y_{11}.

A slightly different estimation approach was taken by the United Kingdom 2001 One Number Census (ONC). Here, the classical DSE (1) was applied at the EA level, and a ratio estimator was used to estimate the size of the entire population.

$$\tilde{N} = \tilde{R}Y_{1+} \text{ where } \tilde{R} = \sum_{i \in U} \hat{N}^i \bigg/ \sum_{i \in U} Y_{1+}^i = \sum_{i \in U} \frac{Y_{1+}^i Y_{+1}^i}{Y_{11}^i} \bigg/ \sum_{i \in U} Y_{1+}^i, \qquad (3)$$

where i indicates EAs, see Brown et al. (1999).

Wolter (1986) derived an expression for the asymptotic expectation and variance of $\tilde{N}$ given in (2) by applying the standard Taylor series method, $E\tilde{N} = N + C$ and Var $\tilde{N} = N \cdot C$, where

$$C = \frac{(1 - p_{1+})(1 - p_{+1})}{p_{1+}p_{+1}} + \frac{1 - f}{f}\frac{1 - p_{1+}}{p_{1+}p_{+1}}, \qquad (4)$$

and $f = m/M$ is the sampling fraction. When other estimation schemes are adopted (e.g., the ONC ratio estimator (3)), the variance is usually estimated by resampling methods.

The estimators described above rely heavily on model assumptions. Some modified versions of those estimators have attempted to deal with the potential failures of these assumptions. Bias resulting from failure of the heterogeneous independence or causal independence assumptions (assumptions (A2) and (A3) in Section 2.1.1, respectively) is referred to as correlation bias. Heterogeneity in counting probabilities is often handled by poststratification. Typical stratification variables include geographic units, age × gender groups, and other socioeconomic variables. Thus, it is assumed that the heterogeneous independence assumption is satisfied within poststrata, and a DSE is computed for each stratum. Huggins (1989) and Alho (1990) further generalize the DSE to cases in which counting probabilities vary for different people (e.g., cases when some heterogeneity still remains within strata). The individual counting probabilities are estimated through logistic regression using relevant explanatory variables. The model predicts the propensity that a person will be counted in the census and in the sample (for further insights on this topic, see also Haines et al., 2000, and a presentation by Bell, 2007).

Causal independence is not satisfied if the act of someone being included in the census affects his or her probability of inclusion in the coverage survey. This can happen, for example, when data collection for the PES and census enumeration are not conducted at completely different times or if some information about the first enumeration is available at the second enumeration. Let $\theta = Y_{11}Y_{00}/Y_{10}Y_{01}$ be the odds ratio in

Table 1, then $E\theta \cong 1$ under the assumptions of the model. However, in the presence of correlation bias, the total population N can be estimated as

$$\hat{N}(\hat{\theta}) = \hat{N} + (\hat{\theta} - 1)Y_{10}Y_{01}/Y_{11}, \tag{5}$$

where $\hat{\theta}$ is a predictor of θ and $\hat{N}$ is the DSE estimator (1) (Bell, 1993). Additional independent (external) data are required to predict θ (e.g., administrative data or demographic estimates). If those data are available, we may have a good demographic total estimate at the national level. Plugging this estimate in place of $\hat{N}(\hat{\theta})$ in (5) yields a prediction of θ. If we are interested in corrected DSEs within poststrata, either external total estimates for these strata are required or some assumptions on θ can be made. Since no accurate external estimates are available at subnational levels in many instances, the second alternative is usually adopted. The simplest assumption is that θ is constant across all poststrata. In that case, a synthetic estimate is obtained using strata-specific census and PES counts combined with a national estimate of θ. Other possible assumptions that yield corrected DSEs have been proposed by Bell (1993, 2001), Elliott and Little (2000), and Brown et al. (2006). Methods based on a third enumeration, known as triple system estimators, are discussed by, for example, Darroch et al. (1993) and Zaslavsky and Wolfgang (1993).

Finally, we note that the dual system model is based on the multinomial distribution. An alternative model for the target population capture process is based on the Poisson distribution (e.g., Cormack and Jupp, 1991). The main advantage of the Poisson model is its amenability to standard maximum likelihood theory. Log-linear models are also discussed extensively in the capture–recapture literature (e.g., Rivest and Levesque, 2001), including models that accommodate heterogeneous capture probabilities. As far as we know, the Poisson and log-linear models have not been used in a census context.

2.1.2. *Principles of sample design and an illustrative example*

To estimate undercount using the DSE, undercoverage postenumeration surveys have been conducted in many countries, including Australia (Australian Bureau of Statistics, 2007), Italy (Cocchi et al., 2003), Turkey (Ayhan and Ekni, 2003), and the United Kingdom (Brown et al., 1999). Typically, two-stage or multistage stratified area samples are used in those surveys. To keep the description simple, we present a two-stage design that has the following features:

Target population. Ideally, the PES target population should be the same as the census target population. In practice, however, some countries exclude population groups such as people living in nonprivate dwellings or in remote areas.

Primary sampling units. These first-stage units are relatively large geographical units such as municipalities.

Secondary sampling units. The primary sampling units (PSUs) are partitioned into smaller plots that typically comprise several dozens of households or dwellings. These may correspond to existing administrative units such as postcode areas, geographical units such as blocks, or often EAs defined specifically for the census. The main rationale for selecting an area-based cluster sample is the lack of reliable sample frames that are independent of the census enumeration, and that can be used to select units such as dwellings, households, or individuals. In addition, an area sample might be preferred for operational reasons.

Sample frames. When address files or building registers exist, they may be used to design and select the secondary sampling unit (SSU) sample. These lists can be on a national scale or can comprise a combination of local lists. As mentioned before, in many situations, only a PSU-level list is available and the second-stage sampling is based purely on geographic area maps. It should be emphasized that the frames should not depend on census information.

Stratification of PSUs. Undercount is generally not homogenous across PSUs. For example, it is expected that undercount will be higher in areas with a high immigration rate or with a high proportion of young people. Therefore, many countries stratify PSUs by characteristics that were found to be important determinants of undercount to attain sample efficiency. For example, a national Hard-to-Count (HtC) score was constructed in the U.K. ONC based on the previous census information. The index distinguished among PSUs by their expected level of census coverage.

Sample size and sample allocation. The overall sample size usually balances accuracy requirements, cost, and other considerations such as using the PES to collect additional socioeconomic data (see Section 5). Some countries aim at sample sizes ranging from 200,000 to 400,000 households (e.g., Italy and United Kingdom) and others have smaller samples ranging from 20,000 to 40,000 households (e.g., Australia and Turkey). Sample allocation to strata usually aims to minimize the variability of the DSE in key areas and population groups. Hence, sampling fractions are generally unequal in different strata. As such, sampling rates are typically expected to be larger in areas with higher undercount rates, as can be seen in Eq. (4). Since counting probabilities are unknown at the time of sample design, the design should be robust to deviations in the hypothesized distributions. Hence, simulation studies (e.g., Brown et al., 1999), sensitivity analyses (e.g., Nirel et al., 2003), and pilot surveys are used to design the sample.

The U.K. 2001 Census Coverage Survey. The 2001 ONC project aimed to identify and adjust for omissions in the 2001 Census. Undercount was evaluated on the basis of a PES known as the Census Coverage Survey (CCS). The objective of the CCS was to provide undercount estimates at the subnational level by age groups and gender and was to allocate the undercount to small areas. The CCS was a stratified two-stage sample, where the primary strata were estimation areas comprising approximately 500,000 people. There were 101 estimation areas in England and Wales, eight in Scotland, and three in Northern Ireland. The PSUs were the 1991 Census Enumeration Districts (EDs).

Using the HtC distribution, three strata corresponding to three levels of enumeration difficulty (lowest 40%, middle 40%, and top 20%) were defined within each estimation area. Within these strata, PSUs were further stratified by size groups of the key age-gender distribution. The size strata were formed on the basis of a design variable, which captured the age-gender structure of the EDs using babies, young males, and elderly female age-sex groups. A sample of EDs was selected within those strata, and postcodes (SSUs) were selected within each selected PSU, with probability related to the mean number of addresses per postcode within the selected ED. A sample of about 19,500 postcodes was selected, which consisted of about 370,000 households (including about 320,000 in England and Wales alone). The sample size aimed at a 1% relative error rate for the EA level estimates, and a relative rate of 0.1% for the national counts. Data was collected by face-to-face interviews as compared to the Census self-completion questionnaire. For more details, see Brown et al. (1999) and Abbot and Marques dos Santos (2007).

2.2. DSE-based estimation of undercount and overcount

As we have pointed out, the problem of census overcount has become as important as undercount in census methodology. Surprisingly, for example, in the U.S. 2001 census, the estimated net undercount was −0.5%, that is, the census data essentially revealed a net overcount rather than a net undercount. In this section, we will describe two approaches to estimating overcount, together with a dual system for estimating undercount. The first approach suggests a design-based estimate of overcount rate. The second approach is based on the multinomial-Poisson model, which extends the dual system multinomial model and includes both undercount and overcount parameters.

2.2.1. Extending the dual system estimator

In the undercount scenario described in Section 2.1, two sources of information were required to estimate undercount—census enumeration and a sample of EAs. We refer to this sample as the undercoverage sample or in short, the U-sample. Consistent with the DSE assumptions, the U-sample is drawn independently from the census list, for example, an area sample. If the census list comprises Z enumerations, of which X are ineligible, then $Z = Y_{1+} + X$. We also assume, as before, that the U-sample list does not include ineligible persons.

To predict X, a second sample is selected, that is, the overcoverage sample or the O-sample. The objective of this sample is to identify erroneous enumerations that result from counting ineligible people in the census. Therefore, the sampling frame is now the census list, which is assumed to consist of eligible and ineligible people. To reduce cost and simplify data collection, it is convenient that the O-sample comprises the same EAs as the U-sample. Specifically, the O-sample consists of all people who are enumerated by the census in the same plots as those that are selected for the U-sample.

Assume, as before, that a simple random sampling scheme of m plots out of M has been selected. The design-based adjustment of the standard DSE for overcount first shrinks the census count, Z, by the O-sample estimate of the share of correct enumerations in the list, $(Z^O - X^O)/Z^O = Y_{1+}^O/Z^O$, where the superscript O indicates O-sample counts. The predicted number of eligible people is then expanded by the U-sample estimate of the counting probability $\tilde{p}_{1+}$ of (2). In sum, we obtain

$$\tilde{N}_D = Z \frac{Y_{1+}^O}{Z^O} \frac{1}{\tilde{p}_{1+}} = \frac{Z \cdot Y_{1+}^O \cdot Y_{+1}^U}{Z^O \cdot Y_{11}^U}. \tag{6}$$

This design-based approach to estimating overcount together with the DSE has been used in U.S. Censuses (e.g., Hogan, 1993, 2003; Mulry, 2007, see below) and in Switzerland (Renaud, 2007a). Variance estimates for $\tilde{N}_D$ can be obtained by resampling methods such as stratified jackknife methods (U.S. Census Bureau, 2004, Section I, Chapters 7–14).

A model-based approach for estimating and adjusting for overcount is based on five basic assumptions. The three assumptions of the classical DSE model (A1)–(A3), plus two additional assumptions: (A4), the number of ineligible people counted by the census in an EA, i is distributed according to the Poisson distribution, with expectation λN^i for $\lambda > 0, i = 1, \ldots, M$; and (A5), all EA counts of eligible and ineligible people are mutually independent. Note that assumption (A4) states that the rates of ineligible counts are homogeneous across EAs. This multinomial-Poisson model was proposed by Glickman et al. (2003) for the 2008 Israeli Integrated Census.

When $Z = Y_{1+} + X$ represents the number of eligible and ineligible people counted in the census list as before, the estimators derived from the model are

$$\tilde{p}_{1+} = \frac{Y_{11}^U}{Y_{+1}^U}, \quad \tilde{p}_{+1} = \frac{Y_{11}^U}{Y_{1+}^U}, \quad \tilde{\lambda} = \frac{X^O}{Y_{1+}^O / \tilde{p}_{1+}} \quad \tilde{N}_M = \frac{Z}{\tilde{p}_{1+} + \tilde{\lambda}}. \tag{7}$$

Thus, the estimate of the undercount parameter p_{1+} is the same as in (2), and the overcount parameter λ is estimated by the share of ineligible persons out of the eligible persons on the census list, with an adjustment for undercount. Note that since p_{1+} is the expected share of eligible persons in the list and λ is the expected share of ineligible persons, their sum amounts to the expected size of the list, Z, divided by the size of the target population. In that way, we obtain the estimator $\tilde{N}_M$ in (7). Note that for the sample design considered here, the census list corresponding to sampled EAs was the same for the U-sample and the O-sample.

The expressions for the asymptotic expectation and variance are similar to those derived by Wolter (1986), for the case with no overcount in the census list. We obtain $E\tilde{N} = N + C$ and $\text{Var}\, \tilde{N} = N \cdot C$, where

$$C = \frac{(1 - p_{1+})(1 - p_{+1})}{p_{1+}p_{+1}} + \frac{1 - f}{f} \frac{1 - p_{1+}}{p_{1+}p_{+1}} + \frac{1 - f}{f} \frac{\lambda}{p_{1+} + \lambda} \left(\frac{p_{1+}}{p_{1+} + \lambda} - \frac{1 - p_{1+}}{p_{1+}} \right). \tag{8}$$

The last term in the right side of Eq. (8) represents the contribution of overcount to the variance of the census population estimator.

It should be noted that the share of eligible persons estimated in (6) is defined with respect to the total number of census enumerations (eligible and ineligible), whereas the overcount parameter λ estimated in (7) is defined only with respect to the number of eligible persons.

2.2.2. Sample design and illustrative examples

The O-sample typically consists of all people who are enumerated by the census in the same EAs selected for the U-sample. Thus, the sampling units and stratification variables are defined according to principles similar to those described in Section 2.1.2. Operationally, data collection for the two samples is linked in the following schematic stages:

(1) U-sample data collection;

(2) Construction of U-sample list;

(3) Matching U-sample and O-sample lists; and

(4) Follow-up of unmatched cases (overcoverage-follow-up, OF).

Hence, the OF fieldwork is limited to people listed in the sampled EAs who were not linked to the U-sample list. These may be ineligible people, eligible people missed by the U-sample enumerators (U-sample undercount), or false nonmatches. Sample size and sample allocation take into account expected overcount patterns in addition to expected undercount patterns. Note that whereas the U-sample unit is a household or a dwelling, the enumeration unit in the OF is typically an individual.

The following is a description of the U.S. 2001 Census Accuracy and Coverage Evaluation and the Israeli Integrated Census paradigm for the 2008 census. Two examples illustrate estimators (6) and (7), respectively.

2.2.2.1. U.S. census 2000 accuracy and coverage evaluation. The ACE consisted of an undercount sample (the Population sample, or P-sample) and an overcount sample (the Enumeration Sample, or E-Sample). The PSU was a block cluster comprising about 30 housing units. A national sample was selected in three phases:

(a) In the first phase, block clusters within states were classified into three size strata (small, medium, and large) and a fourth American Indian Reservation stratum. The size of the block clusters was based on preliminary census files, and an initial sample of about 30,000 block clusters was selected. For these block clusters, lists of housing units were created by field work.

(b) In the second phase, the field-based lists and updated census address lists were used to substratify the first phase sample within the large and medium strata in each state (reduction strata). A subsample of block clusters was selected, with equal selection probabilities within second-phase strata and possible differences in selection probabilities across strata. The second phase sample consisted of 11,303 cluster samples with about 850,000 housing units.

(c) In the third phase, a subsample of housing units was selected in block clusters consisting of 80 or more housing units. This phase elicited a final sample of about 301,000 housing units from the 11,303 block clusters selected in the second phase. The respective E-sample list consisted of 311,000 housing units (approximately 700,000 people).

Data were collected for the P-sample by means of computer-assisted personal interviewing (CAPI). The P-sample list was matched to the census list in the sampled blocks or in adjacent blocks, using computerized or computer-assisted clerical matching procedures. All the unresolved cases in the P- and E-samples (e.g., nonmatches, possible matches) were sent for follow-up interviews. About 50,000 people were included in the P-sample follow-up and about 143,500 people in the E-sample follow-up (for a detailed account, see U.S. Census Bureau, 2004).

The form of the DSE used in each estimation poststratum of the ACE was essentially as follows:

$$\tilde{N} = Z \times \frac{\hat{Y}_{1+}^E}{\hat{Z}^E} \times \frac{\hat{Y}_{+1}^P}{\hat{Y}_{11}^P} \times \phi, \tag{9}$$

where Z is the total number of census enumerations, $\hat{Z}^E$ is the estimated number of census enumerations from the E-sample, $\hat{Y}_{1+}^E$ is the estimated number of eligible enumerations from the E-sample, $\hat{Y}_{+1}^P$ is the estimate of the total population from the P-sample, $\hat{Y}_{11}^P$ is the estimated number of enumerations from the P-sample that match to the census, and ϕ is a correlation bias adjustment factor applied for male adults. All four E-sample and P-sample estimators in (9) are basically expansion estimators adjusted for missing data and other operational problems. The middle expression in (9), $\hat{Y}_{1+}^E/\hat{Z}^E$, is the census overcoverage correction factor, while the last expression, $\hat{Y}_{+1}^P/\hat{Y}_{11}^P$, is the census undercoverage correction factor (see, e.g., Mulry, 2007).

2.2.2.2. The integrated census paradigm planned for the 2008 Israeli census. The basic idea of the Integrated Census (IC) is to replace the traditional nationwide field enumeration with an "enumeration" of the Population Register (PR) augmented by survey data for estimating and adjusting for coverage errors. Note that enumeration of the PR means collecting the data from the PR files. Estimates of the coverage parameters are obtained through two coverage surveys: The U-survey is based on an area sample and provides estimates of undercount rates, and the O-survey is based on a sample of people from the PR and provides estimates of overcount rates. Notably, U-sample enumeration is "blind" to the PR. Thus, the enumerators do not know if and where a person in their area is listed in the PR.

In the Israeli administrative-statistical system, the country is divided into statistical areas (SAs), which comprise 3000–4000 residents on the average. The aim of the IC is to provide population estimates by age and sex subgroups within SAs. In preparation for the IC, SAs are divided into EAs, which include about 50 households each on the average. All PR records are geocoded and clustered into the above-mentioned EAs and a random sample of EAs is then selected within each SA. The U-sample comprises all eligible people who live in the sampled EAs, and the O-sample includes all people who are listed in the PR in the same EAs.

The planned sample for the first IC will comprise about one-fifth of the population (about 400,000 households). However, the sampling fraction within SAs will vary in accordance with accuracy requirements and on the level of the coverage parameters. Sample allocation is basically extracted from the variance estimator (8), and estimates of the coverage parameters are plugged in by matching the previous census with the PR, as well as on the basis of other demographic data (e.g., percentages of children, young people, elderly people, religious people, and new immigrants—for further details, see Nirel et al., 2003).

The U-sample fieldwork will start one day after Census Day and will last from four to six weeks. The U-sample file will then be matched to the PR and a list of the people remaining in the O-sample EAs will be created, which includes the people in the O-sample file but not those in the U-sample file. To complete the O-sample fieldwork, all the people in the remainder list (the overcoverage-follow-up sample) will be traced and interviewed to determine their status. The OF-sample is expected to include approximately 20% of the O-sample on average.

2.3. Other approaches to estimation of coverage errors

Several countries evaluate coverage errors through PESs, which provide direct estimates of overcount and undercount rates by usual weighting methods rather than through DSE-based estimates. A prime example of this approach is the Canadian Coverage Error Measurement Program. A traditional census is conducted in Canada every five years, with modifications in the data collection methodology introduced in each cycle. Thus, in the 2001 census, questionnaires were distributed by enumerators, completed by household members, and returned by mail (mail out). In the 2006 census, an additional online option was offered for completion and return of the census questionnaire. The Canadian evaluation program will be illustrated below on the basis of the 2001 census. The program comprised the following four studies (Statistics Canada, 2004a):

(1) The Dwelling Classification Study (DCS), which focused on undercount due to misclassification of dwellings as unoccupied and due to nonresponse;

(2) The Reverse Record Check (RRC), which estimated total undercount and overcount that was not included in the other studies (AMS and CDS below);

(3) The Automated Match Study (AMS), which focused on overcount of people who were counted more than once within the same area; and

(4) The Collective Dwelling Study (CDS), which dealt with overcount of people who are counted in noninstitutional collective dwellings as well as in private dwellings.

Due to lack of space, we provide here a brief description of these samples (for a detailed account, see Statistics Canada, 2004). The main features of the above samples are highlighted in Table 2. In the DSC, enumerators returned to unoccupied and non-response dwellings in the sampled EAs in an attempt to determine whether or not the dwellings had been occupied on Census Day. The estimates from this survey were the only ones that were fed back into the census database. Approximately 223,000 people were added to the census database.

The RRC was sampled independently from six sampling frames that cover the entire census target population. The total sampling fraction was approximately 0.2% although the sampling fractions varied between and within frames. The main frame was the 1996 census (providing 74% of the total RRC sample), and the other frames included people who had been missed in the 1996 RRC, as well as files of births, immigrants, nonperma-nent residents, and health care beneficiaries. An effort was made to identify people who appeared in more than one frame. The 1996 census frame was stratified by province, gender, age, and marital status. Sample allocation was based on past coverage and tracing rates, which yielded higher sampling fractions for strata with a high percentage of hard-to-count people. An intensive tracing operation provided telephone numbers and other contact details for approximately 50% of the people in the RRC sample, and data were collected primarily by computer-assisted telephone interviewing. Notably, this study was completely independent of the census operation.

Table 2

The main features of the sample design in the Canadian 2001 Census Coverage Error Measurement Program (for a detailed description, see Statistics Canada, 2004)

Study	Target Population	Sample Size	Sample Design
Dwelling Classification Survey (DSC)	Nonresponse and unoccupied dwellings	1399 enumeration areas	Urban: single-stage stratified sample Rural: Two-stage sample
Reverse Record Check (RRC)	People who should be enumerated	60,653 persons	1996 Census: single-stage stratified sample with varying sampling rates
Automated Match Study (AMS)	Matched pairs of households within the same region	17,275 pairs of households	Single-stage stratified sample
Collective Dwelling Study (CDS)	Usual residents in noninstitutional collective dwellings	4500 residents	Single-stage stratified sample. Sample size was related to population size

Although the last two studies did not involve field operations, they are described here to provide a complete presentation of the program. The objective of the AMS was to identify duplicate households within the same geographic region. Automated record linkage identified exact and near matches, and the pairs were then stratified by variables such as geographic proximity and level of similarity. A questionnaire review was conducted for a random sample of pairs. Finally, the CDS reviewed questionnaires completed by a sample of usual residents in noninstitutional collective dwellings who had reported an alternative address on their census form.

For each study, weighted estimates of undercount and/or overcount were calculated. These estimates were combined arithmetically to provide the overall estimates. For instance, the estimated undercount and overcount rates for the 2001 Census were 0.0395 and 0.0096, respectively, yielding a net undercount rate of $\hat{D}/\hat{N} = 0.0299$ (SE = 0.0014).

3. The use of sample surveys to evaluate statistical adjustment of census counts

The coverage sample surveys discussed in the previous section estimate the bias in census enumeration due to undercount and overcount. In many countries, the primary use of these postenumeration surveys (PESs) has been to evaluate the quality of the census and to gain insight into coverage issues for future censuses. In several countries, however, such surveys are considered or even used to improve the accuracy of the census counts by statistical adjustment. In the United Kingdom, for example, the 2001 ONC comprised full enumeration as well as a PES, which was known as the CCS (Section 2.1.2), and covered a sample of approximately 370,000 households. Using DSE, census counts at the local authority district level were adjusted for undercount (Office for National Statistics, 2000).

Another interesting example is the U.S. ACE survey, which was conducted following the 2000 Census. The idea that census counts may be adjusted by sample survey data led to some controversy regarding the usefulness of such a procedure. Freedman and Watcher (2003, 2007) argue that "error rates in the adjustment are comparable to if not larger than errors in the census." In 2003, the Census Bureau decided not to use the ACE results for the population base of the intercensal estimates due to "technical limitations" of the ACE Revision II estimates (Mulry, 2007; U.S. Census Bureau, 2003b).

This section deals with evaluation of adjusted census counts. The primary aim of such evaluation is to assess whether potential errors that may be introduced by a PES are small enough so that its use for adjustment will improve the census accuracy. Some error components can be incorporated in the adjusted estimates. For example, when dual system models are used, correction for correlation bias can be included in the adjusted estimates through a correlation bias factor (see Section 2.1.1). This correction was applied in the U.S. ACE (Bell, 2001; Mulry, 2007), as well as in the U.K. CCS (Brown et al., 2006). Model and sampling errors can be expressed through variance estimations, as in the Israeli IC (Glickman et al., 2003; Section 2.2.1).

However, the adjusted estimates can still be subject to biases resulting from errors in data collection and data processing. Fieldwork in PESs attached to traditional censuses may take place a few weeks or even a few months after census day. This may lead to recall problems, for example, it is possible that people will not remember the exact date they moved from their census day address or that they will even have problems remembering birth or death dates around census day. For other types of censuses such as a register-based censuses combined with coverage surveys, fieldwork for the undercount survey may start on census day. However, differences between register variables and field variables, as well as differences in data collection procedures (administrative versus fieldwork), can introduce new and additional types of errors. In light of that situation, the following section focuses on some of the error components that require additional evaluation.

We start by analyzing known and unknown counts at the end of the data collection process in a census that comprises full enumeration and coverage surveys. We then discuss possible errors in the adjusted estimates and suggest additional data collection with the objective of evaluating the adjusted estimates. We conclude the section by describing evaluations of coverage estimates based on sample surveys that were carried out by the U.S. Census Bureau after the 1990 and 2000 censuses.

3.1. Known and unknown counts in the DSE and extended DSE paradigm

The first step is to analyze the various pieces of information provided by a full enumeration and coverage surveys. This analysis will highlight the missing information and help to design a program for evaluating potential biases in the final estimates. Table 3 summarizes the counts obtained by an extended DSE paradigm. We extend the subscript notation of Section 2 to include a third data source and apply it to eligible counts Y and ineligible counts X. Thus, Y_{101} denotes the number of eligible people counted in the full enumeration, who were not on the undercoverage list but were in the OF. Similarly, X_{100} is the number of ineligible people who were counted in the full enumeration but not in either of the two surveys. In contrast to the previous section, we do not assume here that the undercoverage survey is free from overcount. Let us follow the stages in data accumulation for this paradigm, which are given as follows:

(1) *Full enumeration.* The first step obtains a full enumeration list, either through field collection or administrative "collection." At this stage, we observe Z, which includes eligible and ineligible people. We assume that this list is incomplete and note that $Z = Y_{11} + Y_{101} + Y_{100} + X_{11} + X_{101} + X_{100}$. However, the break down of Z is not known at this stage.

Table 3

Data summary for a sampled area in an extended DSE paradigm, including an undercoverage and an overcoverage-follow-up surveys

		Undercoverage Survey List						Total
		Eligible			Ineligible			
		Counted	Missed		Counted	Missed		
			OF Survey List			OF Survey List		
			Counted	Missed		Counted	Missed	
Census List	Counted	Y_{11}	Y_{101}	Y_{100}	X_{11}	X_{101}	X_{100}	Z
	Missed	Y_{01}	Y_{00}		X_{01}	–		

Note: Shaded cells indicate unknown counts.

(2) *Undercoverage survey.* Once the survey data are collected and matched to the full enumeration, we obtain (a) the count of matches $Y_{11} + X_{11}$; (b) those in the full count and not in the survey $Y_{101} + Y_{100} + X_{101} + X_{100}$; and (c) those who participated in the survey but not in the full list $Y_{01} + X_{01}$. The additional information we seek, namely the omissions of the full enumeration (Y_{01}), is included in the sum $Y_{01} + X_{01}$. However, this information is masked by erroneous survey enumerations (e.g., fabrications) X_{01}. In addition, Y_{00} is unknown. Note that although in theory X_{11} may be positive, that is, both the full enumeration and survey do not identify a person as ineligible, it is believed to be either equal to zero or negligible.

(3) *Overcoverage-follow-up survey.* Data collection and matching of OF data with previous data shed some light on omissions in the undercoverage survey through Y_{101}. Respondents in the OF also provide information on ineligibles through X_{101}, but the remaining uncertainty due to OF survey nonresponse is still included in X_{100}. Actually, the sum $Y_{100} + X_{100}$ is known, but the break down of eligible and ineligible respondents is unknown. To complete the table, we note that by definition, $X_{000} = 0$ because the OF list is extracted from the census list.

In sum, after completing the data collection and matching, the unknown counts that remain are Y_{00} and a break down of $Y_{01} + X_{01}$ and $Y_{100} + X_{100}$ (see shaded cells in Table 3). The other counts in the table are considered "known." Note that a DSE paradigm that does not include an OF survey is a special case, with $Y_{10} = Y_{101} + Y_{100}$ and the appropriate X's equal zero.

3.2. Error components

The second step in developing an evaluation framework is to analyze the sources of missing data, on the one hand, and the sources of errors in the known counts, on the other. Table 4 summarizes typical sources of errors leading to undercount and overcount. It is important to note that the types of errors and their relative weight depend on the country, on the census methodology, and on data collection methods.

We start by discussing errors pertaining to the unknown counts (see Table 3). The main sources for these are nonresponse in the OF survey affecting the break down of $Y_{100} + X_{100}$, nonresponse and inadvert omissions of dwellings in the undercoverage survey affecting Y_{00}, and erroneous enumeration of buildings that do not belong to a sampled area, as well as fabrications that affect the break down of $Y_{01} + X_{01}$. Another group of errors is those caused by model biases, when missing data are imputed using a coverage model (e.g., DSE) and nonreponse imputation models.

Other errors relate to "known" counts in Table 3. Those are subject to measurement and matching errors. One group of measurement errors deals with census eligibility in the national and local dimensions. People who do not belong to the census population (ineligible) may be erroneously counted. Such cases include visitors who came to the country for a limited period (e.g., for less than three months) and happen to be present on Census Day. Similarly, there are eligible people who might not be counted, for example an illegal work emigrant who has been living in the country for more than a certain period (e.g., for over one year). Errors with regard to Census Day Eligibility also include babies born after Census Day who were counted in the census or people who died after census day and were not counted (see Fig. 1).

Table 4
Typical sources of undercount and overcount in adjusted estimates

Error Type	Sources of Undercount	Sources of Overcount
Unknown cells		
Nonresponse	Closed dwellings	
	U-survey and OF survey refusals and noncontact	
Misclassification	Erroneously classified as an empty or nonresidential dwelling	Erroneously classified as a residential or noninstitutional dwelling
Geocoding/address	Buildings or dwellings erroneously deleted by enumerators from area	Buildings or dwellings erroneously added by enumerators in area
Other		Undercoverage survey fabrications
Modeling	DSE and Extended DSE assumptions not met	
	Nonresponse imputation errors	
"Known" cells		
National eligibility	Forgotten	Counted
	Long-term visitors	Short-term visitors and tourists
		Citizens living abroad
	Born before Census Day	Born after Census-Day
	Died after Census Day	Died before Census Day
Local eligibility		Counted more than once
	Students at parents' home (depending on definition)	Students in dormitories and in parents' home
	Not counted	
	Outmovers after Census Day	Inmovers after Census Day
	People with multiple residences	People with multiple residences
	Children in shared custody	Children in shared custody
	No permanent address	
Geocoding/address	Erroneously omitted from maps of area	Erroneously geocoded to area
Other		OF fabrications and multiple response
Matching	False nonmatches due to insufficient data	False matches
	False nonmatches due to geocoding to wrong area	

Census residence is another difficult eligibility issue. At the time of the PES interview, some people may have already moved away from their census address (outmovers) and others may have moved in. Since spatial accuracy is a key issue in censuses, incorrect records of census residency can lead to substantial bias. Part of the problem for outmovers is that information about them is obtained by proxy (e.g., from neighbors) and may thus be unreliable. Other common problems in measuring census residency pertain to short-term visitors, people with vacation homes, college students, and children in shared custody. All these problems can result in erroneous enumerations (people counted in two locations) or in omissions (not counted in any location).

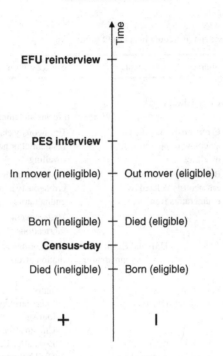

Fig. 1. Schematic illustration of eligibility errors resulting in erroneous enumerations (+) and omissions (−).

Another group of errors relates to measurement of geographical locations. In field operations, enumerators can mistakenly classify a dwelling as empty or nonresidential. Alternatively, there may be errors in the geographic association of a building to an EA (geocoding): buildings can mistakenly be added or omitted from an EA because of erroneous geocodes. Countries that do not have a register of buildings or dwellings are particularly susceptible to geocoding errors. However, address and building registers are also susceptible to risks such as duplication (e.g., a building on the corner of two streets can be entered with both addresses) or omitted addresses. The last group of measurement errors relates to fabrication and multiple response (mail, interviewer, internet), which results in erroneous enumerations.

Finally, record linkage procedures are used to produce the DSE and extended DSE tables (Winkler, Chapter 14). These procedures are applied to identify the people who were enumerated in both the census and the survey, or who were enumerated in one and not in the other. Two types of error can occur when this procedure is used. The first type of error can be made when people enumerated in both the census and the survey are not linked, and results in false nonmatches. The second type of error can be made when people are erroneously matched and results in false matches. Linkage errors can result from insufficient matching information, erroneous data leading to false nonmatches, imputation errors, or geocoding errors.

An important example of error component analysis is the U.S. Census Bureau total error model for PES estimates (Hogan and Wolter, 1988; Mulry and Kostanich, 2006; Mulry and Spencer, 1991). The model attempts to estimate systematic errors remaining after adjustment. It incorporates modeling, sampling, data collection, measurement, and processing errors. The idea is to break the overall bias of the empirical census estimate into major error components and to try and estimate each component separately. Specifically, the model includes components such as matching errors, errors in census address reporting, fabrication errors, errors in measuring erroneous enumerations, correlation bias, and sampling errors. For an interesting analysis of potential problems in a PES, see also Hogan (2003).

3.3. Evaluation follow-up

The last step is to design the required operations. Some evaluation operations, such as expert clerical reviews and analyses based on demographic estimates and administrative data, can be carried out in the office. Evaluation of errors such as Census Day residence and missing data require a reinterview. We focus here on evaluation components that are carried out by means of additional sample surveys, which will be referred to as EFU. Notably, reinterviews are only useful if they provide more accurate data than the census and the PES. Therefore, when we consider such an operation, it is important to balance the need for information with the plausibility of obtaining accurate and consistent data. A key feature is a questionnaire with specific questions, which aim to elicit more accurate recall and understanding of information relevant to determining all dimensions of census eligibility.

Depending on the census methodology and data collection methods, a system of EFU surveys can be designed to evaluate the expected errors. Those surveys may include the following:

Dwellings/households sample. The first proposed operation is to conduct reinterviews in a sample of dwellings that were classified by the PES as empty, closed, or nonresidential, as well as in a sample of dwellings with residents who were not linked to the census enumeration or who were linked with low probability. This sample is intended to add information on PESs omissions, as well as to correct classification errors and false nonmatches.

Buildings/address sample. Another possible operation involves buildings that were deleted from a sampled area or added to it by a PES enumerator, as well as buildings with residents who were interviewed at an address that differs from the census address. This sample can shed light on geocoding errors, as well as on errors in census residence, and errors in address registers.

Sample of individuals. Finally, a sample of individuals from the census list, including OF nonrespondents, nonmatches within a household (especially students and babies) or other nonmatches (e.g., people who reported no change of address), and people in hard-to-count groups such as young male bachelors, residents of institutions, and households with members who have more than one place of residence or households comprising a divorced/separated parent and children. This sample attempts to supplement information on OF missing data, Census Day recall problems, matching errors, and fabrications.

To conclude this section, we will describe some evaluation operations and error component estimation that were carried out by the U.S. Census Bureau after the 1990 and 2000 PESs. These evaluation studies were mostly aimed at assessing errors in particular aspects of the PESs operations and estimation. They were not intended or designed to provide corrections to the undercount estimates. The first net undercount estimate derived from the 1990 PES was 2.1%. A sample of whole households and partial household nonmatches in 919 block clusters and a sample of matches from the same blocks were selected to assess P-sample data collection errors. Moreover, in the E-follow-up survey, interviews were conducted among the same 919 block clusters to assess E-sample errors. The results showed that the net error in the DSE estimate was approximately 0.5% (Mulry and Spencer, 1993). Based on the results presented by Mulry and Spencer (1993, Table 1), the main error components that reduce the undercount estimate are matching error (0.21%), P-sample collection error (0.31%), E-sample operations error (0.25%), and ratio-estimator bias (0.11%). The main components that increase the estimate were E-sample collection error (−0.17%) and model bias (−0.29%). The contribution of fabrication errors, imputation errors, and sampling errors to bias in the 1990 census was negligible. The revision of the 1990 PES estimates involved developing a new poststratification, redoing matching for 104 influential block clusters, and correcting two computer processing errors that affected the estimation of erroneous enumerations (Hogan, 1993).

The ACE survey that followed the 2000 Census provided three estimates of net census undercount. The original net undercount estimate published in March 2001 was approximately 1.2%. Evaluation of this estimate, which involved reinterviews and/or rematching of approximately 1/10 of the P- and E-samples, is known as the EFU. The EFU evaluated Census Day residency by reinterviewing people who were included in the P-sample and people who were included in the person follow-up (PFU) samples with unusual living situations, moving status, etc. The ACE Revision Preliminary Estimate published in October 2001 indicated a net undercount of 0.06% (Thompson et al., 2001). However, the final ACE Revision II Estimate published in March 2003 revealed a net undercount of −0.5% (i.e., a net overcount of 1.3 million, Mulry, 2007). The Revision II evaluation was motivated by the first revision evaluations but involved additional work. It revealed that the reduction of 1.7% in the net undercount was mainly due to census duplications that fell in the E-sample but were not detected as ineligible (−1%), E-sample coding corrections (−0.9%), P-sample duplications (−0.4%), and correlation bias (+0.6%, U.S. Census Bureau, 2003a, p. 31, Table 12). Duplicates were identified by a nationwide search rather than by the limited area search conducted earlier. Coding corrections pertain to conflicting addresses provided by the PFU and the EFU. Other errors included dwellings registered under two different addresses in the address file. In sum, the post-ACE evaluations showed that the main failure of the ACE was inaccurate measurement of the Census Day residence.

4. The use of sample surveys for carrying out a census

The search for new census-taking methods has yielded an entirely different type of census, which is based on continuous cumulative sample designs and utilizes the principles of rolling sample design (see Section 4.1 below). The main advantage of a sample-based census is that it provides more frequent and timely estimates of large national domains. Such estimates supply information on temporal variation, in contrast to the "once in ten years" census. The main drawback of the sample-based census is that it does not provide a detailed geodemographic "snapshot" on a particular date so that comparisons between domains are much more complicated. Furthermore, the issue of coverage becomes much more complicated in a census that is rolled over time because of the population movements in time and space. Thus, the likelihood of coverage errors may increase substantially in a sample-based census (for further discussion of the merits and drawbacks of sample-based censuses, see Office for National Statistics, 2003).

4.1. Rolling samples and some extensions

The concept of a rolling sample was proposed and developed by Leslie Kish in a series of papers (e.g., Kish, 1998). A rolling sample design jointly selects a set of k mutually exclusive (not overlapping) periodic samples, each of which is a probability sample with a fraction $f = 1/F$ of the entire population. One sample is interviewed at each time period, and the accumulation of k periods yields a sample with a fraction $f' = k/F$. The main aim of the rolling sample design is to provide detailed estimates in temporal as well as spatial dimensions. Specifically, Kish emphasizes the need for adequate annual estimates at the national and major regional/domain levels. By keeping the samples mutually exclusive, maximum efficiency of the accumulation is attained and the estimation procedure is simple. Note that the rolling sample design assumes that each sample is a representative sample of the relevant regions and domains. This means that if the PSUs are clusters, they should be smaller than the target domains. Moreover, the PSUs should form a rolling sample themselves. Therefore, a design in which all samples include the same PSUs, with a rolling sample within a PSU, has been referred to by Kish (e.g., 1999) as a *cumulated representative sample* (CRS) design. This design is not strictly a rolling sample.

Kish (e.g., 1990, 1999) further extended the concept of a rolling sample to a *rolling census* by taking $k = F$, that is, a sample with a cumulative fraction $f' = F/F = 1$. Thus, the rolling census replaces the simultaneous and complete enumeration of the population, carried out once in several years, by a continuous cumulative sample survey that covers the entire population over a time period F. For example, a rolling census comprising of 10 annual 1/10 samples yields full coverage of the population after 10 years. A rolling census can provide national and large domain estimates each year based on the latest sample, as well as smaller domain estimates based on appropriate accumulations of a number of samples.

The basic rolling census design is defined by the number of samples F and by the choice of sampling unit. The number of samples is likely to coincide with the "regular" intercensal periods of 5 or 10 years. The natural sampling unit is a small, well-defined geographical area. Some modifications of the basic design can include unequal sampling fractions in different strata as well as some overlap between samples (Kish, 1998). Because of the complexity involved in conducting a census, it is expected that a rolling sample design will be combined with a CRS design or with other panel designs. For example, small local authorities can be sampled by a rolling design, whereas large local authorities can be included in all samples, with a rolling sample of addresses within them. The actual design is clearly determined by the census objectives, as well as by numerous statistical and nonstatistical constraints. A qualitative analysis of alternative designs is presented in the Office for National Statistics (2003).

4.2. An illustrative example: the French rolling census

At the time of the writing of this chapter, France is the only country that decided to carry out a sample-based census (e.g., Dumais et al., 1999; Isnard, 1999). The first sample was interviewed in 2004 and the first census will be completed in 2008. Because this is the only actual example of a sample-based census, we will describe its general design and estimation procedure.

The national statistical agency of France, INSEE, has decided that to comply with the requirement of timeliness on the one hand and budgetary constraints on the other, the census will enumerate 5/7 of the population over a five-year period (Durr, 2004; Durr and Dumais, 2002). The sampling unit is a "commune," which is a subdivision of a French territory and a local authority. The size of a commune varies from several dozen residents to over 300,000 residents. Therefore, the communes are first stratified by size. The *large communes* stratum includes all communes with 10,000 or more residents, and the *small communes* stratum comprises all communes with less than 10,000 residents.

A two-stage annual sample is selected as follows: in the large communes stratum, all communes are included in the sample, and 0.08 of the dwellings are selected for enumeration. In the small communes stratum, approximately 0.20 of the communes are sampled, and all dwellings in the sampled communes are included in the sample. Thus, large communes are surveyed every year and small communes are surveyed once during a five-year period. Because the small communes comprise approximately 50% of the population, the annual sampling fraction is about 0.14 ($0.5 \times 0.08 + 0.5 \times 0.2$) and the accumulated five-year fraction is about 0.70, as required.

The sample design in both strata is controlled by a multiannual rotating scheme. The large communes sample is drawn from the "inventory of located buildings" (RIL) list. The addresses in each large commune are divided into five balanced groups that have a similar distribution with regard to variables such as age, gender, and type of dwelling, based on the 1999 population census data. For year $i, i = 1, \ldots, 5$, a sample of addresses in the ith group is selected. All dwellings within a sampled address are enumerated. The sampling fraction of an address within group i is approximately 0.40 so that 0.08 of the addresses (and hence of the dwellings) in a commune are sampled annually. The RIL is updated every year to account for new and demolished buildings, and the five sample groups are updated accordingly.

Small communes are divided into five representative groups within each of the 22 regions of France based on the same variables as those of the large communes. During year $i, i = 1, \ldots, 5$, the sample of small communes is comprised of the ith group in each region.

In sum, the French design is referred to as a "rolling census" but it covers approximately 70% of the population over a five-year period. The design extends Kish's ideas by combining a CRS design for the large communes with a rolling sample of small communes.

4.3. Estimation

The census design defines a feasible resolution of estimates in time and space. Specifically, estimation for small domains can involve a combination of direct (design-based) estimates and synthetic (model-based) estimates, possibly using auxiliary information from additional sources.

Let Y be the outcome of interest, with annual estimators $\hat{Y}_i i = 1, \ldots, F$ and $\hat{Y}(\mathbf{W}) = \sum_{i=1}^{F} W_i \hat{Y}_i$, a census estimator with $\mathbf{W} = (W_1, \ldots, W_F)$, $\sum_{i=1}^{F} W_i = 1$. Kish (e.g., 1999) considers several basic weighting schemes: (a) using the last year only, where $W_F = 1$ and all other weights equal zero; (b) averaging all years with equal weights, where $W_i = 1/F, i = 1, \ldots, F$; and (c) monotonically nondecreasing weights—$W_1 \le W_2 \le \ldots \le W_F$. Clearly, cases (a) and (b) are special cases of (c).

The example of the French census will be used to illustrate a typical system of estimates that can be produced by a sample-based census. Consider years F, F-1, F-2, F-3, and F-4. Three types of estimates are provided: (a) population counts for year F-2 for every commune; (b) small area estimates based on data collected over the previous five years and pertaining to year F-2; and (c) national and regional estimates for the current year F (Durr, 2004).

4.3.1. Commune population count for year F-2

For every commune, regardless of the year it was surveyed, an estimated population count is provided at the end of year F for the beginning of year F-2. For a large commune, let X_i be an auxiliary variable (e.g., number of dwellings in the RIL) during year i, and $\overline{X} = \sum_{i=F-4}^{F} X_i/5$ the average number of dwellings over the last five years. Let $\hat{Y}_i$ be the expansion estimator for population count of that commune based on the data for year i, and $\overline{\hat{Y}} = \sum_{i=F-4}^{F} \hat{Y}_i/5$ will be the average estimate. Accordingly, the estimated count for year F-2 is the ratio (synthetic) estimate given by

$$\hat{Y}_{F-2} = \overline{\hat{Y}} \frac{X_{F-2}}{\overline{X}}. \tag{10}$$

For a small commune, the estimate depends on the year it was enumerated. Denote by Y_i the population count of a commune that is fully enumerated in year i. For a commune surveyed in year F-2, the estimate is equal to its population count, Y_{F-2}. For communes surveyed prior to year F-2, the estimates are extrapolated using a ratio estimate similar to the one in (10), yielding

$$\hat{Y}_{F-2}^{F-4} = Y_{F-4} \frac{X_{F-2}}{X_{F-4}} \quad \text{and} \quad \hat{Y}_{F-2}^{F-3} = Y_{F-3} \frac{X_{F-2}}{X_{F-3}},$$

where the superscript in $\hat{Y}_{F-2}^{(\cdot)}$ denotes the actual survey year. For communes surveyed after year F-2, the estimates are obtained by interpolation. The first value used for the interpolation is the actual count on the year surveyed. The second value is an estimate for year F-3, obtained from extrapolation of the previous count (five years before). We obtain

$$\hat{Y}_{F-2}^{F-1} = \alpha_{F-1} Y_{F-1} + (1 - \alpha_{F-1}) Y_{F-6} \frac{X_{F-2}}{X_{F-6}} \quad \text{and}$$

$$\hat{Y}_{F-2}^{F} = \alpha_F Y_F + (1 - \alpha_F) Y_{F-5} \frac{X_{F-2}}{X_{F-5}},$$

where $0 \le \alpha_i \le 1$ $i = F - 1, F$, and is typically no smaller than 0.5. These estimates are further calibrated to the national and other large-scale estimates (for other versions, see Durr and Dumais, 2002).

4.3.2. Small-area estimates for year F-2

Every year, a file containing data for the previous five years will be constructed, including a sampling weight for every person and dwelling. This file enables expansion estimation for any geodemographic subgroup, subject to accuracy limitations. These estimates are taken to pertain to the midpoint of the period, for example, at the beginning of year F-2. For large communes, the weight can be extracted from (10) and is equal to $\varphi X_{F-2}/\overline{X}$, where φ is the inverse of the respective inclusion probability. For a small commune, the weight is $\hat{Y}^i_{F-2}/Y_i$, corresponding to the estimates for year F-2.

4.3.3. National and regional estimates for current year

Each annual survey is a representative sample comprising about eight million people. Hence, usual survey methods (e.g., expansion estimates) enable reliable national and regional estimates for the current year. These estimates are used to calibrate the commune and small area estimates.

5. Sample surveys carried out in conjunction with a census

Censuses originally focused on enumeration of people. However, as the demand for more detailed social and economic data increased, there was pressure to include additional variables in the census questionnaire. Nonetheless, there is a delicate balance between the length of the census form on the one hand and data quality and response burden on the other. To solve this problem, many countries use two types of census forms: a *short form* with a few (about 10–20) demographic variables, and a *long form* with comprehensive information on topics such as housing, employment, education, income, immigration, fertility, and disability. The short form is completed for all people in the census population, whereas the long form is typically completed for a random sample of the population. In that way, the response burden is reduced for the population at large, and the census machinery can be used to supplement information based on the detailed data collected from the population that filled out the long form. Because long-form data collection is carried out during the same time period as the census, it is considered to be one of the census outputs (United Nations, 2006).

Overall sampling rates for the long-form sample vary for different countries and range from 5 to 20% of the population. Because of the high sampling rate compared with current annual or panel surveys, the long-form sample provides detailed snapshot information. Another strong link between the long-form sample and the census is that calibration of sample estimates to the census counts is straightforward because the two forms share the same geodemographic data.

The simplest selection method is to sample every lth unit systematically (e.g., dwellings, households), where l is the inverse of the sampling fraction. One in every five and 1-in-10 are common rates. Variable rates are also possible. In Brazil, for example, municipalities with more than 15,000 residents are sampled with a 1-in-10 rate, whereas smaller municipalities are sampled with a 1-in-5 rate. Another sampling scheme is to select geographical areas (clusters) and survey all households within a selected area. An example of such one-stage cluster sample is the undercoverage survey of the Israeli Integrated Census (Section 2.2.2). This survey aims to estimate coverage errors and collects the long-form data at the same time. The EAs are sampled within SAs, and sample allocation is predominantly determined according to a coverage model that typically yields differential sampling rates for different SAs.

5.1. An illustrative example: The U.S. Census 2000 long-form survey

As an illustrative example, we present some main features of the sampling and estimation procedures used in the long-form survey in the U.S. Census 2000. The sampling frame was the Decennial Master Address File, and the target overall sampling fraction was 1/6. This rate was achieved through systematic sampling with variable rates within four size strata. The rates ranged from 1-in-2 for the smallest size stratum with less than 800 housing units to 1-in-4 and 1-in-6 for the 800–1200 and 1200–2000 size strata, respectively, and 1-in-8 in the ≥2000 stratum.

To obtain sampling weights, weighting areas with at least 400 people were defined within the four size strata. These areas were generally similar to the census tabulation areas. The initial weight was the inverse of the sampling fraction, and the final weight was obtained by iterative proportional fitting methodology, otherwise known as raking. The long-form estimates were calibrated to the census counts in dimensions such as household type and size, and race by gender and age groups. Each stage of the raking procedure was adjusted for one dimension and lasted until a predefined stopping criterion was attained. Hefter and Gbur (2002) indicate that the difference in percentages between the census counts and the weighted sample totals ranged from -5.09 to 6.84% for single race groups (see Table 2 in Hefter and Gbur, 2002).

It is interesting to note that the direct variance estimates were calculated by the successive difference replication (SDR) methodology, which takes advantage of the systematic sampling of housing units. For an estimated total $\hat{Y} = \sum w_j y_j$, the basic successive difference estimator (Wolter, 1984) is

$$\text{Var}(\hat{Y}) = (1 - f)\frac{n}{2(n-1)} \sum_{j=2}^{n} (w_j y_j - w_{j-1} y_{j-1})^2,$$

where w_j is the final weight for person j and f is the sampling fraction. The replication version is based on replicate samples and is simple to use

$$\text{Var}(\hat{Y}) = (1 - f)\frac{4}{R} \sum_{r=1}^{R} (\hat{Y}_r(\text{SDR}) - \hat{Y})^2$$

where $\hat{Y}_r(\text{SDR})$ is the estimate for the rth replicate using the appropriate replicate weights (Fay and Train, 1995; Gbur and Fairchild, 2002). The variance estimates produced by the U.S. Census Bureau do not calibrate each replicate to the census counts. Schindler (2005) examined a raked-SDR method and argued that it best reflects the sample design and estimation procedure. According to Schindler, the raked-SDR produces smaller variance estimates than other methods such as SDR and Jackknife. Finally, design factors by state and size strata are calculated by comparing the SDR standard errors to simple random sampling standard errors based on a 1-in-6 sample.

5.2. Rolling the long-form survey

A relatively recent development of the long-form idea has been referred to by the UN as a "traditional census with yearly updates of characteristics" (UNECE, 2006). This type of census is based only on the short-form data, and the long form is replaced by a set of annual surveys, where socioeconomic data are collected continuously during the intercensal years. The census with yearly updates aims to provide more frequent information for small domains.

To date, the United States is the only country that has decided to use a large continuous survey to obtain data on the long-form topics on a regular basis instead of using the traditional census long form. Interest in intercensal information in the United States goes back to the 1940s (see Section 3 in Alexander, 2002) and provided the basis for the ACS. The ACS methodology was tested in nationwide surveys from 2000 to 2004. Full implementation of ACS then began in 2005. These tests were carried out separately from the census and provided data for comparison between the long-form survey and ACS estimates. The main benefits of the ACS compared to the long-form sample are timely and frequent estimates. Another benefit is higher data quality in terms of completeness of response. The main weakness is larger estimation errors due to factors such as smaller sample size and less accurate population controls for adjusting the survey weights (National Research Council, 2007).

The ACS basically uses a monthly rolling sample design (Section 4.1). Approximately 250,000 addresses are surveyed each month, corresponding to an average sampling fraction of $f = 1/F = 1/480$ (a total of approximately 120 million addresses). The survey uses $k = 60$ mutually exclusive monthly samples, yielding 5-year average estimates with approximately 1-in-8 sampling rates, as compared to the 1-in-6 long-form average rate. The ACS uses a systematic sample of addresses, and the sample is selected in two stages. In the first stage, a "super sample" is selected, using a constant rate in all strata, which equals the largest sampling rate required for any one stratum. In the second stage, samples from the first sample are selected to give the desired fraction for each stratum. This design simplifies the handling of stratum-specific sampling rates and their dynamics over time.

Five-year accumulations of ACS data provide products similar to those obtained by the census long form. Average annual estimates are provided for areas with over 65,000 residents, and average three-year accumulations are provided for areas with over 20,000 residents. More frequent estimates for small domains can be obtained using small area methods, see Malec (2005). For multiyear estimates, the ACS accumulation of samples over the years approximately averages the annual estimates with equal weights, contrary to Kish's (1998) inclination to increase w_i with i. For a comprehensive description of the ACS methodology, see U.S. Census Bureau (2006b), and for a discussion on alternative estimation approaches, see Breidt (2007).

6. Concluding remarks

Developments in survey methodology used for censuses can largely be attributed to technological progress, which influences data quality, timeliness, and the cost of direct data collection. The CAPI, use of personal computers, handheld devices, and global positioning systems (GPS) can greatly improve the accuracy of key census variables, in particular, eligibility topics. Advances in internet technologies and in encryption also enable online self-interviewing, which can improve response rates in the long run. It is beyond the scope of this chapter to address this topic and the interested reader can find an overview in UNECE (2006, Chapter II). Regarding data processing, we have mentioned that progress in record linkage capabilities allows census surveys to be matched with PES on a national level and, thus, reduces matching errors and duplication.

Another important topic that has not been covered in this chapter concerns estimation procedures for small subgroups. Although estimators of coverage errors for national and subnational counts were described in Section 2, of this chapter, PESs cannot provide direct coverage estimates for small groups due to sample-size limitations. To derive adjusted counts for small areas, estimates obtained by small-area techniques can be used. Suppose that the coverage factor for a given poststrata is $1/(\tilde{p}_{1+} + \tilde{\lambda})$ (Eq. (7)). Assuming that there is homogeneous coverage within poststrata, a synthetic adjusted census count for a subgroup g is given by $Z_g/(\tilde{p}_{1+} + \tilde{\lambda})$, where Z_g is the census count for subgroup g. The small-area estimates are generally calibrated to the national and subnational direct estimates using methods such as the generalized regression (GREG) or prediction regression (PREG) (Bell et al., 2007). For further details on these procedures, see Brown et al. (1999), Dick and You (2003), Office for National Statistics (2000); see also U.S. Census Bureau (2004, Chapter 8).

Finally, we mention the issues of missing data on the one hand and multiple responses on the other. A unique feature of census-related surveys is the need to impute eligibility status at the national and local levels. In Israel, for example, 5–10% of the listings in the Population Register are emigrants, many of whom are not traced by the O-survey. Imputation models based on administrative data as well as on sample data estimate the propensity of a missing case to be an emigrant. At the other extreme, the census and PES might provide two different census addresses. In such cases, decision rules should determine which address is more likely to be the correct one. Another important census issue is household counts. Entire households can be missed or individuals within households can be missed. The ONC project, for example, developed an imputation methodology which provided a fully imputed census file that is consistent with the adjusted census counts. The methodology involved three steps: missed household imputation, missed persons within households imputation, and "pruning and grafting" of imputations to the adjusted census counts (Office for National Statistics, 2002; Steele et al., 2002; see also U.S. Census Bureau, 2004, Chapter 6).

We have presented two directions in the development of sample survey methodology for censuses. The first direction tightens the relationship between censuses and sample surveys and focuses on evaluation of the census counts, with or without adjustment. The second direction moves away from the "snapshot" census and involves large-scale continuous sample surveys that either replace the census altogether or collect detailed and timely socioeconomic data to supplement the census data. A third direction that may be pursued in the future is to shorten the census cycle from 5–10 years to 2–5 years by basing full enumeration on administrative data and adjustment through coverage surveys.

Essential Methods for Design Based Sample Surveys
ISSN: 0169-7161
© 2009 Elsevier B.V. All rights reserved
DOI: 10.1016/B978-0-444-53734-8.00011-X

11

Opinion and Election Polls*

*Kathleen A. Frankovic, Costas Panagopoulos and
Robert Y. Shapiro*

1. Introduction: the reasons for public opinion and election polling

Public opinion polls are widely used to learn about the political attitudes, voting, and other behavior of individuals, by asking questions about opinions, activities, and individuals' personal characteristics (e.g., Abramson et al., 2007; Asher, 2007; Erikson and Tedin, 2007; Glynn et al., 2004; Traugott and Lavrakas, 2004). Responses to these questions are then counted, statistically analyzed, and interpreted.

Historically, academicians and government researchers in the United States engaged in opinion polling have called themselves "survey researchers," many with interests in psychologically oriented attitude research (Converse, 1987). There is also a separate and more visible group of "pollsters," originally involved in commercial research and in journalism, whose poll results on political and social matters are reported widely in the news media (Converse, 1987; Frankovic, 1998; Moore, 1992; Rogers, 1949). Others conduct proprietary polling for political candidates, political parties, or other clients (Eisinger, 2003, 2005; Jacobs and Shapiro, 1995; Stonecash, 2003). Today, survey researchers and pollsters have become synonymous, although the polls they do can vary in their purpose, type, scope, and quality (see Chapter 10).

This chapter describes these aspects of opinion and election polls, focusing largely on the United States, but providing some international comparisons and discussions of similar aspects and uses of polling. It begins by reviewing the public and private uses of opinion and election polling. Next, it summarizes the general methodological issues in polling that require attention in doing public opinion research. It then examines the cases of preelection polls, "exit polls" on election days, and, briefly, postelection polling. The last sections consider other methods of interview-based opinion measurement and the challenges ahead for opinion and election surveys as interest in public opinion and polling continues.

---

* The authors are listed alphabetically. This article is dedicated to our late colleague and friend, Warren J. Mitofsky, who would not have been shy about critiquing what we take full responsibility for here.

1.1. Polling for public consumption

Polling has been motivated by wide interest and curiosity about public opinion in general and especially about voting during election periods. This is clear in the history of "straw polls" described later. Ordinary citizens are curious about this; in turn, journalists who write about politics try to appeal to their audiences' interests in candidates and issues, especially in the latest elections. This has been apparent from the early days of informal "straw votes" or "straw polls" in the United States to the big expansion of national polling in the 1970s as valid and reliable surveys could be done by telephone (Blankenship, 1977; Nathan, 2001).

Contemporary opinion and election polls, especially those done for public consumption, have many sources. By the end of the 19th century, politicians, academics, market researchers, journalists, and government all had begun serious data collection. In many states local leaders kept "poll books" registering the preferences of every registered voter. These "poll books" had their greatest value in times of political movement and uncertainty. An 1880 Republican poll of 26,000 Indiana Civil War veterans showed that 69% would vote Republican that fall; by 1888, only 30% of those same individuals would (Jensen, 1971, p. 26). In 1886, the social survey movement began in England, and social welfare workers collected data on poverty, housing, and crime. Sociologists developed tools for measuring opinion. In the United States, newspaper market research departments were established and national advertizing campaigns for brands like Ivory soap became prominent in the 1880s. Political advertising on a mass scale soon followed. In 1888, the Republican National Committee placed presidential campaign ads on New York City streetcars for the first time, changing campaign tactics from persuasion to marketing (Jensen, 1971, p. 159). The U.S. Census first used an early version of the Hollerith computer card for storing and analyzing data in 1890.

Opinion and election polls also date from the 19th century in the United States. In 1824, straw poll counts appeared in partisan newspapers – along with suggestions that the public might not agree with political leaders in their choice of presidential candidates (Smith, 1990). In some cases, counts of candidate support were taken at public meetings. In others, books were opened for people to register their preference. Some newspapers praised the technique. The *Niles Weekly Register* said of a count taken at a public meeting, "This is something new; but an excellent plan of obtaining the sense of the people" (*Niles Weekly Register*, May, 1824).

In 1896, the *Chicago Record* sent postcard ballots to every registered Chicago voter, and to a sample of 1 in 10 voters in eight surrounding states. The *Record* mailed a total of 833,277 postcard ballots, at a cost of $60,000; 240,000 of those sample ballots were returned. The *Record* poll found Republican William McKinley far ahead of the Democrat, William Jennings Bryan. McKinley won; and in the city of Chicago, the *Records* preelection poll results came within four one-hundredths of a percent of the actual election-day tally.

By the 1920s, even papers whose editorial pages were clearly partisan were apparently comfortable reporting straw polls that indicated their editorial choice in an election was losing the contest for voters. Between 1900 and 1920 there were nearly 20 separate news "straw polls" in the United States. The *Literary Digest* established a poll in 1916. In 1920, it mailed ballot cards to 11,000,000 potential voters, selected predominantly from telephone lists. In later years, car registration lists and some voter registration lists were added to the sampling frame. Although the *Digest* touted its polling as impartial and accurate, it was also tied to an attempt to increase the magazine's subscriber base.

These news straw polling operations involved outreach to as many groups as possible and included huge numbers of interviews (often conducted on street corners). In its 1923 mayoral election poll, the *Chicago Tribune* tabulated more than 85,000 ballots. In the month before the election, interviews were conducted throughout Chicago, and results published reporting preferences by ethnic group (including the "colored" vote), with special samples of street car drivers, moviegoers (noting the differences between first and second show attendees), and white collar workers in the Loop.

But the real emergence of preelection polls as we now know them came in the 1930s. In 1935, both George Gallup and Elmo Roper began conducting a new kind of news poll: Gallup for a consortium of newspapers, and Roper for *Fortune* magazine. The stated goals were both democratic and journalistic. Gallup co-authored a book called *The Pulse of Democracy* (Gallup and Rae, 1940), while *Fortune*'s editors in their very first poll report in June 1935 explicitly linked impartial journalism and polls: "For the journalist and particularly such journalists of fact as the editors of *Fortune* conceive themselves to be, has no preferences as to the facts he hopes to uncover. . . . He is quite as willing to publish the answers that upset his apple cart of preconceptions as to publish the answers that bear him out" (*Fortune*, 1935).

Gallup and Roper (as well as Archibald Crossley, whose polls for the Hearst newspapers began in 1936) interviewed only a few thousand adults, unlike the tens of thousands in the *Tribune*'s canvass or the millions answering the *Literary Digest* polls. However, samples were selected to ensure that regions, city sizes, and economic classes were properly represented.

Although not a true probability sample, they were far more representative than the larger street corner or postcard polls (cf. Chapters 1, 5, and 9). And in that first test in 1936, the so-called scientific polls successfully predicted a Roosevelt victory. In fact, Gallup not only predicted a Roosevelt victory, but also predicted the *Literary Digest's* mistake. The flaws in the *Literary Digest*'s procedures (using lists that in nearly every state were biased in favor of the economically better-off during an economic depression) are fairly obvious today. First, by almost always limiting the sampling frame to those owning telephones and automobiles, lower-income voters were excluded, even though by 1936 social class would matter more than state or region in a person's presidential choice. Second, only about two million of the *Digest*'s ten million or so postcard ballots were returned, limiting the polling count to those who both received a questionnaire and bothered to respond. This led to a "selection bias" and the *Digest* overestimated support for Roosevelt's opponent, Republican Alfred Landon (see Squire, 1988). Although the new pollsters embarrassed the *Digest*, they too underestimated Roosevelt's margin of victory, suggesting biases in their methods as well. Those biases recurred in 1940 and 1944, foreshadowing the polling debacle of 1948 (see below and Converse, 1987).

By May 1940, 118 newspapers subscribed to the Gallup Poll. *Fortune* surveys appeared monthly. Between 1943 and 1948, at least 14 state organizations were conducting their own polls, using methods approximating those of the national pollsters (Frankovic, 1998).

The very first question asked in the first Gallup Poll was: "Do you think expenditures by the government for relief and recovery are too little, too great, or about right?"

Totally 60% of respondents said "too great," which was similar to what Americans thought about spending on "welfare" in the National Opinion Research Center's General Social Survey more than 50 years later (see Davis et al., 2005; Gallup, 1972, p. 1; Page and Shapiro, 1992, Chapter 3). By 1940, Gallup asked if the public approved of how President Franklin Roosevelt was handling his job, as well as which problems facing the country Americans believed were most important. These questions continue to be asked and are widely cited today.

Preelection polling began in other western democracies in the period between the wars and expanded afterwards (early European survey data are cited in Cantril with Strunk, 1951). There was a Gallup Institute in Great Britain beginning in 1938, and one in Canada starting in 1935. In Britain, "Mass Observation," which collected information about everyday life in Britain from 1937, used a different approach, creating a national panel of volunteers to reply to regular questionnaires (Hubble, 2005).

As democracy spread, so did polling and preelection polls. After the surrender of Japan, the U.S. occupying forces instituted public opinion polling, and the techniques established in the United States were adapted to accommodate at least some Japanese traditions (Worcester, 1983).

Wendell Willkie, an American businessman, used market researchers in his unsuccessful 1940 presidential campaign. Franklin D. Roosevelt was the first president to receive ongoing polling information – though from a distance, compared to later presidents – from Hadley Cantril, whose Office of Public Opinion Research at Princeton University had both an academic base and collaborated with the Gallup Organization. Roosevelt's use of polls during the period leading to the United States' entry into the war showed how polls provide strategic information for leading – or manipulating – public opinion (Page and Shapiro, 1992, Chapters 5 and 9).

At the same time, government agencies expanded their use of survey methods and large-scale polling beyond normal Census operations, beginning with the Department of Agriculture in 1939 (see Chapter 7). Survey research was used for public policy and public administration purposes, in attempts to improve wartime agricultural policies, control prices, understand race relations, stimulate war bond drives, measure the morale of civilians, and the well-being, outlook, and opinions of members of the armed forces (Converse, 1987; Gosnell and David, 1949; Stouffer et al., 1965 [1949]; Truman, 1945; Wallace and McCamy, 1940). The State Department, unknown to Congress, had the National Opinion Research Center (NORC, the first independent national survey research center, see below) survey the public's opinions toward American foreign policy during the Cold War (Eisinger, 2003; Foster, 1983; Page and Shapiro, 1992).

University-based surveys had their beginnings in 1939–1940, when Columbia sociologist Paul F. Lazarsfeld founded the Office of Radio Research, later renamed the Bureau of Applied Social Research. Lazarsfeld et al. conducted the first sophisticated survey study of decision-making in presidential elections in 1940 in Erie County, Ohio, followed by a study of the 1948 election in Elmira, New York (Lazarsfeld et al., 1944; Berelson et al., 1954). Two large and widely known research centers developed the long-term capacity to do national surveys: NORC, established in 1941 by Harry Field at the University of Denver and moved to the University of Chicago in 1947; and the University of Michigan's Survey Research Center (SRC, Institute for Social Research [ISR]), established in 1946 under the direction of Rensis Likert and Angus Campbell, who had been involved in government survey research.

Later, many other universities established their own survey research centers, especially after telephone surveying became more common (Sudman and Bradburn, 1987). The localized election studies begun at Columbia were subsequently continued on a larger national scale elsewhere. In 1948, Angus Campbell and Robert Kahn at the University of Michigan conducted the first national election study in the United States – or anywhere. This small study led to the long-term series of election studies conducted first by the Survey Research Center and the Center for Political Studies at Michigan, formalized as the American National Election Studies (ANES or NES) in 1977 with National Science Foundation (NSF) support, expanding control to a larger academic community. In 1972, NORC became the home of the ongoing, NSF-funded General Social Surveys (Davis et al., 2005).

At the outset, interest in polling was related to the original "democratic" philosophy associated with it – that individuals' opinions could be added up to constitute the will of the people that ought to have influence on political leaders and governing (Gallup and Rae, 1940). Critics argued that this simple definition of public opinion was misguided: public opinion as a concept and influence involved processes of purposive group interactions and communications that reached the attention of government, and in this process not all opinions counted equally (Blumer, 1948; Rogers, 1949). Ironically, the results of the early academic studies supported the views of the critics of polling. These studies showed that the public was not very knowledgeable and well informed about politics, was not influenced in expected ways by mass communication, and voted in ways related to seemingly mindless social characteristics and interpersonal influences. Partisan attachments were no more than equivalent to team loyalties (Berelson et al., 1954; Campbell et al., 1954, 1960; Lazarsfeld et al., 1944). The electorate voted neither on the basis of issues nor awareness of the competing political ideologies of the day (Converse, 1964). This spurred further debates about the capability, "rationality," and overall "democratic competence" of public opinion and the American electorate (cf. Althaus, 2003; Glynn et al., 2004, Chapter 8; Page and Shapiro, 1992; Zaller, 1992).

By 1948, pollsters had gained a national audience far beyond that of their press releases and print media subscribers. Elmo Roper gave weekly Sunday radio talks on CBS, and both Roper and Gallup were televised. The front-page headline of the *Washington Post* on Election morning read, "Dewey Deemed Sure Winner Today," although the last preelection Gallup Poll had given the Republican Thomas E. Dewey only a five-point lead over the incumbent Democrat Harry Truman (49.5–44.5%).

Poll predictions of an easy win for Dewey made Truman's victory all the more devastating for the polling community. Gallup lost some subscribers. In one study that interviewed 47 editors who had used polls before the election, half said they would no longer do so (Merton and Hatt, 1949).

There were selection biases in the 1948 polls. Interviewers used their own judgment in interviewing quota samples of the public. An academic investigation conducted under the aegis of the Social Science Research Council resulted in a greater emphasis on probability selection of respondents (see Mosteller et al., 1949, and Section 3 later). In addition, the final 1948 published polls were conducted weeks before the election, and as a result, they were not in the field to pick up any shifts in the electorate up until the day people voted. Beginning in 1950, Gallup made its sampling methodology more rigorous, providing interviewers with more strict instructions regarding the selection of respondents.

But polling was now embedded in American politics. As an extension of its campaign activities, the Republican Party provided poll results to the White House after Eisenhower's election. George Gallup and later Louis Harris routinely reported their latest polls to presidents. Harris worked as John F. Kennedy's political consultant and pollster. These president–pollster relationships became regularized with Lyndon Johnson and his pollster, Oliver Quayle; Nixon and his polling operation run by H.R. Haldeman, with pollsters David Derge, Robert Teeter, and members of the Opinion Research Corporation; and Gerald Ford continuing with Teeter (see Eisinger, 2003; Jacobs and Shapiro, 1995, 1995–1996).

Although their wartime survey operations were disbanded, government agencies' interest in survey and Census data continued after the war (Converse, 1987; Sudman and Bradburn, 1987). United States government sponsored political polling was devoted, interestingly and lawfully (as there were prohibitions against polling that could be construed as partisan), only to long-term surveying in foreign countries through the United States Information Agency (USIA; see Crespi, 1999). The tradition of public administration and policy-related survey research continued, including state and local survey research to improve policy formation and implementation, and to facilitate feedback on the performance and effectiveness of government bureaucratic agencies, policies, and programs (e.g., see Desario and Langton, 1984; Kweit and Kweit, 1984, Van Ryzin et al., 2004a, 2004b).

Polling in the United States, especially telephone surveys, expanded substantially in the 1970s with the development of Random Digit Dialing (see Chapter 3) and of computer-assisted telephone interviewing (CATI) allowing polls to be done quickly and relatively cheaply. Household telephone penetration reached 94%, easing interviewing over the phone, which cost less and enabled centralized control over interviewers. Sampling by random digit dialing was also easier and cheaper than the enumeration and sampling procedures used for in-person interviewing. Concerns about interviewer access to respondents in urban areas also motivated the use of phones (see Blankenship, 1977; Nathan, 2001). In other countries, data collection also moved from in-person to telephone when possible.

Journalistic organizations that wanted to poll could centralize data collection with phone banks, occasionally utilizing advertizing or classified ad department offices at night and on weekends. Improved newsroom computer technology, especially the arrival of PCs in the 1980s, meant that polls in response to breaking news could be conducted and reported within a half hour of an event's occurrence. The same expansion occurred for polling by political parties, candidates, and advocacy and interest groups.

But Vietnam and Watergate, by increasing cynicism among journalists and the public, were as important as technology in the development of news surveys during the 1970s. Extending the trend that began in the Progressive era of the 1890s, there was an even further shift away from partisan influence over the press. Journalists again believed it important to bypass party leaders and go directly to the public for its opinions. Media polling made journalists less vulnerable to manipulation by political parties, candidates, organized groups, and others who used poll data for their own purposes (see Gollin, 1980a, 1980b, 1987; Jacobs and Shapiro 1995–1996). The news media could now verify claims by others about the state of or changes in public opinion.

CBS News and NBC News had conducted occasional polls for many years, and had used computers and some statistical modeling in election coverage as early as

1952 (Bohn, 1980), but serious polling units were organized only in time for the 1976 election. CBS News formed a contractual arrangement with *The New York Times* in 1975. NBC News worked on its own in the 1976 election, but joined with the Associated Press from 1977 to 1983, and later with the *Wall Street Journal*. ABC News joined forces with the *Washington Post* in the early 1980s; that partnership continues. CNN first joined with *Time* magazine in 1989, and worked with *USA Today* and the Gallup Organization from 1992 to 2005. Partnerships by different media outlet have occurred in state and local polling as well.

As the ease of fielding surveys increased and costs dwindled, public pollsters increasingly probed the public's vote intentions and preferences on candidates for lower offices. State-level polls routinely asked respondents' views about candidates for statewide office, such as gubernatorial and U.S. Senate candidates. Some survey organizations, such as SurveyUSA, which use automated polling, Interactive Voice Response (IVR, see below), on a large scale, have also fielded congressional-district and even municipal-level surveys that ask respondents about candidates in those races.

The historical literature on polling in other countries is less extensive and cumulative than in the United States (but see Worcester, 1983, 1987, 1991). Polling spread to Mexico and Latin America after World War II through the work of Joe Belden and others. After the fall of Communism, polling in Russia and Eastern Europe came through sociological institutes and partnerships. There are now polling companies in nearly every country, with preelection polls conducted wherever there are elections. This suggests that polling has expanded its reach in tandem with democracy and democratization worldwide. There are also many collaborations, including a European Social Survey, modeled after the NORC General Social Survey, and the broader International Social Survey Programme (ISSP). The Eurobarometer, established in 1973, has been used by the European Commission to measure public opinion in all EU member countries. The Eurobarometer concept has been extended to other continents: the Latinobarometro, established in 1995, the Afrobarometer, established in 1999, and the Asian Barometer Survey (formerly the East Asian Barometer Survey), begun in 2000. The broadest international survey collaboration to study elections is the Comparative Study of Electoral Systems (CSES).

In countries other than the United States, there have been wide variations in direct government involvement in political polling. No systematic comparison across countries has been made to date, although comparative research has slowly begun (e.g., Nacos et al., 2000; Worcester, 1983). In Canada, like the United States, polling for the government originated during World War II. In addition to consulting available Gallup polls, government officials initiated regular opinion polling of their own to study public morale, attitudes toward living costs and rationing, and opinions about Canada's involvement in the war and expectations about the postwar economy (Page, 2006). This began a tradition of government polling and politics in which polls are used for purposes of determining what issues should be high on the agenda and for determining communication and policymaking strategies for promoting policies (see Jacobs and Shapiro, 2000). In the former Soviet Union, the collection of data on media use and other "sociological" topics led to the collection of other opinion data (Mickiewicz, 1981).

As public polling became institutionalized, there was a concurrent increase in professional survey organizations. The American Association for Public Opinion Research (AAPOR) was founded in 1947, followed soon after by the World Association for Public Opinion Research (WAPOR). These organizations provided meeting places where both pollsters and academic researchers could discuss and debate issues in survey research (Sheatsley and Mitofsky, 1992). The National Council on Public Polls (NCPP), an association of public polling organizations with a particular interest in the polls produced for public consumption in the United States, was founded in 1969. The European Survey Research Association was founded in 2005 to provide communication among European survey researchers. There are also organizations that are more concerned about the market research industry, including CASRO (Council of American Survey Research Organizations) and CMOR (Council for Marketing and Opinion Research), now merged with the Marketing Research Association, in the United States, and ESOMAR (originally known as the European Society of Opinion and Marketing Research and now branded as a "world organization").

In addition, archives collect and make available polling results beyond the original data collection and publication. The Roper Center (now at the University of Connecticut) was established in 1947, and pollsters like Roper and Gallup deposited data sets there for academic and public use. Currently, nearly every major public polling organization does the same, including some international organizations. The Inter-University Consortium for Political and Social Research (ICPSR, at the University of Michigan) was established in 1962 and has data sets from academic social science studies, from the United States and elsewhere, as well as public polls. A list of some international archives can be found at: http://www.ropercenter.uconn.edu/data_access/data/data_archives.html.

1.2. Polling for private consumption: candidates and campaign strategy

Public opinion and polling research increasingly became a major part of political campaigns in the United States in the 20th century, with the use of focus groups and other means of message testing, initial benchmark polls to provide basic public opinion information to candidates, periodic trend polls to compare to the benchmarks, and even more frequent – weekly or even daily – "tracking polls" (Eisinger, 2005; Jacobs and Shapiro, 2000; Stonecash, 2003). In this century, political campaigns have attempted to "microtarget" voters and engage in vote-getting activities in increasingly precise ways with the aid of both their own proprietary polling and other publicly available data (Hamburger and Wallsten, 2006; Jacobs and Shapiro, 2005). These polls have been used to find out how specific types of voters might be influenced by emphasizing particular issues or otherwise finding messages that will resonate with them, ultimately affecting their voting decisions.[1]

In the 1992 congressional campaigns, candidates perceived public opinion surveys to be a crucial source of information for gauging public opinion in House campaigns. On a scale that ranged from 1 (not important or not used) to 5 (extremely important), the mean score for the importance of polling was 3.5, second only to personal

---

[1] The ways in which polls are used for this purpose are sometimes mistaken with a campaign tactic known as "push polls," that under the guise of an opinion survey is an attempt to talk to large numbers of voters (tens of thousands, in contrast to under 1500) in an effort to spread often misleading information designed to directly influence voters (see Asher, 2007).

candidate contact with voters as a source of information (mean score = 4.4). Candidates apparently perceived polling information to be more reliable than information obtained from newspaper, radio or television sources, local party activists, mail from voters, and national party leaders (Herrnson, 2004). Moreover, polling is widespread even in low-salience, local-level elections; 41% of candidates in municipal elections conducted polls (Strachan, 2003).

Presidents' pollsters have been visible figures beginning with Patrick Caddell for President Jimmy Carter, and continuing with Richard Wirthlin for Ronald Reagan, Robert Teeter for George H.W. Bush, and Stanley Greenberg followed by Dick Morris for Bill Clinton (Eisinger, 2003; Heith, 2004; Jacobs and Shapiro, 2000; Murray, 2006). George W. Bush, despite his disparaging claims about political leaders using polls, appointed an academic public opinion expert, Peter Feavre, as a member of his national security team dealing with public support for the Iraq war.

American-style campaign politics, including heavy use of polling, has spread to other countries, and American consultants have routinely worked in both developed and developing democracies. One study found that nearly 60% of U.S.-based political consultants had worked abroad, especially in Latin America, postcommunist countries, and Western Europe (Plasser and Plasser, 2002). For example, Greenberg Quinlan Rosner, whose principal Stan Greenberg polled for Bill Clinton's 1992 presidential victory, worked on the campaigns of Britain's Tony Blair, Germany's Gerhard Schroeder, South Africa's Nelson Mandela, Israel's Ehud Barak, as well as in Bolivia, Honduras, Poland, and Mexico. However, the extent to which American electioneering styles have been adopted abroad has depended on the political, social, cultural, institutional, and regulatory environments in each country.

Polls also have been conducted by interest groups and other organizations. These groups also have polled their own members and supporters, as a way of maintaining contact with them. They have publicized their polling to promote their organizations' goals, just as candidates and parties have promoted their political objectives. Some organized groups have engaged in fundraising or membership solicitation in the context of mail or other political surveys – "soliciting under the guise" of survey research (SUGing) or "fund raising under the guise" (FRUGing) – which survey professional associations such as the American Association for Public Opinion Research (AAPOR) and the National Council of Public Polls (NCPP) have considered unethical uses of surveys (Traugott and Lavrakas, Chapter 3, 2004).

2. General methodological issues in public opinion and election polls

Methodological issues in public opinion and election polling can perhaps best be summarized under the general rubrics of *errors in sampling, measurement error, and errors in conceptualizing and specifying* what's being studied (Brady and Orren, 1992; see especially the full guide provided by Weisberg, 2005). Errors that are part of sampling go beyond the estimation of the margin of sampling error and include defining and adequately covering the population that is to be studied. Then comes the problems of nonresponse and the need to weight the data appropriately. Measurement errors include the potentially substantial effects of the *mode* of surveying used, question wording, question order or context effect (including the order in which response categories are offered), coding and data processing errors, problems in imputing missing data, and the effects that interviewers can have on responses to questions (see Asher, 2007; Schuman and Presser, 1981; Weisberg, 2005). Conceptual and specification problems bear on the ultimate validity of the opinion measures used. In some cases researchers assume respondents have genuine attitudes or that particular issues and attitudes have a high degree of salience, which may not in fact be the case. The researcher must consider the possibility of "nonattitudes" and the transitory nature of survey responses (see Asher, 2007; Bishop, 2004; Converse, 1964), and they should make use of insights to be gained from "cognitive interviewing" methods and the results of survey pretests (see Beatty and Willis, 2007; Tourangeau et al., 2000; Sudman et al., 1996). Even in cases where researchers are confident that survey responses reveal real attitudes, they may have difficulty.[2]

The different sources of survey error can overlap or interact with each other, and they can vary by the mode in which the survey is administered. Modes (or ways) of conducting surveys include in-person/face-to-face interviewing, mail and other paper surveys, telephone surveys, and most recently, on-line polling. Each survey mode has advantages and disadvantages in the way interviewers or data collectors interact with respondents (Asher, 2007; Dillman, 1978; Fricker et al., 2005; Dillman, 2007; Weisberg, 2005). Nonresponse is a persistent and growing problem across modes, but it is more prevalent in those selected for lower cost (nonresponse error includes noncontacts and refusals, as well as problems of noncoverage of the desired population because of a limited sampling frame). If the attitudes and opinions of nonrespondents differ significantly from those of respondents in ways not easily corrected through demographic weights, nonresponse bias is introduced. In-person surveys normally have higher response rates. Their larger costs insure the close contact with respondents that typically lead to higher response rates.

But even when different surveys use the same mode there can be survey "house effects," the effects of the particular procedures, instructions to interviewers, and other rules for interviewing that are specific to individual survey organizations (Smith, 1978).

Survey respondents often appear to conform to "social desirability" pressures: self-reported rates of voting in elections are substantially (and inaccurately) higher than actual turnout rates (cf. Belli et al., 2001; McDonald, 2003a). Complications and bias may also arise from respondents' reactions to interviewer characteristics, including race, gender, age, and ethnicity. Evidence of interviewer effects dates back to the 1940s (Katz, 1942). Most notably, there have been significant race-of-interviewer effects (cf. Finkel et al., 1999). Black respondents may report more favorable attitudes about white candidates to white interviewers than they do to black interviewers (Anderson et al., 1988; cf. Finkel et al., 1999). Black respondents also demonstrate higher levels of political

---

[2] One illustrative case arose in the 2004 U.S. presidential election in which responses to the exit polls indicated that more than 20% of voters chose "moral values" from a list of issues as the most important influence on how they voted. This led to a lively debate about how the exit poll respondents interpreted the meaning of "moral values" – whether this phrase referred to issues like abortion or the rights and behavior of homosexuals, whether it was a statement about the candidates' morality, or whether it was a fall-back answer if respondents did not think the other issues on the list were sufficiently salient to their vote choice (see Langer and Cohen, 2005).

knowledge in telephone surveys when interviewed by black interviewers than when they are interviewed by whites (Davis and Silver, 2003). The race-of-interviewer effect on expressed voter preferences in preelection polls has been especially acute in elections with a white candidate running against a black candidate. In one study, white respondents were 8 to 11 points more likely to express support for the black candidate to black interviewers than to white interviewers (Finkel et al., 1999). In some past elections, white respondents have been far more likely to report an intention to vote for the black candidate compared to their behavior on Election Day, although there has been less evidence of this in more recent elections and no evidence that it happened in the 2008 election of Barack Obama (Hopkins, 2008). Gender and ethnicity-based interviewer effects have also been reported (Hurtado, 1994; Kane and Maccaulay, 1993). Interviewer effects have also been raised as a significant problem in conducting exit polls (discussed below; cf. Bischoping and Schuman, 1992).

Different modes of interviewing can mitigate or exacerbate survey problems.

2.1. Mail surveys

Mail surveys have been an appealing mode of data collection, partly due to lower cost. Because mail surveys are self-administered, interviewers are not necessary, thus eliminating them as a source of bias. The assurance of anonymity and confidentiality may also encourage respondents to be more forthright, especially when being probed on sensitive or controversial issues (Asher, 2007; Dillman, 1978; Dillman, 2007). Mail surveys can be an effective way to gather information from smaller, specialized population groups. The main drawback of mail surveys has been their high rate of nonresponse, which has tended to exceed the nonresponse rates for telephone and personal interviews. General population mail surveys have frequently achieved response rates under 5% (Dillman, 1978; Dillman, 2007; Glynn et al., 2004). Monetary incentives, personalized notification letters that arrive before the survey, multiple correspondences with nonresponders, and prepaid return postage can boost response rates for mail surveys (Asher, 2007). Other problems have included having little assurance about who actually completed the survey, no opportunity for question clarification, and some uncertainty as to what order questions were answered in (pertaining to possible context effects). Mail surveys also tend to require long field periods for follow-up mailings.

2.2. Telephone surveys

Most surveys in the United States since the 1970s have been conducted by telephone. Random Digit Dialing (RDD) techniques allowed researchers to select a random sample of households, helping to resolve the sampling challenges often inherent with other survey modes. Refinements to RDD made the process even more efficient.

One early innovation was the Mitofsky–Waksberg procedure developed by Warren Mitofsky (Mitofsky, 1970) and later refined by Joseph Waksberg (Waksberg, 1978). Residential phone numbers represented only about 20% of phone exchanges nationwide but tended to be highly clustered within working blocks of 100 consecutive numbers. Mitofsky–Waksberg, a form of area probability sampling, took advantage of this fact to exclude attempts to banks of phone numbers that were unlikely to be residences. The procedure randomly generates 10-digit phone numbers that can be called to identify working blocks. Random samples can be generated from those blocks identified as working. This procedure increased considerably the efficiency of sampling for telephone surveys while preserving the scientific integrity of the sample.

Telephone sampling was further refined to take advantage of list-assisted sampling techniques, using telephone book listings as a seed to generate random household numbers (Brick and Waksberg, 1991). Important refinements were made, involving the statistical theory underlying list-assisted methods and the effects of telephone system changes, including the lower proportion of telephone numbers assigned to residential units (see Tucker et al., 2007, 2002).

Registration-based sampling (RBS) allows survey researchers to draw random samples of registered voters from publicly available voter files rather than relying on the respondent's self-reported registration status. Pollsters for political campaigns have used RBS sampling extensively, despite the fact that registration-based samples have frequently suffered from incomplete coverage (due to unavailable telephone numbers) of the population of registered voters and inaccuracies in the voter file.

An advantage of sampling from registration lists has been that many lists contain additional information from public records that can be used to forecast voter turnout. Although the quality of information available has varied across jurisdictions, typical registration files have contained dates of birth, dates of registration, and (where relevant) party registration. Data on past voter turnout has been furnished by registrars or acquired by private vendors, and researchers can use this information to assign voting propensity scores to individuals on the list and thus draw samples of voters most likely to participate in the election. Some have found that registration-based sampling can improve the accuracy of election forecasts over RDD (Green and Gerber, 2006).

Computer-assisted telephone interviewing (CATI) has allowed polling to be faster, more efficient, and more accurate (Asher, 2007). CATI has also permitted randomizing the order of questions and even response categories for each question, diminishing the possibility of order effects. CATI has helped facilitate and expand experiments in public opinion research, through the National Science Foundation-funded Time-sharing Experiments for the Social Sciences Program, using both telephone and internet surveys (see TESS).

Telephone surveys have had several disadvantages, however, including growing rates of nonresponse, partly due to innovations like caller ID, answering machines, privacy managers, and increasing cell phone coverage (see below). Interviews have tended to last no more than 15 or 20 minutes, as respondents become fatigued faster than in other modes. Telephone surveys have also been vulnerable to interviewer effects, as the characteristics of the interviewer and their levels of training and motivation can influence the quality of the data collected.

2.3. In-person/face-to-face surveys

Although in-person surveys became less common in public opinion and election polls well before the end of the last century, primarily due to cost and time considerations, they have remained an effective way to collect rich and complete public opinion information. The National Opinion Center's General Social Survey (GSS), for example, continued to be administered in-person. Any type of simple, stratified or systematic sampling of respondents dispersed over a vast geographic area would be impractical, so in-person surveys have typically used area probability samples that are essentially stratified multistage cluster samples (MCS).

MCS requires the country to be divided into geographic regions (four in the United States). Within each, a set of counties and standard metropolitan statistical areas have been randomly selected. Within these, primary sampling units (PSUs) of four or five city blocks are randomly selected and then four or five households randomly selected from each block. Random selection has sometimes been abandoned at the household level because complete lists of all household members may not be available, but interviewers have generally relied on some systematic method to select respondents within the household (Erikson and Tedin, 2007). The General Social Survey and the National Election Study continued to select respondents randomly at the household level. The response rate for the 2006 GSS was 71%; the preelection response rate for the 2004 NES was 66%.

One of the main disadvantages of in-person surveys is their higher cost, but these costs are necessary because these surveys are designed to achieve (and they are usually successful in achieving) higher response rates than other modes of data collection. On the plus side, nonresponse rates tend to be lower, and respondents are usually more engaged and forthcoming in the interview. Interviewers are also able to be more personal and interactive, use visual aids and clarify questions, and monitor nonverbal behavior. However, the potential for interviewer effects described earlier is maximized with surveys administered in-person.

2.4. Online surveys

Online surveys offer many advantages, including cost efficiency and speed. Researchers can also develop and administer complex questionnaires and they inform respondents how far they have proceeded through the survey (something they can directly see in mail surveys). More important, researchers can include visual aids and they can conduct experiments with random assignment of "treatments," as well as provide graphic or multimedia enhancements. Item nonresponse can be reduced because respondents can be reminded and motivated to revisit incomplete items. Web-based surveys are self-administered, so interviewer effects can be avoided.

On the other hand, web-based surveys present serious methodological problems that may limit researchers' ability to make valid statistical inferences from them, including challenges in assembling sampling frames for probability sampling, coverage issues, and selection bias. Despite rapid penetration of the Internet into households, coverage issues remain especially acute in the United States, because access is restricted to about 60% of the population. Web-based surveys are also not immune to nonresponse. Other disadvantages include questionable security and authentication procedures and differences in format and presentation across computing systems and browsers (Dillman, 2007).

Since as yet there is no available listing of electronic mail (e-mail addresses) and not everyone has had online computer access, one approach in the United States has been to use RDD or in-person survey probability sampling methods to draw random samples of adults, giving respondents without internet access such access to form "panels" who participate in multiple surveys, thereby reducing the costs of drawing repeated samples. A more controversial variation of this method has recruited thousands of volunteers to join panels through invitations on internet sites; this approach has assumed that any selection bias can be corrected through sophisticated methods of "propensity score" weighting (Rosenbaum and Rubin, 1983; Taylor, 2000; Taylor and Terhanian, 1999). Online panels have been used in multiple countries in Europe and Asia, even some with relatively low internet penetration. Yougov was founded in 2000 in the United Kingdom, and has maintained an internet panel for research. It was successful in predicting the outcome of the 2001 Parliamentary election, performing better than conventional polls. It was not quite as good in the 2005 Parliamentary election, and it fared poorly in its efforts in the U.S. 2004 presidential election. To date, preelection polls from self-selected internet panels in the United States have done no better than telephone surveys, and often less well. But, given the problems facing telephone surveys, they have received a great deal of interest. There has been a need for more transparency and openness so that these surveys (like IVR, see below) can be evaluated fully (Blumenthal, 2005).

Probability sampling for Internet surveys originated with Willem Saris who developed this method in the Netherlands prior to the development of the Internet (Saris, 1998). It utilized "computerized self-administered questionnaires" (Couper and Nichols, 1998, p. 13) and was implemented by the Telepanel of the Netherlands Institute for Public Opinion (NIPO or Dutch Gallup). Respondents were provided with computers and modems, and were trained to download and fill out the questionnaires by computer. Upon completing the questionnaire, each respondent uploaded it for data collection and processing. The Telepanel idea was adopted in other countries. Most recently, in 2006, the US National Science Foundation funded a trial run of this method using in-person full probability sampling recruitment of respondents. The Dutch government also provided a major grant to support a large-scale Telepanel.

2.5. Mixed or multiple mode surveys

To date, mixed mode methods have not been used extensively in public opinion and election surveys. Mixed mode surveys offer respondents a choice of response modes. Multiple mode surveys have collected information from respondents using several survey modes. Some respondents may participate by mail, for example, while others in the same survey may be interviewed on the telephone. The methodological considerations described earlier for each survey mode continue to apply to mixed mode surveys respectively, and others may arise when combining samples of respondents interviewed using different modes, such as differences in responses depending on the method of data collection. Offering respondents a choice of response modes may increase rates of participation.

3. Preelection polling: methods, impact, and current issues

The earliest scientific preelection polls, in the 1930s, relied on in-person interviewing. Interviewers were sent to selected locations and instructed to interview a specific number of men and women, young and older people, higher and lower-status voters. Completed questionnaires were then returned for tabulation and reporting. The Gallup

Organization developed a procedure to speed up the polling as Election Day drew near, in which interviewers telegraphed back the responses to specific questions.

As already noted, the early Gallup and other polls had significant successes in predicting presidential victors, at least until the 1948 election. Their methodology, particularly the reliance on the quota selection process administered by the interviewers themselves, had been questioned by government statisticians like Louis Bean, of the Department of Agriculture, Philip M. Hauser and Morris Hansen, of the Bureau of the Census, and Rensis Likert, from the Bureau of Agricultural Economics, and some academics. The post-1948 Social Science Research Council Report (Mosteller et al., 1949) also questioned the reliance on quota samples, and it observed that the pollsters had overestimated the capabilities of the public opinion poll.

There was another source of concern in making prediction from preelection polls. What the pollsters (and even some academics, like Paul Lazarsfeld) had discovered from the 1936, 1940, and 1944 presidential elections was that few if any changes could be attributed to the campaign (Lazarsfeld et al., 1944; on the history of "minimal" campaign and especially "media effects," see Klapper, 1960). In 1948, they believed that the lead held by the Republican candidate in the fall could not be affected by anything either candidate could do. Polling stopped several weeks before the election. After 1948, pollsters would poll much closer to Election Day.

Methodological changes tend to follow problems in election prediction internationally as well. For example, in Britain, preelection pollsters wrongly predicted a Labor victory in the 1992 parliamentary election. The Conservative Party won by eight percentage points. The British Market Research Association conducted an investigation, and found similar problems to those in the United States in 1948 (Jowell et al., 1993). Voters changed their minds at the last minute, and there were also problems with the sampling methods. Most British pollsters opted to continue in-person quota sampling for the next national election, although they did change their quotas. But others moved to implement greater changes, including the adoption of telephone polls. Similar issues were reported in 2002 preelection poll errors in France (Durand et al., 2004).

When preelection polls underestimated Ronald Reagan's victory margin in the 1980 election, pollsters decided that in future years they needed to continue interviewing through the night before the election (Hansen, 1981; Kohut, 1981; Mitofsky, 1981).

Differences in methods may apply to the designation of likely voters or allocation of undecided respondents (discussed below), but may also mean framing preference questions differently. Martin et al. (2005) found evidence of these differences in the preelection polls conducted in 2000.

3.1. The allocation of undecided voters

Preelection polls are routinely adjusted from pure probability samples of the adult population. Some of those adjustments include the management of voters who refuse to give a preference when asked. The range of the number of undecided voters in preelection surveys can vary greatly across surveys and across stages of the campaign. Between 1988 and 1996, the proportion of undecided voters reported in polls ranged from 3 to 73% of the sample (Visser et al., 2000). Typically, 15% or more of the electorate may be undecided during a presidential campaign. This figure tends to be even higher in lower-level, less-salient races (Erikson and Tedin, 2007). The proportion of "undecideds" drops substantially in the final days of the election, although 5% of voters in the 2004 exit poll claimed they made up their minds on Election Day.

Some polling organizations treat undecided voters as a separate category and include that percentage in the final preelection estimate; others remove them and recalculate the percentage for each candidate or party, and others have attempted to allocate them. Allocation schemes have varied considerably. One approach has assumed that undecided voters who end up voting do so randomly, so truly undecided respondents should be allocated equally to the main candidates. This procedure can yield more accurate forecasts than eliminating undecided respondents altogether (Erikson and Sigelman, 1995; Visser et al., 2000). Another approach allocates undecided voters disproportionately to the challenger, as preelection surveys may systematically underestimate support for the challenger because of the "spiral of silence" (Noelle-Neumann, 1993 [1984]). In one study of a wide range of statewide, congressional and municipal primary and general election races, undecided respondents disproportionately voted for challengers in 82% of elections (Panagakis, 1989).

Changes in procedures also make comparisons difficult. In 1992, for example, the Gallup Organization changed its allocation method and, as a result, severely overestimated support for Bill Clinton and underestimated support for George W. Bush and independent challenger Ross Perot (Traugott and Lavrakas, 2004). In the weekend before the election, Gallup decided to assign all undecideds in its tracking poll to Clinton, citing the tendency of undecided voters to ultimately choose challengers. Clinton's lead grew from two points on Friday to 12 on Monday. But rather than moving to Clinton, many undecided voted for independent candidate Ross Perot (Erikson and Tedin, 2005).

3.2. Weighting and determining likely voters

The British difficulties in 1992 and the reaction to them underscored a major methodological issue for preelection polls. Should one weight or adjust the results? The adjustments can range from insuring that the original sample reflects the appropriate population parameters and the probabilities of selection to weighting on past voting behaviors. Adjustments may also be required to ensure that the final published results reflect the opinions of actual voters, not all adults.

In countries like the United States, where voter registration is not automatic, a large portion of respondents may not vote. Candidate preferences of nonvoters may be different from those of actual voters.

In the United States, pollsters have almost always first asked respondents whether or not they were registered and then asked those registered a series of questions designed to separate voters from nonvoters, including whether the respondent had voted in the past and would vote in the current election, as well as a measure of political interest. Some screens have included whether respondents knew the location of their polling places. Based on their answers, registered respondents can be assigned a probability of voting which is then used as a weight when tallying the projected vote. A more common solution (used by Gallup and many other pollsters) has been to divide registered respondents into two groups. Respondents who scored beyond a specified cutoff have been designated as "likely" voters, whose choices are then counted in the tally. The choices of those scoring below the cutoff are excluded in the estimation (Asher, 2007; Crespi, 1988; Daves, 2000; Erikson et al., 2004).

Estimates of likely voters in the weeks and months prior to Election Day may reflect transient political interest on the day of the poll, and might have little relationship to behavior on the day of the election. Even though such likely voter samples might well represent the pool of potential voters sufficiently excited to vote if a snap election were to be called on the day of the poll, they may not be the same people voting on Election Day. An analysis of Gallup polls in the 2000 presidential election indicated that the sorting of likely and unlikely voters is volatile and that much of the change (although certainly not all) is an artifact of classification. Pollsters can mistake shifts in the excitement level of the two candidates' core supporters for real, lasting changes in preferences (Erikson et al., 2004).

3.3. *Tracking polls*

Tracking polls, which monitor campaign dynamics on a daily basis, originated with political campaigns, which used them to evaluate campaign events and the impact of political advertizing. Journalists adopted them in the 1980s. Typically, tracking polls contact small samples of respondents (100–350) each day. To update results, a new day's sample of respondents is added to the total sample and the oldest day's sample of respondents is dropped. On their own, these samples are too small to provide precise estimates of preferences, but pollsters have used rolling averages of two or three consecutive days' worth of interviewing. Thus, estimates can be based on 500–600 interviews aggregated across all days (Traugott and Lavrakas, 2004).

Tracking polls may be useful to assess campaign dynamics, but there are shortcuts used in these surveys. Tracking polls have typically been one-night surveys that have not always employed the rigorous sampling and respondent selection procedures that many other polls do (Traugott and Lavrakas, 2004). Respondent call-back appointments have rarely been made and interviewers have not always selected respondents randomly within households. There have been disagreements over whether samples should be weighted each day or over several days. The *Washington Post* tracking poll in the 2004 presidential election, for example, adjusted each day's randomly selected samples of adults to match the voting-age population percentages by age, sex, race, and education, as reported by the Census Bureau's Current Population Survey. The *Post* also adjusted the percentages of self-identified Democrats and Republicans by partially weighting to bring the percentages of those groups to within three percentage points of their proportion of the electorate, as measured by national exit polls of voters in the last three presidential elections (*Washington Post*, 2004). Despite these challenges, the accuracy of tracking polls has been shown to be superior to other polls in some studies (Lau, 1994).

Several companies, in both the United States and Great Britain, have conducted election polls among a sample of individuals recruited to be part of web panels. The resulting interviews conducted online have been adjusted by demographics and politics to reflect a predetermined estimate of the electorate (Taylor, 2000).

For the most part, pollsters have been reluctant to weight by party identification. The main hesitation has been that party identification is not a fixed characteristic of the electorate in the United States, and there has been evidence of significant short-term fluctuation in party ID. But overall partisanship exhibits considerable stability over time. Pollsters can estimate the underlying proportions of Democrats and Republicans in the electorate based on moving averages of results from surveys conducted over several weeks, and these estimates might be used, with caution, to weight samples (see Abramowitz, 2006).

3.4. Preelection poll accuracy

Although lopsided attention has been devoted to notable failures to predict election outcomes, the results of preelection polls conducted at the end of an election cycle have overall tended to come within a few percentage points of the actual outcome (Crespi, 1988). Analyzing presidential races between 1984 and 1992, one study reported an average error of 4.5 percentage points (Gelman and King, 1993). The National Council on Public Polls (NCPP) calculated that the average error of national polls in 2004 and in 2008 was just 0.9 percentage points (see Martin et al., 2005; Traugott, 2005; NCPP, 2004, 2008). In fact, the "trauma" that has often followed inaccurate poll-based predictions of election results has been a testament to the general reliability of polls (Mitofsky, 1998).

4. Exit polling

4.1. Uses

Exit polls are polls of *voters*, interviewed after they have left their places of voting and no later than Election Day. They may include the interviewing before Election Day of postal, absentee and other early voters. Exit poll functions are not mutually exclusive: they can predict election results, describe the patterns of voter support for parties, candidates, and issues; and support extensive academic research efforts with which the results are formulated and disseminated.

Election projections can be made in ways other than by interviewing voters as they exit the polling place. Though most projections are based on exit polls, interviewing voters after having voted at a polling place, other forecasting models may include: CAPI, CATI or other interviews on Election Day with voters after *or before* having cast their votes and counts of official votes in a sample of precincts, often known as quick counts.

A standard use for exit polls in new democracies has been as a check on voting itself. In recent years, exit poll results in Venezuela, the Ukraine, Georgia, Peru, and Serbia have been hailed by some as better indicators of election outcomes than the vote count. Although a well-conducted exit poll can sometimes be a check on fraud, sampling error limits any poll's precision, and operational difficulties, including restrictions on carrying out exit polls, and possible bias due to interviewer–respondent interactions can call into question the accuracy of those, and other, exit poll results.

The first exit poll was perhaps conducted inadvertently in the United States by Ruth Clark in 1964. Clark, a well-known newspaper researcher, began her research career as an interviewer. In 1964 she worked for Louis Harris and was sent to conduct interviews in Maryland on its primary election day. Tired of door-to-door interviewing to look for voters, she decided to talk with them as they left the polling place (Rosenthal, 1998, p. 41).

The exit poll did not become a staple of news election coverage until the 1970s and 1980s. CBS News, under the leadership of Warren Mitofsky, began exit polling in a

1967 gubernatorial race in Kentucky to collect voting data in precincts that did not make their vote available at poll closing. It later expanded its questionnaire to include questions about voter demographics and issue positions. The process was adopted by other American news organizations and then quickly spread to other countries (Frankovic, 2007; Mitofsky, 1999).

The first exit poll in Great Britain was conducted in 1974 for ITN by Humphrey Taylor's United Kingdom company, part of Louis Harris and Associates, followed soon after by exit polling in other Western European countries. Other non-European democracies adopted exit polling soon after – for example, Social Weather Stations in the Philippines conducted its first Election Day poll in 1992, and Mexican researchers did so in 1994. Mitofsky himself did exit polling in the Russian elections starting in 1993, working with the Russian firm CESSI.

To provide a random sample of voters, the exit poll locations (precincts) must be selected using probability sampling, proportionate to precinct size, with some stratification by geographic location and past vote. Interviewers have to be hired and trained, and stationed at the selected poll locations. Voters at the polling locations have to be sampled, either by interviewing every voter or a probability sample of them (every nth, with n determined ahead of time depending on the expected size of the precinct). Records of nonresponse normally should be kept – indicating its size and composition. Results then must be transmitted to a central location for processing, either physically, by telephone, or electronically.

In the United States, estimates of election outcome are made on a state-by-state basis, because of the allocation of electoral votes by states to the presidential candidates. The precinct tallies are weighted by size and their probabilities of selection, a nonresponse adjustment is made following a quality control check, and the results are entered into several estimation models – stratified by geography or past vote, including simple estimates and ratio estimates using the past vote. The models include tests for significance.

There have been different types of exit poll questionnaires. Some, as in Britain, have simply asked which candidate the respondent voted for. In contrast, a typical United States exit poll may contains 25 questions on both sides of a single sheet of paper including the importance of issues and demographic characteristics.

4.2. Problems for exit polls

The most serious methodological issue for exit polls has the level and distribution of nonresponse, as this may result in bias due to differences between those voters willing and those unwilling to respond. In addition, interviewer effects can be great because exit polls are conducted in person, although paper and pencil questionnaires preserve confidentiality and can reduce the impact of this concern on respondents (Bishop and Fisher, 1995). Examples of differential nonresponse have been documented in response rates of voters to interviewers of different races in elections with a racial component (Traugott and Price, 1992), and in other highly intense elections where interviewers may be perceived (correctly or incorrectly) as favoring one or another candidate or party. In the 2004 U.S. presidential election, exit poll overestimates of the vote for Democrat John Kerry were frequently cited as evidence of fraud by some activists; but all analysis indicated the difference was more likely caused by a differential response rate due to the interviewer–respondent interaction (Edison Media Research and

Mitofsky International, 2005; Traugott et al., 2005). Younger interviewers achieved lower response rates than older interviewers. Younger voters in general were more likely than average to be Kerry voters, and that perception (frequently reported before the election from preelection polls) may have influenced potential respondents.

In the United States, some states have passed legislation requiring exit poll interviewers to stand as far as 300 feet (nearly 100 m) from the polling location, effectively making a good sampling of voters impossible. In some countries, the difficulties of interviewing at the polling place (either through legal restrictions or fear of violence) have forced researchers to use different methodologies, such as in-person interviewing at home after people have voted. In the Philippines, day of election surveys at voters' homes are substituted for an exit poll at the polling place (voters there can be identified by an indelible mark on their hands).

In the 2000 United States election, mistaken projections in the state of Florida were attributed to exit polls, although when news organizations first projected that Al Gore would win Florida's electoral votes (7:50 p.m. ET), more than just exit poll results had been received. Twelve of the 120 sample precincts had reported *actual* tabulated results, and six of those precincts were part of the exit poll sample. Four percent of *all* precincts statewide had reported their votes. At 7:50 p.m., all of the estimation models indicated a Gore victory, and the estimates met the tests of significance.

There were several data problems. Precincts selected for the exit poll were not a true reflection of the state results. The difference between the actual precinct vote and the state totals was at the outer edge of sampling error. The ratio estimation model used only one past race for comparison, and that was the 1998 gubernatorial election, which had a 0.91 correlation with the vote for the 2000 Republican candidate, Bush. But using this race, the size of the absentee vote was underestimated – at only 8% of the total. As it turned out, the correlation of the 2000 Bush vote and the 1996 vote for Bob Dole was nearly as high (0.88). In addition, the correlation of the Democratic vote in those races was significantly higher than for the 1998–2000 comparison (0.81 vs. 0.71). Had that race been chosen for use by the ratio estimate, the absentee vote would have not been so grossly underestimated. Accurately estimating the size of the absentee vote is extremely important in states like Florida, where absentee votes historically were more than 20 points more Republican than the in-polling place day of election votes.

There was also differential nonresponse. In comparing exit poll results by precinct with the actual vote in that precinct, one can compute the average *Within Precinct Error*. This differential nonresponse has been attributed to many things, including variations in levels of enthusiasm for each candidate.[3] Early on election night 2000, it appeared the exit poll was understating the vote for Gore, and overstating the vote for Bush. That had been the pattern in Kentucky, the only other state where a *WPE* calculation could be made at the time. However, though the overestimate of the Bush vote in the exit poll remained true for Kentucky at the end of the night, it did not remain true in Florida. (The later projection of Bush as the victor in Florida, which was also withdrawn, was made without any use of exit poll results, only tabulated vote counts).

---

[3] The 1992 Republican Presidential primary in New Hampshire provided an instructive example of this. The exit poll indicated that Pat Buchanan might receive as much as 40% of the total vote against then President George H.W. Bush. He did not. According to the exit poll, Buchanan voters were more enthusiastic about their candidate than Bush voters were.

WPE has been rarely calculated elsewhere in the world, because the results in each precinct are not available in many places. In the United Kingdom, votes are aggregated and released publicly only at the constituency, not the polling place level. And in countries where election day polls have not been conducted at the polling place, comparisons can only be made to larger geographic units.

After the 2000 election, the U.S. exit poll operation was reviewed by RTI-Research Triangle Institute, which suggested a number of improvements that could be made to the methods of the Voter News Service (VNS), the organization that conducted the exit polls in that year. The main suggestions were as follows: improving the methodology for estimating the impact of absentee voters, improving the methodology for estimating outstanding votes in close races, improving the measures of uncertainty for election estimates, improving quality control, developing better decision rules, and exploring new approaches (RTI, 2001).

In the United States, the average within precinct error has been consistently in favor of the Democratic candidate. In three recent elections, 1988, 1996, and 2000, the average error on the difference between the candidates has been about 2 points (2.2 in 1988 and 1996, 1.8 in 2000). But in 1992 and 2000 the errors were 5.0 and 6.6 points, respectively. (This WPE calculation does not include polling places where there are many different precincts voting.) Turnout was higher in 1992 and 2004 than in the other elections, and the level of interest in the campaign was also high. In 1992 and 2004, two-thirds of voters reported paying a lot of attention to the campaign; fewer than half did in 1988, 1996, and 2000.

WPE was higher in larger precincts, in urban precincts, and in precincts where the respondent selection rate was high. It was higher in more competitive states. It was also greater in precincts with more Bush voters. WPE was correlated with interviewer reports of legal or other difficulties with election officials, with the distance an interviewer was forced to stand from the polling place, and with bad weather. But it was also correlated with interviewer characteristics: younger interviewers had higher WPE than older interviewers. The adjustments made in 2006 were the recruitment of a greater number of older interviewers and active attempts to encourage good relations with polling place officials. There was still some evidence for similar problems in the 2006 midterm election exit poll; the early afternoon tabulations compared to official vote returns showed that the Democratic candidates had a margin of vote in the exit poll that was about 4 percentage points too large (Lindeman, 2007).

Exit poll accuracy has also come under scrutiny in other countries. Investigation by a blue ribbon panel of an over-report of the vote in the exit poll for Gloria Macapagal–Arroyo was traced to exceptionally high nonresponse in metropolitan Manila, where many respondents were simply not available during the interview period. The Philippine exit poll was conducted away from the polling place, at respondents' residences.

Changes in the ways elections are conducted will affect exit polls. Absentee voting, vote by mail, and other forms of early voting have been increasingly permitted in the United States, so interviews conducted only at polling places will not include many voters. In two U.S. states (Oregon and Washington), nearly all votes are, at this writing, cast by mail; in more than half, the absentee/early vote has become a quarter or more of the total. Consequently, in the U.S. exit polls must be combined with telephone surveys conducted in the days before the election to see a full portrait of the electorate.

5. Postelection and between-election polls

Postelection surveys are often conducted by academic organizations for scholarly purposes and facilitate the analysis of public attitudes and behavior during election cycles (see discussion of NES above).[4] Media organizations conduct extensive polling to track and monitor public preferences in the postelection period as well as reactions to election outcomes. Perhaps, the most extensive use of postelection polling is by elected officials and party organizations who find it useful to continuously monitor public opinion on issues of public policy. At the presidential level, the polling apparatus has essentially become institutionalized to provide private data about the state of public opinion to the chief executive and his key advisers (Jacobs and Shapiro, 1995; Murray, 2006). There is considerable debate about how this opinion data is used by politicians (cf. Eisinger, 2003; Heith, 2004; Jacobs and Shapiro, 2000), but a general consensus about politicians' growing reliance on postelection private polls in the era of what Sidney Blumenthal has named "the permanent campaign" (Blumenthal, 1982).

6. Other opinion measurements: focus groups, deliberative polls, and the effect of political events

6.1. Focus groups

Most polls through the early 1970s relied on face-to-face interviewing. Ancillary research may have included longer intensive interviews and the use of carefully planned and moderated small group discussions to learn about perceptions and attitudes. These were originally called "focused interviews," but are now widely known as the "focus group" (e.g., Delli Carpini and Williams, 1994; Morgan, 1996). Technically, focus groups are not polls but in-depth interviews with a small number of people (6–12) often selected to represent broad demographic groups (Asher, 2007). Focus groups became widely used in market research and later in political campaigns as ways to learn people's opinions about products and candidates and their perceived sources for these opinions. Focus group participants do not constitute random or purportedly representative samples, because they are rarely, or ever, selected through random sampling, but they can provide useful information and pretest questions being developed for a larger scale survey. Effective ways of framing issues and messages can be explored in focus groups, and conversations among their dozen or fewer participants can provide insights into how individuals' opinions are shaped and change in response to new information, such as candidates' statements, news media reports, and advertizements. Focus group

---

[4] One methodological consideration relevant to postelection studies is inaccurate respondent recall of past behavior. Studies reveal that nontrivial numbers of respondents misreport vote choice. Moreover, there is evidence that retrospective reports of vote choice systematically magnify the support actually received by winning candidates, Using NES data, Wright (1993) reported that the prowinner bias tends to be relatively modest for presidential contests (about 1.5 percentage points) but over-reported for winners of congressional or gubernatorial races average between 4 and 7 percentage points, differences that far exceed amounts we could expect from sampling error. Similar evidence of such "bandwagon effects" has been detected using other data sources. Lindeman (2006) also shows that "false recall" favoring winners in presidential elections often grows over time.

leaders are able to probe participants to react to various stimuli while simultaneously observing other participants' reactions and redirecting the discussion to keep it relevant as necessary.

Generalizations from focus group participants do not apply to broader populations. Still, it is an appropriate methodology to illuminate the process and complexities of preference formation and attitude change. Mainly, focus groups highlight that people regularly and continuously construct views on complex issues through cognitive processes rather than retrieve those views (Glynn et al., 2004).

6.2. Deliberative polling

Deliberative polls have attempted to combine the virtues of focus group studies with those of standard public opinion surveys and to study and improve the quality of survey data by affecting the dynamics and quality of public opinion itself. Their conductors have first drawn a random sample of the public through probability sampling, interviewed respondents, and then brought the sample together to meet and learn about issues and problems through briefing materials, meetings with experts and political leaders, and small-group discussions. This survey method has provided a way of observing how public opinion is transformed, by the time of a later follow-up survey, through information and debate (Fishkin, 1997). The first national deliberative poll "sample" was convened in the United States in Austin, Texas, in 1996, in the context of the upcoming presidential election. Several other deliberative polls have been conducted to date, including British deliberative polls on "Europe 1995" and the "Monarchy;" an "Electric Utility" deliberative poll in Texas; "Australia Deliberates" in 1999 and one on "Aboriginal Reconciliation" there in 2001; one in Denmark on the Euro in 2000; and even a first-of-its-kind deliberative poll sponsored by the local government in Zeguo, China, on local infrastructure projects (Fishkin et al., 2006; see also Luskin et al., 2002). Other survey researchers have attempted to study deliberation through the context of a single survey itself as respondents are asked to react to new information provided in batteries of survey questions (see Kay, 1998). Another variant of the deliberative poll samples participants using a probability sampling method and has them interact in online groups to see whether and how opinions change when participants are interviewed again (see Lindeman, 2002; Price, 2006; Price and Neijens, 1998; on deliberation more generally, see Mendelberg, 2002).

Critics contend that the conclusions of deliberative polls cannot be generalized to the population at large, despite their randomly selected samples, because the public is unlikely to be exposed to information or experiences in the way participants in deliberative polls have been. Moreover, participants' attitudes may be influenced by the heightened sensitivity associated with participation.

6.3. The effect of events, political debates, and changing conditions

One important academic and journalistic use of polling is estimating the effect of events, including crises, political debates, election campaigns, and other changing circumstances and conditions, on short-term changes in public opinion. This has been attempted when surveys have been conducted frequently over short periods of time, or by tracking polls. Impact can also be measured by *panel surveys*, where the same respondents are interviewed again, and individual change (and the reasons for it) can be specified, although the "panel" may be subject to attrition (see Chapter 2).

The effects of presidential debates in the United States have been studied extensively through experiments or small group studies, as well as larger scale surveys conducted before and after those debates. Although there is not a consensus among researchers, there is evidence that debates (and political campaigns generally) increase public knowledge about the candidates and salience of the issues raised. Some studies have reported evidence of effects on candidate support and hence election outcomes (cf. Benoit et al., 2003; Geer, 1988). One found that presidential candidates perceived as victorious in debates against their opponents typically experience a surge in support following the debate: in 1984, Mondale was perceived as the winner of the first debate against Reagan, and experienced a bump of 3 to 4 percentage points (Holbrook, 1996).

U.S. political conventions have offered parties an opportunity to present their candidates and image to voters in a positive and relatively uncontested format. The resulting spike in support for the party's nominee can be substantial and have lasting implication that can carry through to Election Day. On average, presidential contenders between 1964 and 2004 received a 12-point boost in two-party support following their convention (Panagopoulos, 2007).

The impact of events on political attitudes and preferences can also be detected in performance evaluations of incumbent presidents. Major foreign policy actions, scandals, and other events can influence how respondents perceive the president. Short-term surges in presidential support have often followed momentous foreign policy events, for example, such as the attacks of 9/11 or the response to the invasion of Kuwait in 1991. Such rally-around-the-flag effects resulted in a net-positive shift of nearly 30 points in approval for President George W. Bush in 2001 following the 9/11 attacks (Mueller, 1973; see also Erikson et al., 2002). Generally, the impact of events dissipates over time, although traces of event-related effects on opinion have often lingered (Campbell, 2000; cf. more generally Page and Shapiro, 1992, on changes in the public's policy preferences).

7. Present and future challenges in polling

7.1. Response rates and nonresponse bias

The proliferation of telephone surveying, along with the growth of market research, telemarketing, and telephone solicitations in the United States since the 1970s created new challenges for pollsters and for the study of public opinion and voting. These challenges also have begun to occur in other countries as well. Telephone calling became increasingly disruptive, and household members became less willing to participate in such conversations. New technology also gave potential respondents the ability to screen calls through answering machines and caller identification devices. Decreased response rates increased the potential for "nonresponse" or selection bias in polls. In-person surveys are still conducted in some academic studies, such as the General Social Survey and the American National Election Studies, and government surveys including the U.S. Census. These surveys have had larger budgets and could maintain high-response rates through greater public relations and spending efforts. The trend in nonresponse in telephone surveys due to both noncontact and refusals was steeper from 1996 to 2003 than from 1979 to 1996 (Curtin et al., 2005; see also Zukin, 2006).

Despite this increase in nonresponse in the United States, there has been, surprisingly, little significant bias found *thus far* in comparisons of surveys with low (less than 30%) and substantially higher response rates (60% or more). What is still not known is whether there is still bias related to the sizeable hard core portion of the public that never responds (cf. Groves, 2006; Groves and Couper, 1998; Keeter et al., 2006; Singer, 2006; Weisberg, 2005). Exit polls have also faced increasing nonresponse.

7.2. Technological issues

7.2.1. Polls using interactive voice response
Rising costs in telephone polling and the increasing demand for polls has spurred not only the development of on-line surveys, but also the use of interactive voice response (IVR) technology. This methodology is an offshoot of audio Computer Assisted Self Interviewing (audio-CASI or ACASI). IVR polls are also referred to as automated polls or "robo-polls," which use a computer assisted polling method that replaces human interviewers with a prerecorded voice asking a short set of survey questions. Depending on the technology, respondents provide their answers verbally or key in responses on their touch-tone phones.

One advantage that IVR pollsters emphasize is that IVR controls and makes uniform how questions are asked, and how responses are received and data entered (though there can be respondent errors). A major disadvantage is that these surveys work best if limited to no more than 5 minutes of questions, which means less data can be collected. In addition, they are likely to have more break-offs because respondents are not hanging up rudely on a person who has attempted to build rapport, and respondents might have no hesitation to offer flippant or false responses. Consequently, the main use for these polls is to collect specific opinions and the most relevant background characteristics. Unless the initial introduction and screening of respondents is done by a human interviewer, these surveys may interview individuals who are not members of the sample of appropriate age, voter eligibility, or whatever required characteristic (see Couper et al., 2004; Li, 2006).

At this writing, IVR surveys have performed well in a number of election contests in many states and localities, as well as nationally, but there are also examples of pre-election difficulties (see Blumenthal, 2005).

7.2.2. Cellular phones
For telephone surveys, coverage issues in the past were limited to noncoverage of households without telephones and over-coverage of households with multiple phone lines. The latter can be dealt with by statistical weighting, the former by demographic adjustments. There are concerns about the increasing use of cellular phones not only to supplement regular "land-line" phones but also to replace such phones (see Lavrakas, 2007). According to a 2006 study conducted by the Pew Research Center for the People and the Press (2006), an estimated 7–9% of the American public was "cell phone only" in 2006, 53% of the public had access to both a landline and a cell phone, and 37% had a landline only. The remainder had no telephone access. Subsequently,

the January–June 2008 National Health Interview Survey estimated cell phone-only households at 17.5%. Approximately 2% of households had no phone (Blumberg and Luke, 2008).

Thus far, surveys have dealt with this successfully through weighting the data to adjust for the typically young age and other characteristics of individuals who have no telephone or only cell phones at home. These phone issues were not a factor in 2004 preelection poll estimates, based on self-reports of household phone coverage in the 2004 U.S. national election exit poll (Keeter, 2006). The Pew study (2006) discussed earlier also found only minimal differences between cell-only respondents and those reachable by landline on key political questions, once appropriate weighting procedures were implemented. However, in a subsequent study, Keeter et al. (2007) did find that by including a cell phone-only sample with a standard RDD they could produce population estimates that were nearly the same as those from a landline-only sample, they also found evidence that the noncoverage of young adults (fully 25% of whom had only cell phones) in RDD surveying created biased estimates on certain survey measures.

As the proportion of cell phone-only households increases in ways that might lead to greater biases, it is likely to affect further survey response rates and costs. It is already clear that respondents aged 18–34 have become much harder to reach and that the "portability" of phone numbers in the United States has made it increasingly difficult for sampling purposes to identify the geographic residence of cell phone users (see Zukin, 2006). Further studies of cell phone users are under way to determine the feasibility and effectiveness of interviewing respondents on their cell phones, and compensating respondents for any costs incurred in receiving survey calls (Brick et al., 2007). Response rates for a cell phone sample frame are typically lower than a landline sample. The contact rate for the cell phone sample may be higher, although greater accessibility has not lead to a higher rate of cooperation; in one study half of the people reached in the landline sample (50%) cooperated, when compared with 28% of those reached in the cell phone sample. However, interviewers working on the survey reported that cell phone respondents were as focused and cooperative as those reached on a landline telephone (Pew Research Center, 2006; see also Lavrakas, 2007).

Telephone surveying has greater problems in other countries that do not have the extensive availability of land lines. In some places cell phone penetration is very great (Zukin, 2006), and in others land-line expansion has been by-passed by the large-scale introduction of cell phones. Response rates are a major issue everywhere. In some countries it is still necessary to do in-person interviewing, and in others researchers are turning to the Internet and Internet panels, which have become increasingly appealing (especially to interview young adults).

7.3. Threat of government regulation

In contemporary politics, attacking or fending off negative polls are a normal part of campaigns. In 1992, when George H. W. Bush was trailing Bill Clinton, Bush attacked polls in more than 30 speeches: the equivalent of once in every four times he spoke publicly. In 1996, Bob Dole talked about the polls in one-third of all his speeches (Frankovic, 1998). In recent elections, campaigns and news stories frequently describe differences in preelection polls, and raise questions about methods, including queries about how likely voters are defined, question order, weighting, and assumptions about partisanship (Frankovic, 2005).

Most recently, pollsters themselves have come under direct attack – almost literally – for their election polling. There has been continuing debate and sensitivity to what might be call "polling politics" in which local news media and other public pollsters have been accused of partisan biases in their polls, in which normal variations and the occasional outlier in poll results that can occur due to chance are attributed to manipulative polling practices (Daves and Newport, 2005; Jacobs and Shapiro, 2005).

Some governments have attempted to limit the publication of poll results – both during the preelection period and on Election Day. As late as 2002, at least 30 countries had legal restrictions on the publication of preelection poll results. There had been little change in that absolute number since 1996, when at least 31 countries had embargos on the publication of political poll results on or prior to Election Day. Nine of these embargos applied to Election Day only; 46 countries (61%) had no embargo. Nine countries had increased the time restrictions between 1996 and 2002, while 15 others had decreased it, or eliminated it entirely. Countries with limits on the publication of preelection polls in 2002 included Western European countries like Portugal, Spain, and Switzerland, and countries in Asia and Latin America. In Italy, publication was allowed, but required a poll report to be accompanied by an "information note" with several specifications related to the poll, which must be published together with the results of the poll in the media and recorded on a dedicated website (Spangenberg, 2003).

Several nations ban the publication of preelection poll results at certain stages of the campaign. In Canada, for example, poll results cannot be published during the final 3 days of the campaign. Greece, Italy, and Ukraine are even more restrictive, prohibiting poll publication for the final 15 days of the electoral campaign (Plasser and Plasser, 2002). South Africa bans poll publication for the last 6 weeks of a campaign. In Lithuania, poll publication is prohibited for the entire duration of the official campaign period.

The regulatory framework as it applies to public and media preelection polls varies in other meaningful ways cross-nationally. According to data provided by the ACE Electoral Knowledge Network, 16 countries required the sponsor of a poll to be indicated. Disclosure of the sample characteristics is required by law in 17 countries, and the margin of error is legally required to be disclosed in 13 countries including Albania, Portugal, and Russia (ACE Electoral Knowledge Network, 2006).

The U.S. Congress held hearings after the early projection of a Ronald Reagan victory in the 1980 election (as it had after a previous electoral landslide in 1964), and some claimed that projections of a Ronald Reagan victory before all polls had closed affected turnout in Western states (Jackson, 1983). Some states, including Oregon, passed laws restricting those polls. As of 2002, 41 countries restricted publication or broadcast of poll results until after the polling places have closed. In addition, in both the United States and in Hong Kong, there are no government regulations about the release of exit poll information, but pollsters and news organizations have agreed not to report exit poll results until after the polls close (Spangenberg, 2003).

Bans on reporting preelection polls have been circumvented by posting results on the Internet, and restrictions (whether government-imposed or self-imposed) on reporting exit polls before polls close have also been circumvented, as leaks of exit polling results and their reporting on the Internet have become routine. In 2004, early leaks of partial exit poll results (with the overestimate of the Kerry vote) fueled speculation of voter fraud that continued even after the election.

Despite near-universal belief that poll information affects voting, there is minimal supporting evidence. According to one review of studies about the impact of election polling. "The conclusion is that any effects are difficult to prove and in any case are minimal. Opinion polls do provide a form of 'interpretative assistance' which helps undecided voters make up their mind. But the media are full of such interpretative aids, including interviews and commentaries, and in this perspective, election polls are a relatively neutral and rational interpretative aid" (Donsbach, 2001, p. 12; see also, Adams, 2005).

7.4. Issues in reporting polls accurately

Beyond the technical details in conducting and analyzing polling data, how poll results are reported and interpreted is itself an issue. News stories about polls are increasingly common. One recent estimate is that the number of stories reporting on poll results has nearly doubled from the 1992 and 1996 U.S. Presidential election years to 2000 and 2004 (Frankovic, 2005, p. 684–685).

In the United States, problems in reporting have been affected by changes in journalism – cutbacks and 24 hour reporting leading to more reliance on poll results – and the repackaging of releases of poll results – as news. Journalists too often do not have the time or skills to evaluate fully the quality of the polls they report on (Rosenstiel, 2005).

8. Continued interest in public opinion and polling

Public opinion and election polling has been one of the constant features of late 20th century and early 21st century social and political life in the United States. They have been a persistent source of discussion and debate in the press, and the latest opinions of the public toward political issues and candidates are persistent topics of political contentions, and academics and commentators continue to debate the positive role for American democracy that George Gallup saw for public opinion through polling. Many critics doubt that the public is sufficiently knowledgeable, attentive to politics, skilled in interpretation and analysis, and wise enough overall to deserve attention in governing beyond casting votes on Election Day. Rather, the public should defer to political leaders and experts. The defenders of the "rationality" of public opinion argue that the public – as individuals and especially as a *collective* – was sufficiently capable of taking cues or learning from political leaders and other sources. Indeed, the public had defensible reasons for its opinions to warrant ongoing consideration in the political process.

This has raised classic questions: To what extent do political leaders follow or lead public opinion? What are the implications for this for democracy? Some critics have also argued that the existence of polling gives the public the false sense that its voice is amply represented in the political process (see Ginsberg, 1986; Herbst, 1993). Although the common wisdom, beginning with George Gallup, was that polls enabled political leaders to learn about public opinion and, under electoral pressure, to follow the public's wishes, political leaders are not always under such immediate pressure; they have room to maneuver and attempt to lead – or even manipulate – public opinion, using polling information to learn how best to "craft" their messages for this purpose (see Glynn et al., 2004, Chapter 9; Jacobs and Shapiro, 2000).

Polls continue to vary in their type, scope, and quality; innovations may create new problems for pollsters to wrestle with. As survey researchers try to get good estimates of the public's opinions and behavior, the problems cited in the early days of polling remain the same: sampling coverage; identifying and representing an identifiable population and obtaining a sufficiently high response rate; statistical sampling error, assuming there is acceptable sampling coverage; the effects of questions wording; the use of fixed choice versus open-ended questions; the treatment of "don't know," "no opinion," "undecided" and similar types of responses; effects of question order ("context effects"); lack of clarity in the research questions being studied and assumptions about respondents' familiarity with a particular issue or topic; whether reported behavior validly represents actual behavior (in the present, past, or future); interviewer effects; and the effects of the type of survey method used ("mode effects").

Given such attention to public opinion, the accuracy of poll results and how these results are reported have become increasingly important academic and political issues. Journalists have been widely criticized for their shortcomings in reporting about poll results, often accepting them uncritically without researching the quality of polls and the questions asked in them. The opportunities first offered by polling have led to challenges for the pollsters and democratic politics on a number of fronts in the United States and (increasingly) worldwide, as the reach of democracy and survey research has expanded to more and more countries.

Acknowledgments

We thank Kirk Wolter, one anonymous reviewer and the editor for excellent advice and helpful comments. Robert Shapiro is grateful to the Russell Sage Foundation where he was a 2006–2007 Visiting Scholar and to Columbia University's Institute for Social and Economic Research and Policy (ISERP).

Subject Index

A

AAPOR, *see* American Association for Public Opinion Research

Accuracy and coverage evaluation (ACE), 239

Across-stratum estimation, 21

Activity status, 132

Administrative data, 136–140
- advantages and disadvantages, 138
- calendarization of, 138–140
- categories, 137
- survey evaluation, 138

Administrative files, 132, 134

Agricultural enterprises, 167

Agricultural surveys, sampling methods, 163

Allocation schemes, sample size, 142–143

American Association for Public Opinion Research (AAPOR), 69, 70, 263, 265

American Community Survey (ACS), 97, 254
- objective of, 111

American National Election Studies (ANES), 261

AMS, *see* Automated Match Study

Ancillary information in design, 190–192

ANES, *see* American National Election Studies

Annual tax data, 137

Answering machine systems, 60

Anticipated variance, HTE, 26

Apartment frames, 99

Approximation approach, 123–124

Area codes, 53

Area frames, 99, 163
- for business surveys, 134–135
- use of, 99

Attribute disclosure, 82, 84

Australia Labour Force Survey (LFS), rotation designs for, 112

Australian Bureau of Statistics, 149

Automated dialers, 61

Automated Match Study (AMS), 241

Automated polls, *see* Interactive voice response (IVR)

Auto-Regressive Moving Average (ARMA) process, 140

Auxiliary data for business surveys, 145–147, 157

B

Base weights, 72–73

Bayesian approach, 196

Benchmarking model for calendarization, 139

Bernoulli sampling, 91, 148, 157, 160
- variants, 149

Bias
- conditioning, 222–223
- household dissolutions, 222
- household formations, 222
- household moves, 222
- initial refusals, 220–222
- panel dropouts, 221–222

Births and population relationship, 2

Bowley's Bayesian question, 9, 10

British Market Research Association, 271

Buildings/address sample, 247

Business, 132
- changes in, 135
- complex, 133, 134 f
- register, 132, 134
- simple, 133, 133 f
- survey, 131, 170
- – sample selection in, 141, 147–150
- – sampling frames for, 132–136

C

Calendarization
- of administrative data, 138–140
- for monthly series, 139 f

Calibration estimation in household surveys, 119–121

Caller id, 60

Call-outcome code, 66, 67

Canada Revenue Agency, 134, 140

Canadian Business Register, 132

Canadian evaluation program, 240–241

Printed in the United States
by Bookmasters

Printed in the United States
By Bookmasters